Performance and Stability of Aircraft

Performance and Stability of Aircraft

J. B. Russell MSc, MRAeS, CEng
Centre for Aeronautics
City University
London

BUTTERWORTH
HEINEMANN

OXFORD AMSTERDAM BOSTON LONDON NEW YORK PARIS
SAN DIEGO SAN FRANCISCO SINGAPORE SYDNEY TOKYO

Butterworth-Heinemann
An imprint of Elsevier Science
Linacre House, Jordan Hill, Oxford OX2 8DP
200 Wheeler Road, Burlington, MA 01803

First published 1996
Transferred to digital printing 2003

British Library Cataloguing in Publication Data
A catalogue record for this book is available from the British Library

Library of Congress Cataloguing in Publication Data
A catalogue record for this book is available from the Library of Congress

ISBN 0 340 63170 8

Contents

Preface

The title of this book should really have been 'Performance, stability, control and response of aircraft for undergraduate aeronautical engineers', but that is obviously too cumbersome. The material of the book should probably be spread over the three years of the normal British aeronautical degree, with Chapters 1 and 2 being dealt with in the first year and Chapters 3 to 7 in the second year. The remainder, or most of it, would form a course for the final year; in this way the course will normally keep ahead of the mathematics being taught. The mathematical background expected of students include matrices, differential and integral calculus, the Laplace transform, transfer functions and frequency response. All of these are normally included in a degree course and it was therefore not thought necessary to include much purely mathematical material as have some earlier books on this subject. Equally it has not been thought necessary to include detailed information on the determination of aerodynamic parameters because this information is readily available in the publications of the Engineering Sciences Data Unit (ESDU) and elsewhere.

Many students find this subject fairly difficult. Amongst the reasons are (a) the mathematics involved, (b) the need to 'think in three dimensions', and (c) the very large number of symbols required. Attempts to assist the student in thinking three-dimensionally have been made by discussing a number of simple situations in which the aircraft is considered from different viewpoints, and by means of an illustrative example. In these simple situations the student is introduced to some of the dynamic stability notation rather before it is strictly necessary. There is also a gradual progression from one degree of freedom cases to the six degrees finally required Other aids for the student are a number of worked examples and of examples for the student to attempt with answers. On the subject of notation for the dynamic stability work, a subset of that proposed by Hopkin (reference 8.2) has been preferred; in spite of its age this work is still worthy of study and has received less use than it deserves. This notation has the advantages that (a) it is used by ESDU, (b) it encompasses both the case in which the dimensional stability derivatives are divided by mass or inertia, as appropriate, and the case in which full non-dimensionalization is used, and (c) it is about as compact as may be devised. Conversion to the American notation is also covered.

I must place on record my thanks to my friends and colleagues Mike Freestone, Ranjan Banerjee, Peter Lush and Trevor Nettleton who read parts of early versions and who made valuable comments and suggestions; the mistakes are all my own work. I would also like to record my gratitude to Dick Cox, 'Bram' Bramwell and Malcolm Wright, now sadly all passed on, for their friendship, help and encouragement long before this project began but who contributed indirectly. My thanks must also go to the City University for granting me sabbatical leave without which this book would never have been started and for giving me permission to publish the examples, many of which are based on examination paper questions. Finally my greatest thanks go to my wife, Joy, for her support, tolerance and love.

List of symbols and abbreviations

The following is a list of the symbols used in the text except for a few which are used only in one section and are defined there. A number in brackets following a definition indicates the chapter for which that definition only applies; no number implies general application of the definition elsewhere. If a symbol is defined mathematically the general definitions are to be assumed for the quantities used.

Information on the corresponding American symbols is given in Sections 4.3.1, 8.6.3, 9.2.8 and 11.2.7.

A	Aspect ratio, b^2/S (1)
A_1	Coefficient of λ^4 in longitudinal quartic
A_2	Coefficient of λ^4 in lateral quartic
$\mathbf{A}$	Angular velocity vector
$\mathscr{A}$	Temporary lumped constant
a	Constant term in parabolic drag law (1, 2, 3)
a	Lift curve slope (4)
a	Aircraft-less-tail lift curve slope (5, 9)
a_1	Tailplane lift curve slope
a_1	Coefficient of λ in quadratic representing SPPO mode (9, 13)
$\bar{a}_1$	Tailplane lift curve slope with free elevator, $a_1 - \frac{a_2}{b_2} b_1$
a_1^F	Fin lift curve slope
a_2	Elevator lift curve slope
a_2	Constant term in quadratic representing phugoid mode (9, 13)
a_2	Rudder lift curve slope
a_3	Tab lift curve slope
$\bar{a}_3$	Tab lift curve slope with free elevator, $a_3 - \frac{a_2}{b_2} b_3$
$\mathscr{a}$	Acceleration
B_1	Coefficient of λ^3 in longitudinal quartic
B_2	Coefficient of λ^3 in lateral quartic
$\mathscr{B}$	Temporary lumped constant
b	Wing span
b	Coefficient of C_L^2 in parabolic drag law (1, 2, 3)
b_0	Constant term in equation for hinge moment coefficient
b_1	Slope of hinge moment curve with incidence
b_1	Constant term in quadratic representing SPPO mode (9, 13)
$\bar{b}_1$	$b_1 - \frac{a_1}{a_2} b_2$
b_2	Slope of hinge moment curve with elevator angle
b_2	Constant term in quadratic representing phugoid mode (9, 13)
b_3	Slope of hinge moment curve with tab angle

$\bar{b}_3$	$b_3 - \dfrac{a_3}{a_2} b_2$
C	Damper constant (7, 8)
C_1	Coefficient of λ^2 in longitudinal quartic
C_2	Coefficient of λ^2 in lateral quartic
C_D	Drag coefficient, $D/\frac{1}{2}\rho V^2 S$
C_{Dtv}	Trailing vortex (induced) drag coefficient
C_H	Hinge moment coefficient, $H/\frac{1}{2}\rho V^2 S_\eta C_\eta$
$C_{H_{trim}}$	Trimmed value of hinge moment
C_L	Lift coefficient, $L/\frac{1}{2}\rho V^2 S$
C_{L_1}	Lift coefficient in level flight (5)
C_L^F	Foreplane lift coefficient, $L^F/\frac{1}{2}\rho V^2 S_F$
C_L^T	Tailplane lift coefficient, $L^T/\frac{1}{2}\rho V^2 S_T$
C_L^W	Wing lift coefficient
C_{Lmax}	Maximum lift coefficient
C_{Lmd}	Lift coefficient for minimum drag
C_{Lmp}	Lift coefficient for minimum power required
C_M	Pitching moment coefficient, $M/\frac{1}{2}\rho V^2 S \bar{\bar{c}}$
C_{M0}	Pitching moment coefficient at zero lift
C_{Mfix}	Aircraft pitching moment coefficient with elevator under control
C_{Mfree}	Aircraft pitching moment coefficient with elevator free to float
$\mathscr{C}$	Temporary lumped constant
c	Wing (section or local wing) chord
$\bar{c}$	Mean wing chord
$\bar{\bar{c}}$	Aerodynamic mean chord, see (5.8)
c_s	Speed of sound
D	Drag (1, 2, 3)
D	d/dt
$\hat{D}$	$d/d\hat{t}$
D_1	Coefficient of λ in longitudinal quartic
D_2	Coefficient of λ in lateral quartic
D_{min}	Minimum drag
E	Constant term in inertial coupling quartic (12)
E	Endurance (2)
E_1	Constant term in longitudinal quartic
E_2	Constant term in lateral quartic
e	Specific endurance (2)
$\bar{e}$	$1 - d\varepsilon/d\alpha$
e_x	$-I_{xz}/I_x$
e_z	$-I_{xz}/I_z$
$F(\lambda)$	Characteristic equation
$\mathbf{F}$	Vector of forces
$\mathbf{F}_a$	Vector of aerodynamic forces on aircraft
$\mathbf{F}_g$	Vector of components of aircraft weight
f	Specific fuel consumption, propeller-engined aircraft (2)
f	Frequency (10)
f	$\mathring{N}_\xi/\mathring{L}_\xi$ (12)
f_j	Specific fuel consumption, jet-engined aircraft (2)

$G_1, G_1', G_1'', G_2, G_3$ Response polynomials
G_η Elevator gearing, (elevator deflection)/(stick movement)
$\mathbf{G}$ Linear momentum vector
g Acceleration due to gravity
g_1 $g \cos \Theta_e$
g_2 $g \sin \Theta_e$
$\mathbf{g}_1$ Normalized value of $g_1 = C_L$
$\mathbf{g}_2$ Normalized value of $g_2 = C_L \tan \Theta_e$
H Hinge moment
H_m Manoeuvre margin, 'stick fixed'
H_m' Manoeuvre margin, 'stick free'
H_n cg margin, 'stick fixed'
H_n' cg margin, 'stick free'
$H_{u\eta}, H_{w\eta}, H_{n\eta}, H_{ww_g}, H_{nw_g}$, Aircraft transfer functions
$\mathbf{H}$ Angular momentum vector
h Height, altitude
h Maximum height of mean line above chord line (1)
h_1, h_2, h_3 Components of angular momentum along Ox, Oy and Oz
h_e Equilibrium height
h_{en} Energy height, $h + V^2/2g$
h_s Screen height
$h\bar{\bar{c}}$ Distance of cg aft of leading edge of amc
$h_0\bar{\bar{c}}$ Distance of aircraft-less-tail aerodynamic centre aft of leading edge of amc
$h_m\bar{\bar{c}}$ Distance of manoeuvre point, 'stick fixed', aft of leading edge of amc
$h_m'\bar{\bar{c}}$ Distance of manoeuvre point, 'stick free', aft of leading edge of amc
$h_n\bar{\bar{c}}$ Distance of neutral point, 'stick fixed', aft of leading edge of amc
$h_n'\bar{\bar{c}}$ Distance of neutral point, 'stick free', aft of leading edge of amc
$\mathbf{h}$ Angular momentum vector relative to cm
I_x Moment of inertia about Ox axis (rolling axis)
I_y Moment of inertia about Oy axis (pitching axis)
I_z Moment of inertia about Oz axis (yawing axis)
I_{xy} Product of inertia about Ox and Oy
I_{yz} Product of inertia about Oy and Oz
I_{zx} Product of inertia about Oz and Ox
i $\sqrt{-1}$
i_1 $(I_z - I_x)/I_y$
i_2 $(I_y - I_x)/I_z$
i_x I_x/mb^2
i_y $I_y/m\bar{\bar{c}}^2$
i_z I_z/mb^2
$\mathbf{i}, \mathbf{i}_0, \mathbf{i}_1, \mathbf{i}_2$ Unit vectors along Ox, Ox_0, Ox_1 and Ox_2
$\mathbf{J}$ General vector quantity
$\mathbf{j}, \mathbf{j}_0, \mathbf{j}_1, \mathbf{j}_2$ Unit vectors along Oy, Oy_0, Oy_1 and Oy_2
k_1, k_2, k_3 Magnitudes of eigenvectors
k_0 Proportional control constant (13)
k_1 Rate control constant (13)

Symbol	Meaning
$\mathbf{k}, \mathbf{k}_0, \mathbf{k}_1, \mathbf{k}_2$	Unit vectors along Oz, Oz_0, Oz_1 and Oz_2
L	Lift (1, 2, 3, 4, 5)
L	Rolling moment (6, 7, 8, 11, 12, 13)
L	Integral scale length (10)
L_1	Lift in level flight (5)
L_A	Rolling moment due to aileron deflection
L^F	Foreplane lift
L_R	Rolling moment due to rudder
L^T	Tailplane lift
L^W	Wing lift
$\mathring{L}_p, \mathring{L}_r, \mathring{L}_v, \mathring{L}_\xi, \mathring{L}_\zeta$	Rolling moment derivatives, $\partial L/\partial p$, $\partial L/\partial r$, $\partial L/\partial v$, $\partial L/\partial \xi$ and $\partial L/\partial \zeta$
L_p	Non-dimensional rolling moment derivative due to rate of roll, $\mathring{L}_p/\frac{1}{2}\rho V_e S b^2$
L_r	Non-dimensional rolling moment derivative due to rate of yaw, $\mathring{L}_r/\frac{1}{2}\rho V_e S b^2$
L_v	Non-dimensional rolling moment derivative due to sideslip velocity, $\mathring{L}_v/\frac{1}{2}\rho V_e S b$
L_ξ	Non-dimensional rolling moment derivative due to aileron angle, $\mathring{L}_\xi/\frac{1}{2}\rho V_e^2 S b$
L_ζ	Non-dimensional rolling moment derivative due to rudder angle, $\mathring{L}_\zeta/\frac{1}{2}\rho V_e^2 S b$
l	Characteristic length
l_F	Fin moment arm
l_F	Foreplane moment arm (5)
l_T	Tailplane moment arm
l_R	Rudder moment arm
l_p	$-L_p/i_x$
l_r	$-L_r/i_x$
l_v	$-\mu_2 L_v/i_x$
l_ξ	$-\mu_2 L_\xi/i_x$
l_ζ	$-\mu_2 L_\zeta/i_x$
M	Pitching moment
M_0	Pitching moment at zero lift
M_0^{wb}	Aircraft-less-tail pitching moment at zero lift
M_0^T	Tailplane pitching moment at zero lift
$\mathring{M}_q, \mathring{M}_u, \mathring{M}_w, \mathring{M}_{\dot{w}}, \mathring{M}_\eta$	Pitching moment derivatives, $\partial M/\partial q$, $\partial M/\partial u$, $\partial M/\partial w$, $\partial M/\partial \dot{w}$ and $\partial M/\partial \eta$
M_q	Non-dimensional pitching moment derivative due to rate of pitch, $\mathring{M}_q/\frac{1}{2}\rho V_e S \bar{\bar{c}}$
M_u	Non-dimensional pitching moment derivative due to forward velocity increment, $\mathring{M}_u/\frac{1}{2}\rho V_e S \bar{\bar{c}}$
M_w	Non-dimensional pitching moment derivative due to vertical velocity increment, $\mathring{M}_w/\frac{1}{2}\rho V_e S \bar{\bar{c}}$
$M_{\dot{w}}$	Non-dimensional pitching moment derivative due to vertical acceleration, $\mathring{M}_{\dot{w}}/\frac{1}{2}\rho S \bar{\bar{c}}^2$
M_η	Non-dimensional pitching moment derivative due to elevator angle, $\mathring{M}_\eta/\frac{1}{2}\rho V_e^2 S \bar{\bar{c}}^2$
$\mathbf{M}$	Mach number, V/c_s
$\mathbf{M}_{Dcrit}$	Drag critical Mach number
m	Mass
m_0	Initial aircraft mass
m_1	Final aircraft mass
m_F	Total fuel used in climb
$\dot{m}_f$	Fuel mass flow rate

N	Yawing moment
N	Load factor, lift/weight (3)
N_R	Yawing moment due to rudder angle
$\mathring{N}_p$, $\mathring{N}_r$, $\mathring{N}_v$, $\mathring{N}_\xi$, $\mathring{N}_\zeta$	Yawing moment derivatives, $\partial N/\partial p$, $\partial N/\partial r$, $\partial N/\partial v$, $\partial N/\partial \xi$ and $\partial N/\partial \zeta$
N_p	Non-dimensional yawing moment derivative due to rate of roll, $\mathring{N}_p/\frac{1}{2}\rho V_e S b^2$
N_r	Non-dimensional yawing moment derivative due to rate of yaw, $\mathring{N}_r/\frac{1}{2}\rho V_e S b^2$
N_v	Non-dimensional yawing moment derivative due to sideslip velocity, $\mathring{N}_v/\frac{1}{2}\rho V_e S b$
N_ξ	Non-dimensional yawing moment derivative due to aileron angle, $\mathring{N}_\xi/\frac{1}{2}\rho V_e^2 S b$
N_ζ	Non-dimensional yawing moment derivative due to rudder angle, $\mathring{N}_\zeta/\frac{1}{2}\rho V_e^2 S b$
n	Normal acceleration factor
n_p	$-N_p/i_x$
n_r	$-N_r/i_x$
n_v	$-\mu_2 N_v/i_x$
n_ξ	$-\mu_2 N_\xi/i_x$
n_ζ	$-\mu_2 N_\zeta/i_x$
$Oxyz$	Aircraft reference axes: Ox in forward direction (for wind axes in direction of undisturbed flight), Oy towards starboard wing tip, and Oz pointing downwards, completing the orthogonal set
P	Engine power
P_e	Equivalent shaft power, $P + (T_j V/_\eta)$
P_r	Power required for level flight
P_η	Backward force exerted by pilot
p	Rate of roll
p	Atmospheric pressure (1, 2, 3)
p_0	Atmospheric pressure at sea level
$\hat{p}$	$p\tau$
$\mathbf{Q}$	Torque vector
$\mathbf{Q}_a$	Vector of aerodynamic moments acting on the aircraft
q	Rate of pitch
$\hat{q}$	$q\tau$
R	Gas constant (1)
R	Range (2)
R	Radius of turn (3), of pullout (5)
R	Routh's discriminant (9, 11, 13)
R_e	Reynolds number, Vl/ν
r	Rate of yaw
$\hat{r}$	$r\tau$
$\mathbf{r}$	Position vector
$\bar{\mathbf{r}}$	Position vector of cm
$\mathbf{r}'$	Position vector relative to cm
S	Gross wing area
S	Spectral density (10)
S_F	Foreplane area (5)
S_F	Fin area (6)
S_T	Tailplane area
s	Wing semispan, b/2
T	Thrust
T_j	Jet thrust
t	Time

t_c	Characteristic time
t_D	Time to double initial amplitude of a disturbance
t_H	Time to halve initial amplitude of a disturbance
t_p	Periodic time
t_s	Servomotor time constant
$\hat{t}_c$	t_c/τ
$\hat{t}_D$	t_D/τ
$\hat{t}_H$	t_H/τ
$\hat{t}_p$	t_p/τ
$\hat{t}_s$	t_s/τ
U	Aircraft velocity along Ox in disturbed flight, $U_e + u$
U_e	Aircraft velocity along Ox in datum flight
u	Increment in velocity along Ox
$\hat{u}$	u/V_e
V	True airspeed
V	Aircraft velocity along Oy in disturbed flight (7, 8)
V_E	Equivalent airspeed, $V\sqrt{\sigma}$
V_{Emd}	Equivalent airspeed for minimum drag
V_{Emp}	Equivalent airspeed for minimum power
V_{MCa}	Minimum control speed in air
V_{MCg}	Minimum control speed on ground
V_e	Datum (true) airspeed, resultant of U_e and W_e
$\bar{V}_F$	Foreplane volume coefficient, $S_F l_F/S\bar{\bar{c}}$
$\bar{V}_T$	Tailplane volume coefficient, $S_T l_T/S\bar{\bar{c}}$
v	Aircraft velocity along Oy in disturbed flight, sideslip velocity
$\hat{v}$	v/V_e (= β for small angles)
v_c	Rate of climb
v_{co}	Rate of climb at constant forward speed
v_s	Rate of descent
$\mathbf{v}$	Velocity vector
$\bar{\mathbf{v}}$	Vector velocity of cm
$\mathbf{v}'$	vector velocity relative to cm
W	Aircraft velocity along Oz in disturbed flight, $W_e + w$
W_e	Aircraft velocity along Oz in datum flight
w	Increment in velocity along Oz
w	Rate of gain of energy height (3)
$\hat{w}$	w/V_e
w_g	Upgust velocity
$\hat{w}_g$	w_g/V_e
X	Component of force along Ox
X_a	Component of aerodynamic force along Ox
X_g	Component of weight along Ox
$\mathring{X}_q$, $\mathring{X}_u$, $\mathring{X}_w$, $\mathring{X}_\eta$	Forward force derivatives, $\partial X/\partial q$, $\partial X/\partial u$, $\partial X/\partial w$ and $\partial X/\partial \eta$
X_q	Non-dimensional forward force derivative due to pitching velocity, $\mathring{X}_q/\frac{1}{2}\rho V_e S\bar{\bar{c}}$
X_u	Non-dimensional forward force derivative due to velocity increment along Ox, $\mathring{X}_u/\frac{1}{2}\rho V_e S$
X_w	Non-dimensional forward force derivative due to velocity increment along Oz, $\mathring{X}_w/\frac{1}{2}\rho V_e S$
X_η	Non-dimensional forward force derivative due to elevator angle, $\mathring{X}_\eta/\frac{1}{2}\rho V_e^2 S$
x_q	$-X_q/\mu_1$

x_u $-X_u$
x_w $-X_w$
x_η $-X_\eta$
Y Component of force along Oy, sideforce
Y_a Component of aerodynamic force along Oy
Y_g Component of weight along Oy
$\mathring{Y}_p$, $\mathring{Y}_r$, $\mathring{Y}_v$, $\mathring{Y}_\xi$, $\mathring{Y}_\zeta$ Yawing moment derivatives, $\partial Y/\partial p$, $\partial Y/\partial r$, $\partial Y/\partial v$, $\partial Y/\partial \xi$ and $\partial Y/\partial \zeta$
Y_p Non-dimensional sideforce derivative due to rate of roll, $\mathring{Y}_p/\frac{1}{2}\rho V_e S b$
Y_r Non-dimensional sideforce derivative due to rate of yaw, $\mathring{Y}_r/\frac{1}{2}\rho V_e S b$
Y_v Non-dimensional sideforce derivative due to sideslip velocity, $\mathring{Y}_v/\frac{1}{2}\rho V_e S$
Y_ξ Non-dimensional sideforce derivative due to aileron angle, $\mathring{Y}_\xi/\frac{1}{2}\rho V_e^2 S$
Y_ζ Non-dimensional sideforce derivative due to rudder angle, $\mathring{Y}_\zeta/\frac{1}{2}\rho V_e^2 S$
y Spanwise distance (5, 7)
$\dot{y}$ Lateral velocity of cg relative to mean flight path
y_p $-Y_p/\mu_2$
y_r $-Y_r/\mu_2$
y_v $-Y_v$
y_ξ $-Y_\xi$
y_ζ $-Y_\zeta$
Z Component of force along Oz
Z_q Non-dimensional vertical force derivative due to pitching velocity, $\mathring{Z}_q/\frac{1}{2}\rho V_e S\bar{\bar{c}}$
Z_u Non-dimensional vertical force derivative due to velocity increment along Ox, $\mathring{Z}_u/\frac{1}{2}\rho V_e S$
Z_w Non-dimensional vertical force derivative due to velocity increment along Oz, $\mathring{Z}_w/\frac{1}{2}\rho V_e S$
$Z_{\dot{w}}$ Non-dimensional vertical force derivative due to acceleration along Oz, $\mathring{Z}_{\dot{w}}/\frac{1}{2}\rho V_e S$
Z_η Non-dimensional vertical force derivative due to elevator angle, $\mathring{Z}_\eta/\frac{1}{2}\rho V_e^2 S$
z_q $-Z_q/\mu_1$
z_u $-Z_u$
z_w $-Z_w$
$z_{\dot{w}}$ $-Z_{\dot{w}}/\mu_1$
z_η $-Z_\eta$
α Incidence, angle between chord line and free stream wind direction
α Incidence of aircraft-less-tail no-lift-line (5)
α_0 Incidence at zero lift
α_e Angle between Ox and flight path in datum conditions ($\alpha_e = 0$ for wind axes)
α^T Tailplane incidence
β Sideslip angle
β' $\sqrt{1-\mathbf{M}^2}$
Γ Dihedral angle
γ Angle of descent
δ Relative pressure, p/p_0 (1)
δ Unspecified control angle
ε Downwash angle
ζ Rudder angle
ζ Relative damping, (actual damping)/(critical damping)
ζ' Change of rudder angle from trimmed position

η	Elevator angle
η	Propeller efficiency (2, 3)
η'	Change of elevator angle from trimmed position
η'	Elevator angle for zero hinge moment (5)
η_T	Tailplane setting angle
η_{trim}	Elevator angle for trim
Θ	Climb angle
Θ_e	Angle between Ox and horizontal in datum conditions
θ	Angle of pitch
ϑ	Temperature
ϑ_0	Temperature at standard sea level conditions
Λ	Sweepback angle
λ	Eigenvalue, root of characteristic equation
λ	Temperature lapse rate in atmosphere (1, 3)
μ	Friction coefficient (3)
μ	Aircraft relative density parameter, $m/\rho S l_T$ (5)
μ_1	Longitudinal relative density parameter, $m/\frac{1}{2}\rho S\bar{\bar{c}}$
μ_2	Lateral relative density parameter, $m/\frac{1}{2}\rho Sb$
ν	Kinematic viscosity
ξ	Aileron angle
ξ'	Change of aileron angle from trimmed position
ρ	Air density
ρ_0	Sea level air density
σ	Relative density, ρ/ρ_0
τ	Magnitude of time unit, $m/\frac{1}{2}\rho V_e S$
ϕ	Angle of bank
φ	Phase angle
χ	$\dot{y}/V_e$
ψ	Angle of yaw
Ω	Wave number, spectral frequency, 2π/(wavelength)
ω	Rate of turn (3, 6)
ω_θ	Approximate circular frequency of SPPO mode
$\hat{\omega}_\theta$	$\omega_\theta\tau$
ω_ψ	Circular frequency of directional oscillation
$\hat{\omega}_\psi$	$\omega_\psi\tau$

Abbreviations

AFCS	Automatic Flight Control System
amc	Aerodynamic mean chord
cg	Centre of gravity
cm	Centre of mass
eas	Equivalent airspeed
ESDU	Engineering Sciences Data Unit
PSD	Power Spectral Density
sfc	Specific fuel consumption
SPPO	Short Period Pitching Oscillation
tas	True airspeed

Note to undergraduate students

This book was written with you in mind and my hope (probably forlorn) is that it is possible to read it from start to finish as one would a novel, except that it might take much longer. However, a textbook is a tool to be used in a variety of ways; use the index to find an alternative treatment of a concept you have been given in a lecture, follow through a worked example or just 'browse'. It is my sad duty to inform you that the **secret of passing examinations** in this subject (and others) is to solve as big a variety of problems as possible. You will find some at the end of most chapters; they are arranged roughly in order of increasing difficulty. Answers, partial or complete, are given for those with (A) at the end, but before looking to see if your answer is correct please ask yourself if it is a reasonable result. You should not ignore the problems for which no answer is given; answers are not given on examination papers and still less are they given out there in the real world. Lastly, decimal numbers in parentheses, such as (3.56), are equation numbers with the first number giving the chapter.

1
Introduction

1.1 The travelling species

From the time of our emergence as a separate species *Homo sapiens* has been a traveller, firstly on foot, then using animals and finally developing vehicles. Originally the journey was a daily search simply for food; later it was for new pastures for his animals or better land to grow crops on. This has led to the spread of the species to almost every part of the globe and to the present situation where journeys are made for every imaginable purpose. In spite of the development of telecommunications it appears that every year more people travel greater and greater distances, mostly by air. The vehicles have developed from sledges and carts to aircraft and spacecraft. Two broad characteristics of the vehicles have concerned us from the beginning: how far and how fast they can go and their control and stability. The load a horse can be expected to pull in a cart and how far in a day was of interest; there must be a means to stop, start and steer, and even a cart can overturn if overloaded and a corner is taken too fast.

This book then is concerned with one of mankind's most productive forms of transport and its performance, stability and control characteristics. The later sections of this chapter are intended to be an introduction to the characteristics of aircraft that determine the performance, to engine performance and to the relevant properties of the atmosphere. Chapters 2 and 3 deal with aircraft performance, defined not only as how far and how fast it can fly, but also such things as the ability to climb, turn, take off and land. The performance of the aircraft is, of course, the reason for its existence and the most important starting point for design. The rest of the book is concerned with the stability and control of aircraft to which Chapter 4 forms an introduction.

The safety of the occupants and of the aircraft is one basic driving force in what we choose to study. The design and operation of aircraft is highly circumscribed by government safety regulations and we shall make occasional references to various airworthiness requirements but in no way will they be covered in detail.

1.2 General assumptions

The feature which characterizes all the topics dealt with in this book is that we are dealing with the interaction between the dynamics of the aircraft and the aerodynamic forces and moments generated on its surfaces by the motion. Other factors such as gravity have also to be included. However, the real situation is far more complex than we can reasonably hope to analyze completely. The atmosphere is a variable mixture of gases and vapours; it is never completely at rest and its properties such as density, pressure and temperature vary with position and time. The acceleration due to gravity varies slightly with latitude and height. The aircraft is an elastic body, distorting with every load on it, and losing mass as it burns fuel and uses other consumables. We therefore have to make a number of general assumptions.

- The aircraft is flying in a stationary atmosphere having constant properties.
- The aircraft does not deflect due to the loads placed on it.

- The aircraft is of constant mass.
- The acceleration due to gravity is constant.
- Accelerations of the aircraft due to motion about a curved rotating Earth are negligible.

These assumptions will apply throughout unless it is specifically stated otherwise. Probably the least justifiable assumption is the second which can have serious consequences if its effects are totally ignored.

For the purposes of determining aerodynamic forces and moments it does not matter if we consider the aircraft or some component of it to be flying at velocity $\mathbf{V}$ (a vector) through stationary air, or the aircraft or component to be stationary in a uniform, unbounded airstream of steady velocity $-\mathbf{V}$ at a large distance ahead. We shall use whichever point of view is the more convenient at the time.

Throughout this book we will keep strictly to the use of consistent units (e.g. SI) for simplicity. The practising aeronautical engineer, however, uses the most convenient units (such as dN, hours and knots) correcting the equations with suitable numerical constants.

1.3 Basic properties of major aircraft components

Before we begin to discuss the performance, stability and control of aircraft we need to have some general information on the components of the aircraft, a basic idea of their function and how their characteristics depend on such parameters as Mach number and geometry.

1.3.1 Functions of major aircraft components and some definitions

We must first be clear on the main functions of each component of the aircraft. These are summarized in figure 1.1 which shows an aircraft in level flight.

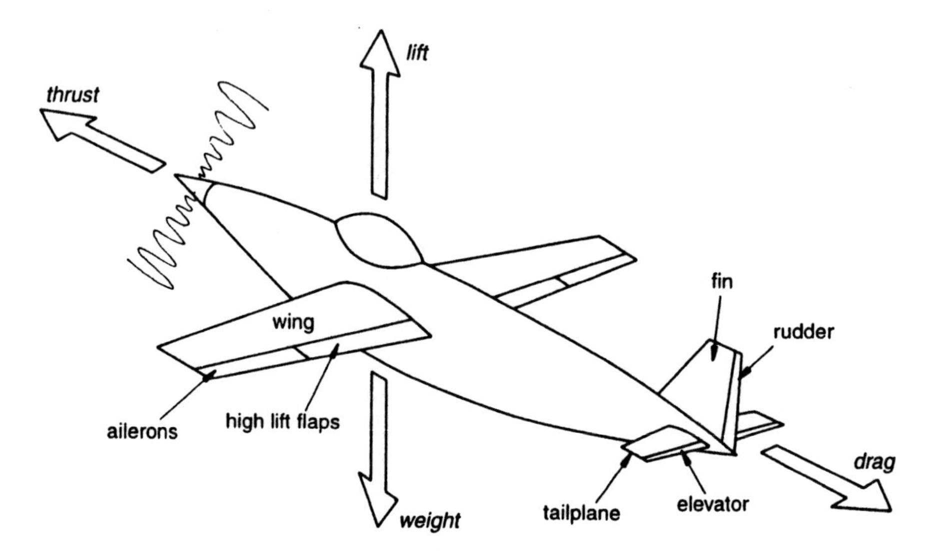

Fig. 1.1 Main components of an aircraft and primary forces

The primary function of the wings is to provide lift, which is defined as the aerodynamic force at right angles to both the direction of motion and the wing surface and therefore in this case vertically upwards. The lift at a constant speed and height can be varied in at least two ways. The usual method is to vary the attitude of the aircraft and therefore that of the wings to the direction of motion. A common secondary method, normally used only at low speeds, is to deploy what are known as high lift devices. The wings also carry the ailerons which can provide a moment about the direction of flight to provide control in roll. The function of the fuselage is to carry and protect the crew, much of the equipment and the payload; in the case of an airliner the latter are the passengers and freight. It also transmits loads from the tailplane and fin at its rear. The fin provides directional stability by generating sideways lift if it becomes inclined to the local airstream. The rudder is used to provide a moment about a vertical axis through the cg for control purposes; it is considered to be part of the fin. Similarly the tailplane provides stability about a spanwise axis and carries the elevator which provides control about that axis. The engine or engines, if present, are to provide a forward force to overcome the drag of the remainder of the aircraft and to enable the aircraft to climb and accelerate.

Collectively the wing, tailplane and fin are the 'lifting surfaces' of the aircraft and to discuss their characteristics we must first set out some definitions. A typical wing section is shown in figure 1.2.

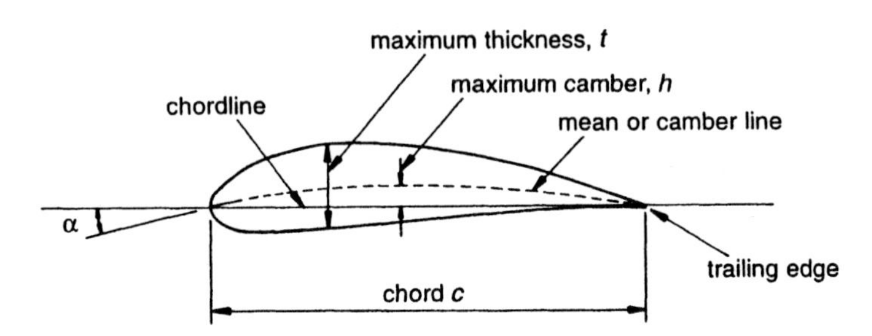

Fig. 1.2 Definition of terms used in describing a wing section

The chord line is a straight line drawn through the centres of curvature of the leading and trailing edges, and the chord, c, is the length of the chord line between the leading and trailing edges. For the present we will define the wing section incidence angle, α,[1] as the angle between the chord line and the direction of the oncoming airstream. Two ratios are frequently used to characterize the section. These are the 'thickness chord ratio' defined as $\tau = t/c$ and the 'camber ratio' defined as $\gamma = h/c$, where t and h are indicated in figure 1.2.

We turn now to definitions relating to wing planforms; these are often trapezoidal or nearly so, as shown in figure 1.3.

It is usual to define the planform by continuing the leading and trailing edges through the fuselage to the centreline. The 'gross wing area', S, is then the plan area of the wing including the part within the fuselage, which for a trapezoidal wing is

$$S = \text{mean chord}(\bar{c}) \times span = \frac{c_0 + c_t}{2} b \tag{1.1}$$

where b is the span. The gross wing area is used chiefly to define coefficients for the whole aircraft. The aspect ratio, A, is defined as (span)/(mean chord) and can be variously expressed as

$$A = b/\bar{c} = b^2/b\bar{c} = b^2/S = S/\bar{c}^2 \tag{1.2}$$

The wing taper is described by the taper ratio λ, defined as $\lambda = c_t/c_0$, where c_0 and c_t are the centreline and tip chords respectively. Sweepback is measured by the angle Λ between a line at a constant fraction, k, of the chord and a line perpendicular to the centreline in plan view as shown in figure 1.3. The fraction of the chord used is indicated by using k as a subscript, thus the sweepback angle of the quarter chord line is written $\Lambda_{1/4}$. These definitions can be applied equally to tailplanes and to fins but in the latter case the chord at the base of the fin is usually taken as c_0.

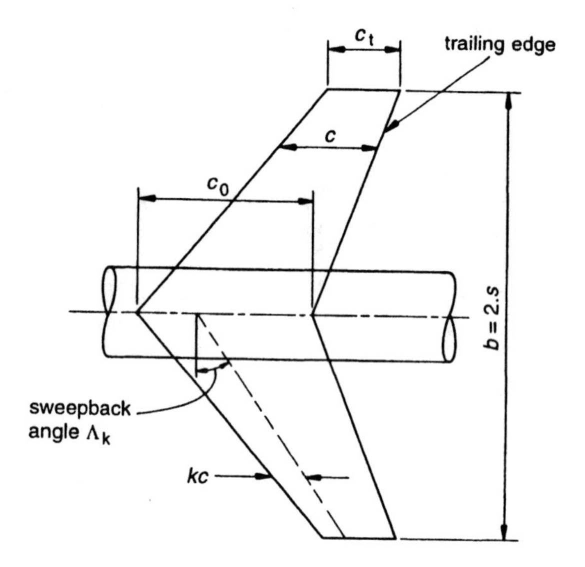

Fig. 1.3 Definition of terms used in describing a wing planform

1.3.2 Lift characteristics of wing sections and wings

In this section we will describe the lifting characteristics of wings. Although modern computer based methods can calculate the complete flow around a wing directly and hence find all the forces on it, the simplest way to approach the lift characteristics of a wing is to first consider those of the wing section. The characteristics of a wing section are those found, for instance, using an accurate theory applied to a wing of infinite span having the same section. Alternatively the characteristics are those found in an experiment in which the wing spans the width of the windtunnel and accurate corrections have been made for the presence of the tunnel walls and their boundary layers. A real wing may have a spanwise variation in its

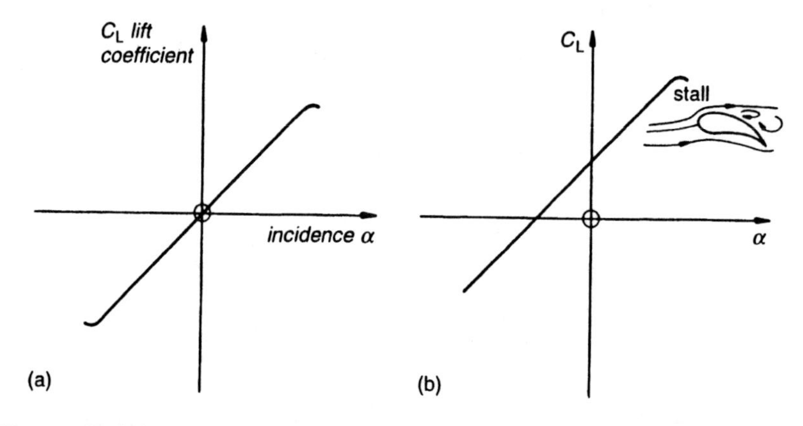

Fig. 1.4 Variation of C_L with incidence: (a) symmetrical section, (b) cambered section

section or be twisted; we can, however, ignore these complications for the present. The lift of a wing is best expressed in terms of the lift coefficient C_L, defined as

$$C_L = \frac{L}{\frac{1}{2}\rho V^2 S} \tag{1.3}$$

where L is the lift and ρ the air density. A typical curve showing the dependence of the lift of a symmetrical section with incidence is shown in figure 1.4(a).

It can be seen that the variation is a linear one for moderate angles of incidence, but when the incidence is about 15° the flow separates from the upper surface and the magnitude of the lift decreases again. This flow separation is referred to as 'stalling'. A symmetrical section is satisfactory for tailplanes or fins which are required to produce lift in both directions. Wings, however, are mostly required to produce positive lift and so we camber the section which has the effect of raising the curve as shown in figure 1.4(b). This also raises the positive stalling value of the lift coefficient, $C_{L\max}$, and causes the lift to become zero at a negative angle of incidence, known as the no-lift angle α_0. The main characteristics of the lift curve of interest are then the lift curve slope, $a_\infty = dC_L/d\alpha$, α_0 and $C_{L\max}$.

The lift curve slope depends on the thickness chord ratio, the Reynolds and Mach numbers and the angle between the upper and lower surfaces at the trailing edge. At low Mach numbers ($\mathbf{M} < 0.4$) and typical Reynolds numbers for an aircraft, the lift curve slope is given approximately by the semi-empirical expression

$$a_\infty = 5.65(1 + 0.8\tau) \tag{1.4}$$

where the incidence is measured in radians. Most subsonic aircraft have thickness chord ratios of about 0.14 so that a_∞ is approximately 6.3. The effect of Mach number is to produce a distinct peak in the lift curve slope near to a Mach number of one as shown in figure 1.5.

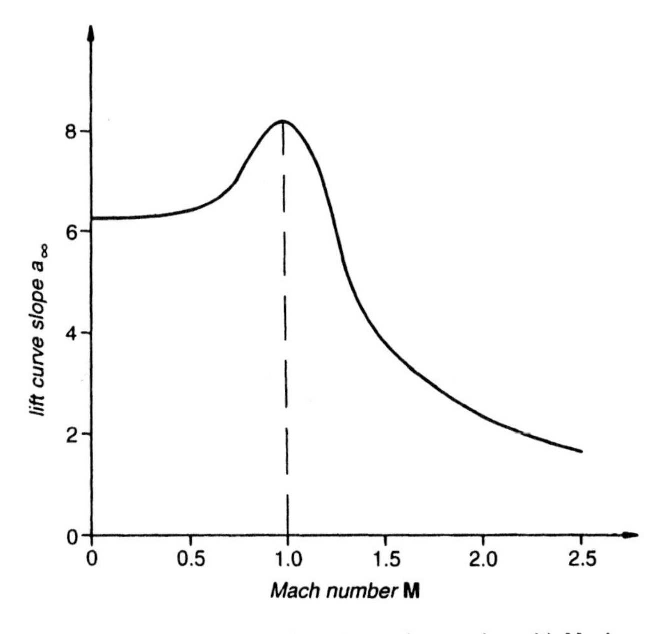

Fig. 1.5 Variation of lift curve slope for a wing section with Mach number

The effects of Reynolds number and trailing edge angle are much smaller; the lift curve slope increasing with increase of both factors.

More accurate data can, of course, be obtained from careful experiment or from computational fluid-dynamics. When these are not available the 'Data Items' of the Engineering Sciences Data Unit (ESDU) will often provide the information. These are based on theory, where applicable, correlated with experimental data. In the rest of this book these will be referenced by quoting the number of the relevant Data Item in brackets without further explanation.[2] Another similar source of information is reference (1.1). In the case of sectional lift curve slope much better data than (1.4) is available (Aero W.01.01.05).

Turning to the lift curve slope of a finite wing, a, this depends on the aspect ratio, sweepback angle, taper ratio, Reynolds and Mach numbers and for subsonic wings the sectional lift curve slope. For unswept wings at low Mach numbers we can use the approximate relation

$$a = \frac{\pi a_\infty A}{a_\infty + \pi\sqrt{A^2 + 4}} \tag{1.5}$$

which is shown plotted in figure 1.6.

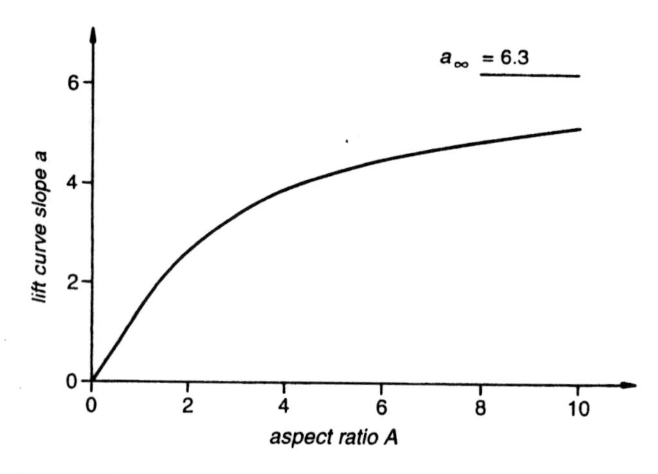

Fig. 1.6 Variation of lift curve slope for a wing section with aspect ratio

The dependence on Reynolds number in this case is expressed through the value of a_∞. The effect of Mach number is to produce a variation similar to figure 1.5 but with lower values. Sweepback reduces the lift curve slope at all Mach numbers; a crude theory suggests that it is proportional to the cosine of the sweep angle. The effect of taper ratio is slight. Better values are available for subsonic speeds (70011) and for supersonic speeds (70012).

The no lift angle of wing sections at subsonic speeds is given roughly by

$$\alpha_0 = -2\gamma \tag{1.6}$$

where α_0 is measured in radians and is the same for plane wings. For more accurate results the shape of the whole camber line must also be taken into account (72024); twisting a wing also changes the no lift angle (87031). At supersonic speeds the no lift angle is close to zero.

1.3.3 Maximum lift and the characteristics of flaps

The maximum lift coefficient in two-dimensional flow depends on the aerofoil section geometry, the surface condition (rough or smooth) and the Reynolds and Mach numbers. There are basically two types of stall, those in which the separation starts predominantly from just behind the leading edge and those which start from the trailing edge. The boundary between the two depends on the leading edge radius, and the maximum lift for the first type varies with the same parameter. This parameter is often not easily available and it is usually substituted for by a parameter such as the section thickness or upper surface ordinate just behind the leading edge. The maximum lift of aerofoils which have rear separation depends on the geometry of the rear part of the section; the camber is also a significant parameter for both types as we saw in the previous section. Thin ($\tau < 0.08$), smooth, symmetrical aerofoils, which inevitably have small leading edge radii, have C_{Lmax} values of 0.9 or less. The highest values are achieved by aerofoils with fairly large thicknesses and leading edge radii and so have rear separations. Values of the order of 1.6 for conventional sections and up to about 2.0 for modern aerofoils which have been specifically designed for high lift can be achieved. Roughness of the surface causes a considerable reduction in these values. Increase of Reynolds number increases maximum lift rapidly up to Reynolds numbers of the order of 10^7 for smooth aerofoils (84026). Compressibility effects reduce the maximum lift, starting in some cases at Mach numbers as low as 0.2; however, maximum lift recovers to some extent at around $\mathbf{M} = 1$.

Finite wings generally have rather lower values of maximum lift as once the wing has stalled at one spanwise station the separated flow region spreads rapidly. Sweepback causes wings to stall earlier due to unfavourable effects on the boundary layer. Low aspect ratio delta wings and wings of similar planforms with thin sections are a special case and have higher than expected values of C_{Lmax} due to the appearance of leading edge vortices. Stalling angles are also much increased, values of 30–35° being normal.

Buffeting is defined as the more or less regular oscillation of a part of an aircraft caused by the wake from some other part; often it is oscillation of the tailplane due to flow separation from the wing, aggravated by compressibility effects. A common cause is separation of flow in the wing–body junction. The effect of buffeting is to limit the usable C_L to rather less than the true C_{Lmax} and a 'buffet boundary' appears on a plot of C_L against $\mathbf{M}$. A typical buffet boundary is shown in figure 1.7.

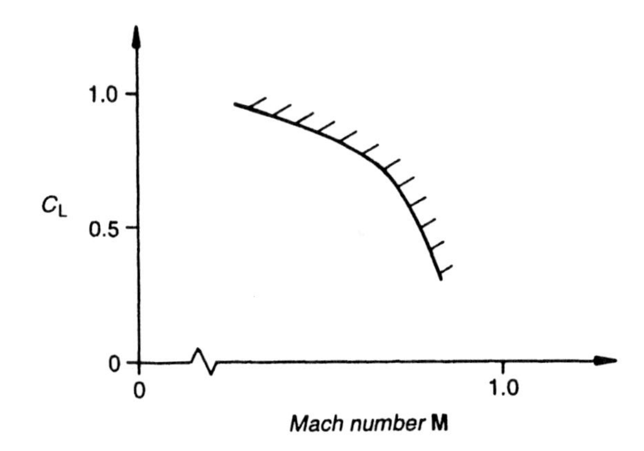

Fig. 1.7 Typical buffet boundary

Flaps were originally front or rear parts of wing sections which were hinged and could move up or down; in doing so the effective camber of the section was changed, thus changing the lift. Leading edge flaps and control flaps such as the aileron, elevator and rudder are still basically of this form: see figure 1.8(a).

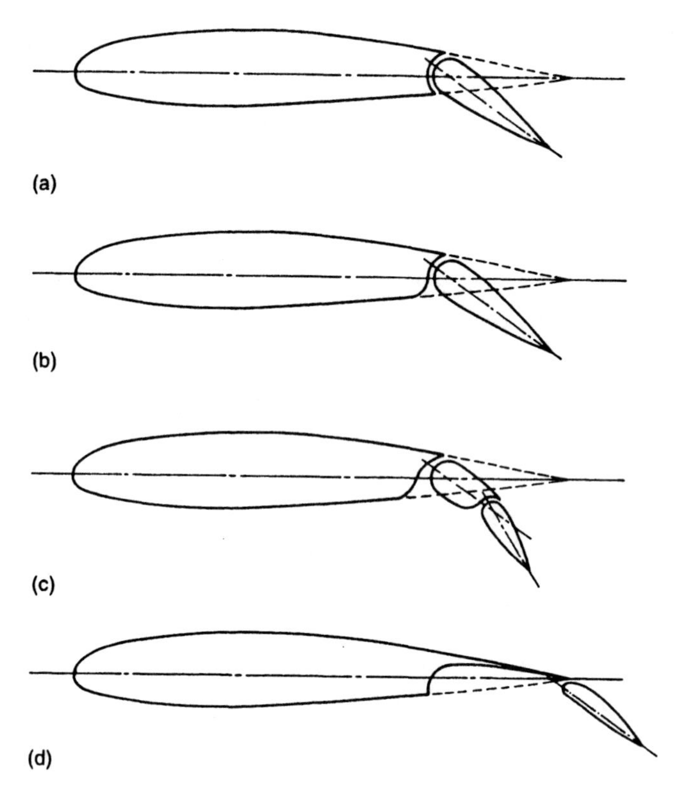

Fig. 1.8 Various types of high lift flap: (a) simple flap, (b) slotted flap, (c) double slotted flap, (d) Fowler flap

However, trailing edge high lift flaps have undergone considerable development and have additional methods of increasing the lift. Many configurations of high lift flaps have been tested, but modern versions can be seen as developments of simple flaps along two lines. The first is the single slotted flap, which has a carefully shaped gap between the flap and the main surface, and is shown in figure 1.8(b). Air is allowed to flow from the high pressure area on the underside to the upper which re-energizes the boundary layer on the flap, increasing the maximum deflection possible and hence the lift. Sectional lift coefficient increments of the order of 1.0 are possible with flap chords of 25 per cent of the wing chord and flap deflections of 50°. A development of this form is the double slotted flap as shown in figure 1.8(c); lift coefficient increments of the order of 1.5 can be obtained in this case. The wing area is increased to a small degree with these types of flap as the elements are moved rearwards as well as rotated for optimum results. This second method of increasing the lift is carried to much greater lengths with the Fowler flap illustrated in figure 1.8(d). Similar lift increments as for the double slotted flap are obtainable; more accurate values for all these types are available (Aero F.01.01.08 and .09). Flaps normally only extend over part of the span, reducing the increment in the practical case (74012). Deployment of high lift flaps also increases the maximum lift coefficient available but this increment is generally less than that at an incidence well below the stall (85033). Both types of flap can be seen in figure 1.9. The high lift flaps

Fig. 1.9 Airbus showing both types of flap. (Courtesy British Aerospace)

occupy the inboard portion of the wings; the ailerons can be seen further outboard on the wings and the rudder on the fin.

Leading edge devices can also be used to increase the maximum lift. In this case there is a small loss of lift at zero incidence but the slope is slightly increased and the curve extended so that C_{Lmax} is considerably increased, as shown in figure 1.10(a).

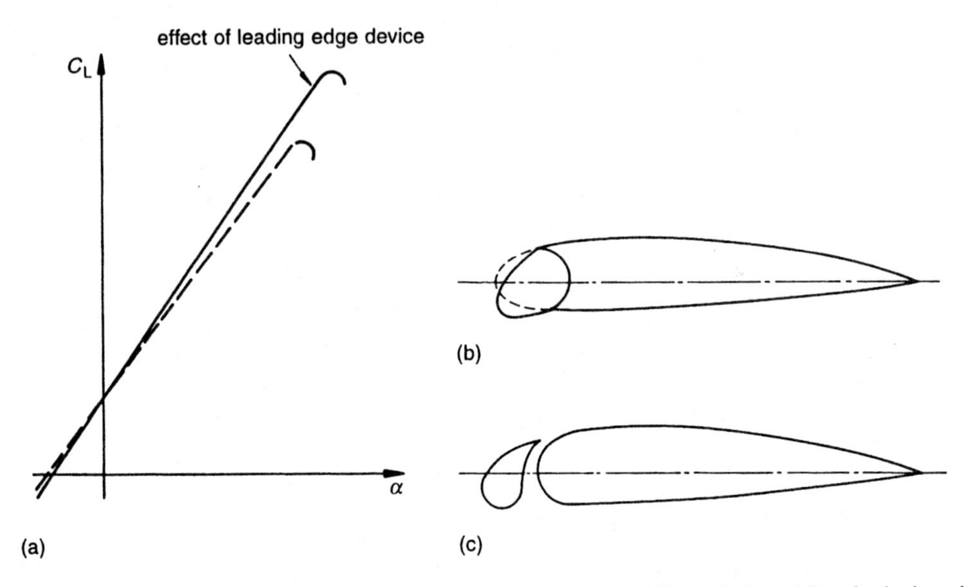

Fig. 1.10 Leading edge flaps: (a) typical lift curve, (b) simple hinged flap, (c) slotted flap, in deployed position

By contrast with trailing edge flaps the extra lift is obtained at increased incidence, which is a slight disadvantage, and these flaps are usually only employed in conjunction with trailing edge flaps. The simplest device is the leading edge flap shown in figure 1.10(b) and is just the forward part of the section hinged to move downwards. On a thin wing section a C_{Lmax} increment of about 0.5 can be achieved using a 15 per cent chord flap; somewhat less will be obtained in the case of a finite wing (85033, 92031). The slot principle is used in the leading edge slat shown in figure 1.10(c), and gives a slightly greater lift increment.

Other devices such as vortex generators, suction and blowing can also be used to augment the lift.

1.3.4 Estimation of drag

The normal method of estimating the drag of an aircraft is to estimate the drags of the components separately and then add them together. This usually underestimates the drag, and since the boundary layers on the components will interact at their junctions, causing further drag, it is usual to factor the sum by a constant for 'interference drag'. In fact gaps, leaks, excrescences and roughness also contribute and their effects are often included in the factor. We can then write the drag of the aircraft as

$$D = k\left(D_{\text{wing}} + D_{\text{fuse}} + D_{\text{nacelles}} + \text{etc.}\right)$$

where k is the 'interference factor'. The wing drag can be expressed in the form $\frac{1}{2}\rho V^2 S_N$, where S_N is the nett or exposed area of the wing, since that part of the gross wing area falling inside the fuselage can have no drag. The drags of the other components are expressed similarly and, dividing through by $\frac{1}{2}\rho V^2 S$, we find the drag coefficient for the whole aircraft as

$$C_D = k\left(C_{D\text{wing}}\frac{S_N}{S} + C_{D\text{fuse}}\frac{S_{\text{fuse}}}{S} + C_{D\text{nacelles}}\frac{S_{\text{nacelles}}}{S} + \text{etc.}\right) \tag{1.7}$$

where $C_{D\text{fuse}}$ is the drag coefficient of the fuselage based on the area S_{fuse}, and so on. Fuselage drag is usually based on the wetted area, and that of bluff components such as the undercarriage on the frontal area.

There are many mechanisms for the production of drag and we can draw a tree diagram to illustrate the breakdown of the drag according to the various mechanisms: see figure 1.11.

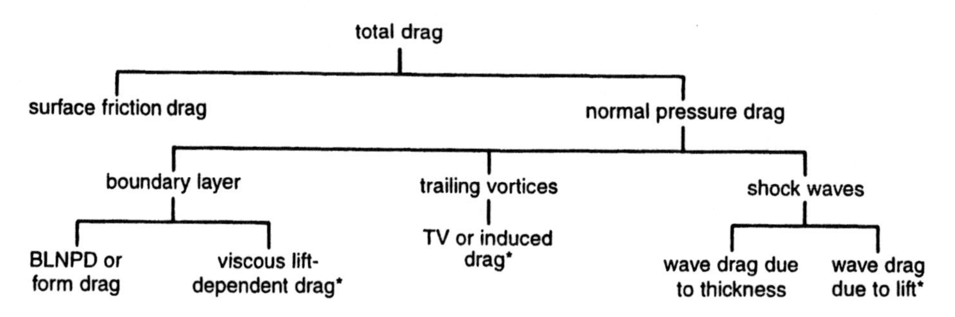

Fig. 1.11 Breakdown of drag by mechanisms. Asterisks identify lift-dependent drags

Drag appears on the components of the aircraft in two fundamentally different ways; either as shear stresses on the surface or as the streamwise components of normal pressures. Although the viscosity of air is very small, the velocity gradients at the surface of a body are large and the resulting 'surface friction drag' is significant. All other forms of drag appear as a result of normal pressure effects, the 'normal pressure drags'.

The normal pressure drag of a two-dimensional aerofoil in inviscid, incompressible flow can be shown to be zero. Three main effects in real fluids give rise to drags of this type. In a viscous fluid the presence of a boundary layer modifies the flow pattern and hence the pressure distribution on the wing and this produces a nett drag. This is correctly known as 'boundary layer normal pressure drag' (BLNPD), but is more conveniently referred to as 'form drag'; for low speed aircraft the sum of the surface friction and form drags is referred to as the 'profile drag'. Boundary layer effects also reduce the lift curve slope of the wing and hence its surface is more inclined to the direction of motion for a given lift, giving rise to 'viscous lift-dependent drag'. Secondly, the wing has a finite span and this gives rise to wing tip vortices if the wing is producing lift; the energy required to generate these continuously appears as a normal pressure drag. This is known as 'trailing vortex drag', 'TV drag' or 'induced drag'. If the flight speed is high enough for shock waves to appear then the energy loss they represent appears as a normal pressure drag. At zero lift this is known as 'wave drag due to thickness'; incidence effects modify the shock waves giving an additional drag, which is known as 'wave drag due to lift'.

Several forms of drag have been omitted from the figure. First, there is 'spillage drag' which arises from the interaction of the airflow into the engine with that around other components of the aircraft; it is usually only important at supersonic speeds. Second, there is 'trim drag' which arises from the need to balance aerodynamic moments about the centre of gravity in steady flight. Another effect not explicitly appearing in the figure is that due to the propeller slipstream; the increased airspeed in the slipstream raises the drag of parts of the aircraft within it. Each component of the aircraft may have a contribution to its total drag from each of the mechanisms that have been described and it is convenient to collect all these drags into two terms, the 'no-lift drag' or 'datum drag' and the 'lift-dependent drag'. The latter consists of the items marked with an asterisk in figure 1.11. A typical value for the interference factor is 1.15, although if we were confident that some types of drag were accurately estimated, for instance the lift-dependent drag, a value nearer unity could be employed for them.

The estimation of the no-lift drag of the wing, tailplane and fin follows similar lines. Conventionally the surface friction drag of a wing is taken to be the same as for a thin, flat plate under the same conditions. This is a function of the Reynolds number (based on the chord) and the boundary layer transition position. At typical aircraft cruising speeds the natural position for transition on a smooth wing is at about 10 per cent of the chord, and figure 1.12 shows the variation of the surface friction coefficient with Reynolds number for this case.

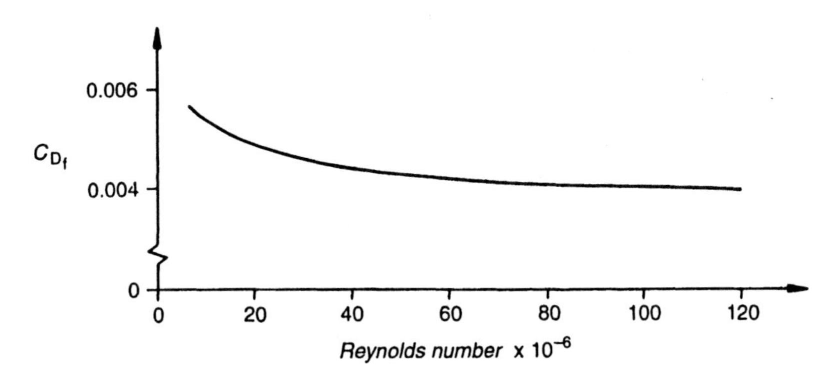

Fig. 1.12 Variation of friction drag coefficient on a flat plate with Reynolds number

More accurate data are available (68019, 68020 and Aero W.02.04.01). The profile drag is then estimated by multiplying the surface fraction by a factor λ given approximately by

$$\lambda = 1 + 3.5\tau \tag{1.8}$$

assuming transition is at about 10 per cent. The profile drag can be estimated more accurately by allowing for its dependence on transition position and aerofoil section shape (Aero W.02.04.02 and .03). In practice the transition position varies with incidence, giving a small variation of profile drag; at incidences approaching the stall, short regions of separated flow may appear, further increasing the drag. When the wing stalls there is a large increase in profile drag. Wing drag is a very significant part of the total aircraft no-lift drag and special low drag aerofoil sections have been developed to reduce it. These have a pressure distribution on the upper surface which is favourable for the maintenance of a laminar boundary layer on a larger part of the upper surface over a range of lift coefficients. The resulting profile drag coefficient variation with lift coefficient has a 'bucket' in it as shown in figure 1.13.

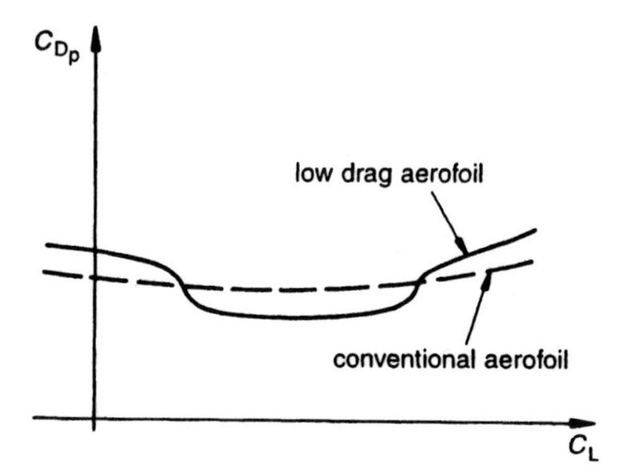

Fig. 1.13 Variation of profile drag for conventional and low drag wing sections with lift coefficient

Trailing vortex drag depends on the aspect ratio, A, the sweepback angle and the taper ratio. Theoretically, about the minimum is achieved by an untwisted wing with an elliptic planform, and has a value given by

$$C_{\mathrm{Dtv}} = \frac{C_L^2}{\pi A} \tag{1.9}$$

The trailing vortex drag for other planforms is obtained by multiplying this value by a factor (74035); for unswept wings it is approximately 1.1. The viscous lift-dependent drag is normally about 10 per cent of the trailing vortex drag but can be estimated more accurately (66032).

1.3.4.1 Effect of compressibility on drag

We need to discuss some aspects of the effect of compressibility on the drag of aircraft. As the Mach number is increased from a low value the speed of the air over the wing increases steadily until at some point the speed reaches the local speed of sound; this Mach number is known as the 'critical Mach number'. At a slightly higher Mach number a shock wave appears on the upper surface and the drag rises rapidly. The Mach number at which the no-lift drag coefficient has risen by a small fixed percentage above the low Mach number value is known as the 'drag critical Mach number', $\mathbf{M}_{\mathrm{Dcrit}}$. The variation of no-lift drag coefficient with Mach number is shown in figure 1.14; as can be seen $\mathbf{M}_{\mathrm{Dcrit}}$ varies with lift coefficient.

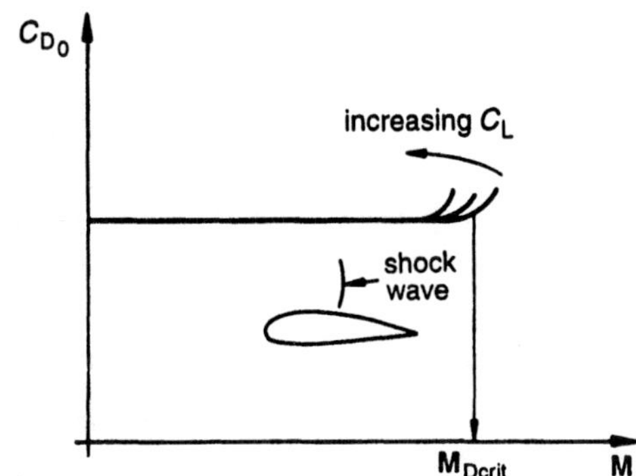

Fig. 1.14 Definition of M_{Dcrit}

M_{Dcrit} depends on wing aspect ratio and sweep and most critically on the wing section (67011); in recent years a great deal of research effort has been expended on developing wing sections with improved performance in this respect.

The no-lift drag coefficient reaches a maximum close to a Mach number of unity and falls off thereafter; a typical variation is sketched in figure 1.15. Also shown (to a different scale) is a typical variation of (lift dependent drag)/C_L^2.

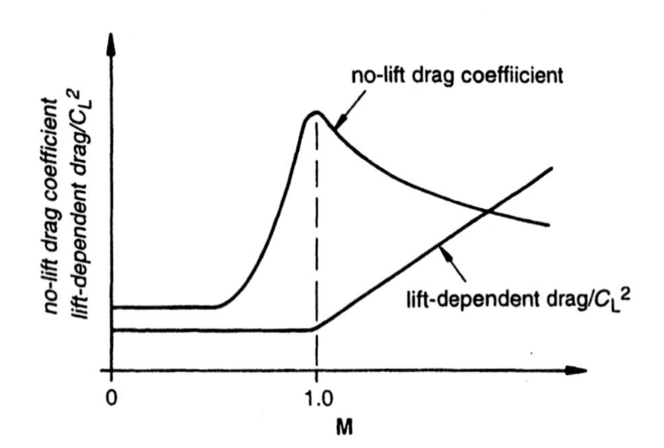

Fig. 1.15 Variation of aircraft no-lift drag coefficient and lift-dependent drag coefficient with Mach number

1.3.4.2 Drag polars

A drag polar is the variation of C_D as a function of C_L, rather than with incidence; the former is more useful as C_L is related to many other variables. We have already noted that the lift-dependent drags are directly proportional to C_L^2; we can often ignore the slight variation of the other forms of drag with incidence, or assume the variation is also proportional to C_L^2, and write as an approximation

$$C_D = a + bC_L^2 \tag{1.10}$$

where a and b are constant at Mach numbers below M_{Dcrit}.[3] This form also suits best-fit lines to experimental determinations of the drag polar. Some authors use the symbols C_{D_0} or C_{D_z} and k for these quantities.

1.4 Engine characteristics

There are a very large number of types of engine which have been, or could be, used to power an aircraft. A short list would consist of petroleum fuelled piston engines, turboprops, pure turbojets, bypass turbojets or turbofans, ramjets and rockets. The first two types need a propeller to convert shaft power into thrust; the turboprop has most of its output in the form of shaft power but a small jet thrust is also produced from the turbine exhaust.

The type of engine chosen to power an aircraft depends on which type gives the most efficient and convenient operation for its function; a balance needs to be struck between the weight of the engine and that of the fuel. A relatively heavy engine with low fuel consumption may be optimum for a long range aircraft, while a lighter engine with heavier fuel consumption might be chosen for a short range aircraft. Propellers are not suitable for use at and above transonic speeds in general. The result is that generally aircraft of a given type will have the same type of engine. Thus light aircraft have piston engines and propellers, small airliners and business aircraft have turboprops, medium and large civil transport aircraft have high bypass-ratio turbofan engines, and combat aircraft have low bypass turbofans. Concorde has pure turbojet engines but future supersonic airliners are likely to have hybrid engines which function as low bypass turbofans subsonically but change their internal configuration to a pure turbojet for supersonic operation.

The most interesting characteristics of an engine for our purposes are the shaft power, if it is a piston or turboprop engine, or the thrust for the other types, and the fuel consumption. For a turbojet engine the specific fuel consumption (sfc) is defined as the rate of consuming fuel mass divided by the thrust. The units will be of the form grams/second/newton ($g\ s^{-1}\ N^{-1}$) or pounds/hour/pound ($lb\ h^{-1}\ lb^{-1}$). Figure 1.16 shows the variation of sfc with thrust for a range of Mach numbers and heights for a typical turbofan engine in the 70 000 lb class with a bypass ratio of 5.

The 70 000 lb is approximately the maximum sea level thrust and it can be seen that the thrust falls off rapidly with height. The sfc increases steadily with Mach number and is weakly dependent on thrust and height.

For the sake of comparison with the turbojet engine the specific consumption of a turboprop engine and propeller combination may be defined as above on the basis of the sum of the propeller thrust and residual jet thrust. Figure 1.17 shows the variation of shaft power, residual thrust, propeller thrust and sfc for a typical turboprop engine in the 1800 shaft hp class equipped with a propeller of an efficiency of about 79 per cent.

Turboprop engines are usually designed to run at a constant rotational speed, the power absorbed by the propeller being varied by varying the pitch of the blades. We see that the shaft power falls with altitude and increases with Mach number, while the specific fuel consumption (as defined in this case) is constant with height but also increases with Mach number. Comparing the two types of engine the turbojet has a sfc in the range 0.49–0.66 whilst that of the turboprop combination has a range of 0.36–0.56.

More usually the specific fuel consumption is defined for the engine alone as the rate of consuming fuel mass divided by the power produced; because some useful jet thrust is produced, the power it represents has to be included by defining an equivalent power (ehp). The form of the units in this case is pounds/hour/ehp ($lb\ h^{-1}\ ehp^{-1}$) or milligrams/second/watt ($mg\ s^{-1}\ W^{-1}$) which equates to milligrams/joule ($mg\ J^{-1}$).

1.5 Standard atmospheres

To deal with the problem of the variability of the atmosphere a number of standardized atmospheres have been agreed internationally, the chief of which is known as the International Standard Atmosphere (ISA). This has the standard temperature of 15°C (288.15 K) as the sea level value and so corresponds to the average atmosphere in temperate zones. Temperature falls linearly with height up to a height of 11 km; this region is known as the 'troposphere'. Above this height, known as the 'tropopause', the temperature is constant up to a height of 20 km and this region is known as the 'stratosphere': see figure 1.18.

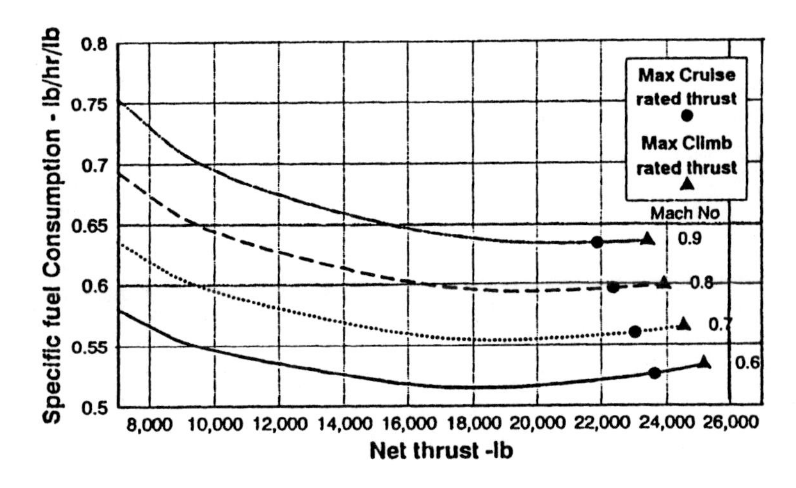

(a) Altitude 20 000 ft

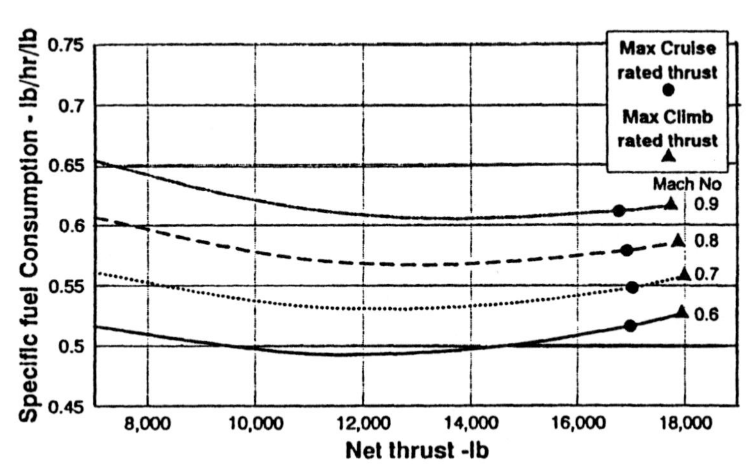

(b) Altitude 30 000 ft

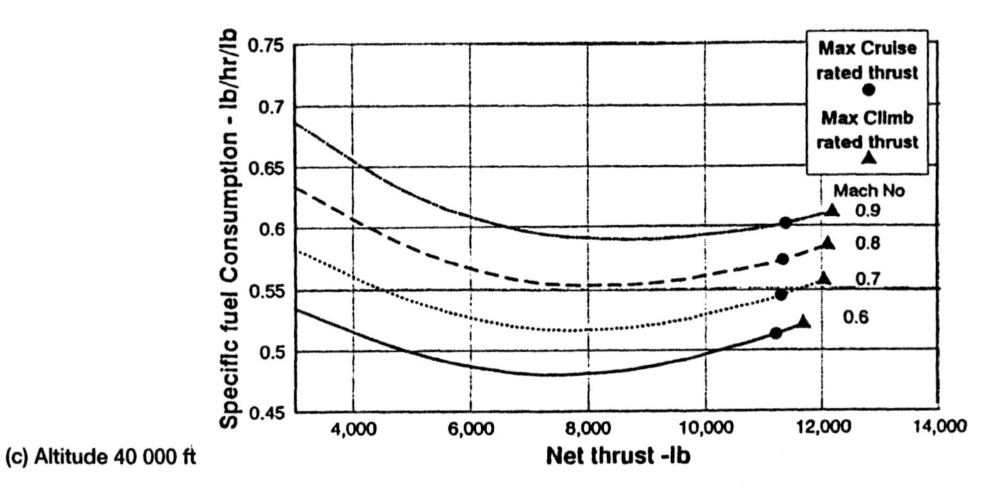

(c) Altitude 40 000 ft

Fig. 1.16 Characteristics of a typical 70 000 lb thrust turbofan engine (ISA) fitted with a propeller. Courtesy of Rolls-Royce Plc

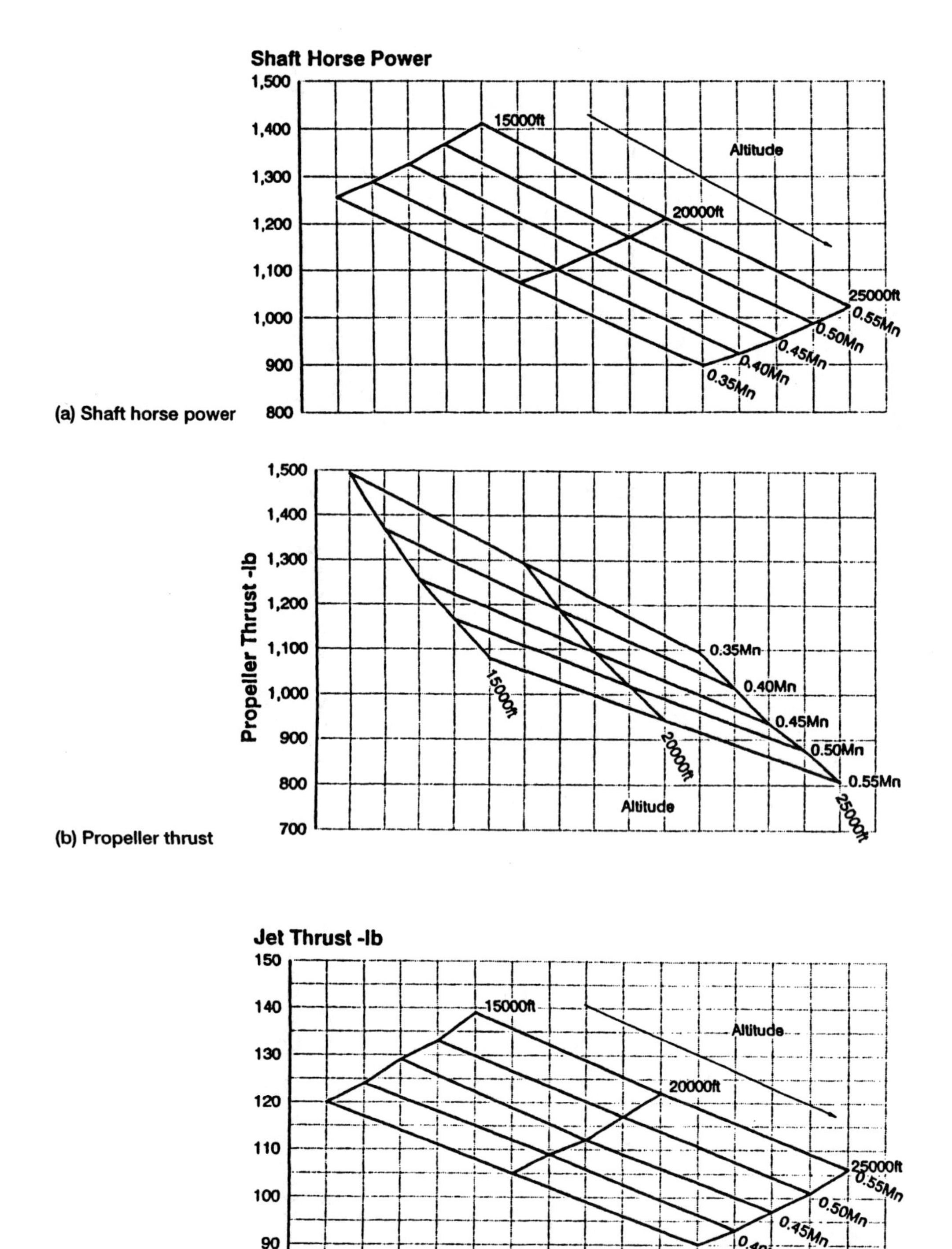

Fig. 1.17 Characteristics of a typical turboprop engine fitted with a propeller. Courtesy of Rolls-Royce Plc

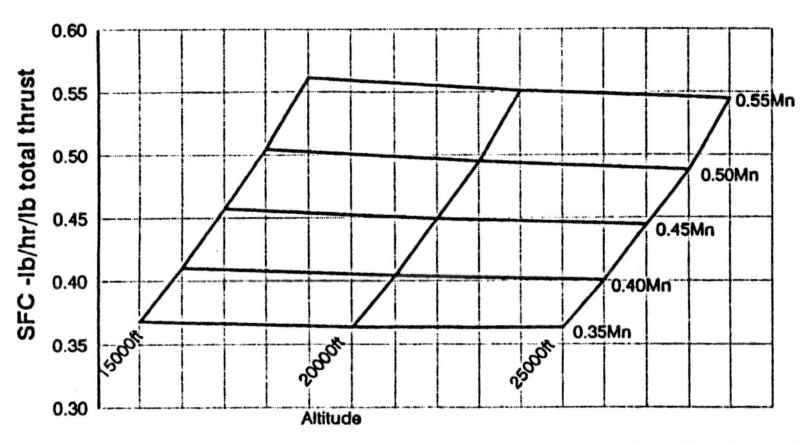

(d) SFC

Fig. 1.17 Characteristics of a typical turboprop engine fitted with a propeller. Courtesy of Rolls-Royce Plc

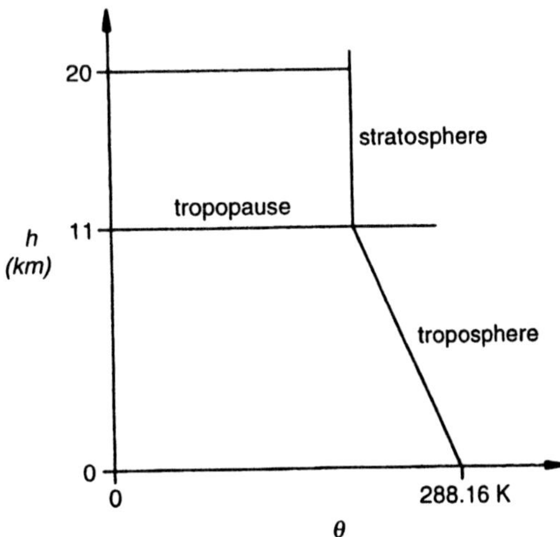

Fig. 1.18 Variation of temperature in the ISA

The rate of decrease of temperature with height, known as the 'lapse rate' λ, has a value of 6.5 K km^{-1}, giving a temperature of 216.15 K in the stratosphere. These temperatures are mean values based on observation; theory is then used to deduce the variation of pressure and density and other properties with height.

The first piece of theory required is the 'buoyancy relation'; consider a cylindrical element of the atmosphere with end area A, as shown in figure 1.19. The pressure at any point in the atmosphere is due to the weight of air above it, and therefore decreases with height. However, as we measure height upwards, we will take the pressure as p on the lower end of the cylinder and $p + \mathrm{d}p$ on the upper. Hence there is a nett upward force due to the pressure of $pA - (p + \mathrm{d}p)A = -A\mathrm{d}p$. This is balanced by the weight of the element ρg.(volume) $= \rho g A\mathrm{d}h$, where ρ is the (mass) density of the air. Equating these we find

$$\frac{\mathrm{d}p}{\mathrm{d}h} = -\rho g \tag{1.11}$$

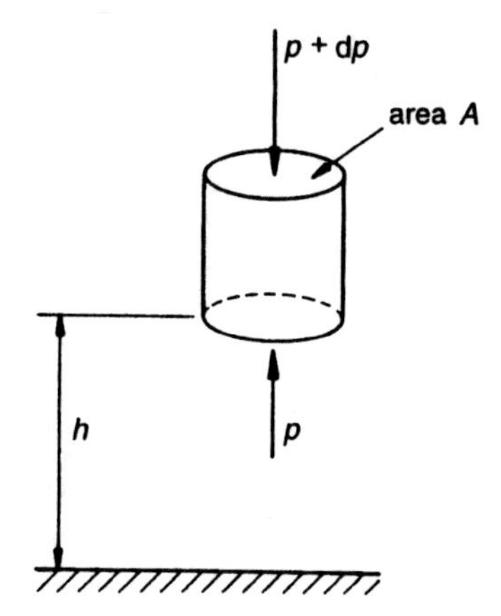

Fig. 1.19 Forces on an isolated cylinder in the atmosphere

1.5.1 Pressure and density in the troposphere

If we write the temperature at height h as ϑ and its value at sea level as ϑ_0 we have the temperature variation with height in the form

$$\vartheta = \vartheta_0 - \lambda h \qquad (1.12)$$

so that if we find pressure in terms of temperature we can find it in terms of height. We also have

$$\frac{\mathrm{d}\vartheta}{\mathrm{d}h} = -\lambda$$

and combining this with (1.11) gives

$$\frac{\mathrm{d}p}{\mathrm{d}\vartheta} = \frac{\rho g}{\lambda}$$

We now substitute for ρ from the equation of state in the form $p = \rho R\vartheta$ to give

$$\frac{\mathrm{d}p}{\mathrm{d}\vartheta} = \frac{pg}{\lambda R\vartheta}$$

or rearranging

$$\frac{\mathrm{d}p}{p} = \frac{g}{\lambda R}\frac{\mathrm{d}\vartheta}{\vartheta}$$

Finally, integrating gives

$$\ln(p) = \frac{g}{\lambda R}\ln(k\vartheta) \qquad (1.13)$$

where k is a constant of integration. To eliminate this we assume that the pressure at sea level has the value p_0, then from (1.13)

$$\ln(p_0) = \frac{g}{\lambda R}\ln(k\vartheta_0)$$

We subtract this expression from (1.13) and use the fact that $\ln(p) - \ln(p_0) = \ln(p/p_0)$. Taking exponentials we find

$$\delta = \frac{p}{p_0} = \left(\frac{\vartheta}{\vartheta_0}\right)^{g/\lambda R} \tag{1.14}$$

where δ is known as the 'relative pressure'. Using (1.12) and the numerical data for the ISA we find

$$\delta = \left(1 - \frac{\lambda h}{\vartheta_0}\right)^{g/\lambda R} = (1 - 0.002\,256h)^{5.256} \tag{1.15}$$

where the value for the R has been taken as 287.053 N m $(\text{kg K})^{-1}$ and h is given in km. Similarly the 'relative density' is defined as $\sigma = \rho/\rho_0$, where ρ_0 is the sea level density, and from the equation of state we have

$$\frac{\rho}{\rho_0} = \frac{p\vartheta_0}{p_0\vartheta} = \delta\frac{\vartheta_0}{\vartheta}$$

hence

$$\sigma = \left(1 - \frac{\lambda h}{\vartheta_0}\right)^{(g/\lambda R)-1} = (1 - 0.002\,256h)^{4.256} \tag{1.16}$$

If we eliminate the bracket $\left(1 - \frac{\lambda h}{\vartheta_0}\right)$ between (1.15) and (1.16) we find

$$\delta = \sigma^n, \text{ where } n = \frac{g/\lambda R}{(g/\lambda R) - 1} = 1.235 \tag{1.17}$$

n is referred to as the 'polytropic index' for the atmosphere.

1.5.2 Pressure and density in the stratosphere

In this case the temperature is constant at ϑ_s say; then again using the equation of state to eliminate ρ from (1.11) we have

$$\frac{dp}{dh} = -\frac{g}{R\vartheta_s}p$$

Rearranging gives

$$\frac{dp}{p} = -\frac{g}{R\vartheta_s}dh$$

and integrating

$$\ln(p) = -\frac{g}{R\vartheta_s}(h + k) \tag{1.18}$$

where k is another constant of integration. The initial conditions at the height of the tropopause, h_t, are $p = p_t$ and $\rho = \rho_t$, so that (1.18) gives

$$\ln(p_t) = -\frac{g}{R\vartheta_s}(h_t + k)$$

Subtracting this from (1.18) and taking exponentials gives

$$\frac{p}{p_t} = \exp\left(-\frac{g}{R\vartheta_s}(h - k_t)\right)$$

Now $\delta = \frac{p}{p_0} = \frac{p}{p_t}\cdot\frac{p_t}{p_0} = \delta_t\frac{p}{p_t}$, and the relative pressure in the stratosphere is

$$\delta = \delta_t \exp\left(-\frac{g}{R\vartheta_s}(h - h_t)\right) = 0.223\,36 \exp\left(-0.015\,805(h - h_t)\right) \qquad (1.19)$$

using numerical data for ISA. For the relative density, noting that in an isothermal atmosphere $\rho \propto p$, we have

$$\sigma = \frac{\rho}{\rho_0} = \frac{\rho}{\rho_t}\cdot\frac{\rho_t}{\rho_0} = \sigma_t\frac{p}{p_t}$$

then

$$\sigma = \sigma_t \exp\left(-\frac{g}{R\vartheta_s}(h - h_t)\right) = 0.297\,08 \exp\left(-0.015\,805(h - h_t)\right) \qquad (1.20)$$

Tables of the variation of the characteristics of the atmosphere with height are available in many aerodynamics textbooks (68046, 77021, 78008). For other parts of the world there are other standard atmospheres, ranging from Tropical Maximum to Arctic Minimum, each with suitable temperature profiles with height (78008). There are also off-standard versions of the ISA in which the temperature in the troposphere is raised or lowered by a constant amount. Such atmospheres are denoted by putting the increment (in Celsius) after the letters ISA, for example ISA+10 (68046).

Student problems

1.1 A wing of zero sweep has an area of 108 m^2, a span of 28 m and a thickness/chord ratio of 0.135. Using (1.4) and (1.5) find the lift curve slope. (A)

1.2 An aircraft, equipped with an engine whose characteristics are given in figure 1.17, is flying at a height of 20 000 ft at a speed of 142 m s^{-1}. Find the fuel flow rate. The speed of sound may be taken as $20.08\sqrt{\vartheta}$ and 1 km = 3281 ft. (A)

1.3 The Arctic Minimum Atmosphere is defined by the following table of temperatures and heights:

Height (m)	0	1524	3048	10 668	20 000
Temperature (K)	223.15	238.15	238.15	203.15	203.15

The pressure at sea-level is taken as the standard ISA value. Find the ratio of the air density at each of these heights to the standard density in the ISA at sea-level. (A)

1.4 Show that the quantity $(1/\sigma^2)\mathrm{d}\sigma/\mathrm{d}h$ (used in Section 3.3.2) in the troposphere is given by

$$-\frac{\lambda}{\delta\vartheta_0}\left(\frac{g}{\lambda R}-1\right)$$

1.5 Use a spreadsheet to produce a table of relative temperature, density and pressure as a function of height in the ISA. Use steps of height 0.5 km up to 11 km and steps of height 1 km from there on. Check the results against a published table. Extend the table to cover other quantities such as relative speed of sound and relative kinematic viscosity using Sutherland's formula for the dynamic viscosity:

$$\mu \,\alpha\, \vartheta^{1.5}/(\vartheta + C)$$

where for air, $C = 114$.

Notes

1. The term 'angle of attack' is also used. The term 'angle of incidence' originally meant the angle that the wing is set relative to the fuselage, but this meaning has completely fallen into disuse in Great Britain.
2. The references are to the main Data Items on any topic and are not intended to refer to all the information available.
3. The symbol b is also used for the wing span; the meaning intended should, however, be clear from the context.

Background reading

Barnard, R. H. and Philpott, D. R. 1989: *Aircraft flight.* Harlow: Longman Scientific and Education.
Jane's all the world's aircraft, published yearly by Jane's Information Group Ltd, London.
Houghton, E. L. and Carpenter, P. W. 1993: *Aerodynamics for engineering students,* 4th edition. London: Edward Arnold.
Katz, J. and Plotkin, A. 1991: *Low speed aerodynamics.* New York: McGraw-Hill.

2
Performance in level flight

2.1 Introduction

In the air an aircraft spends the vast majority of the time in level flight, which makes this a logical topic with which to start our discussion on performance and enables the introduction of some unfamiliar ideas in a simple context. In this chapter we will discuss the maximum level speed, level acceleration, range and endurance. The latter is the time for which an aircraft can fly and has applications for such operations as surveillance. We also include cruise-climb techniques although obviously they are not strictly speaking level flight.

In this chapter and the next we will treat the aircraft as if it were simply a point with mass acted on by forces. We are therefore assuming that all moments are continuously in balance.

2.2 The balance of forces

In level accelerated flight the forces on an aircraft are the lift, L, normal to direction of flight; the drag, D, along the flight direction; the engine thrust, T, which we assume for simplicity is in the direction of flight, and the weight mg.[1] The situation is illustrated in figure 2.1.

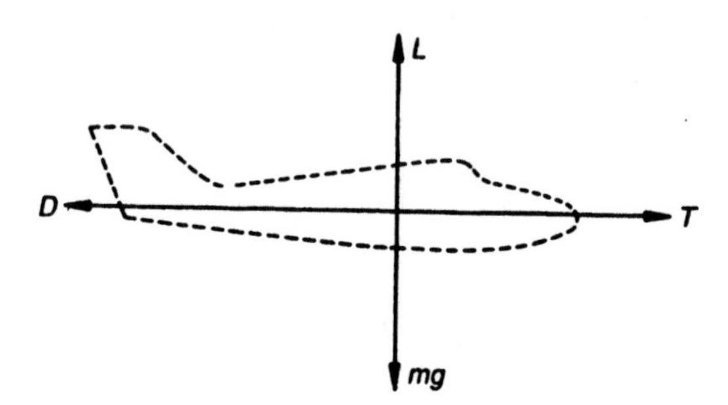

Fig. 2.1 Forces on an aircraft treated as a point

The balance of forces then leads to the equations

$$T - D = m\frac{dV}{dt} \tag{2.1}$$

or in steady flight

$$T - D = 0 \tag{2.2}$$

and

$$L - mg = 0 \tag{2.3}$$

Taking the last equation, substituting for the lift using $L = \frac{1}{2}\rho V^2 S C_L$ and solving for the speed V we find

$$V = \sqrt{\frac{2mg}{\rho S C_L}} \tag{2.4}$$

In level flight m, g, ρ and S are constant so that $V \propto C_L^{-1/2}$, so if we wish to discuss the effect of speed we can use C_L as an alternative variable. This has advantages as other aerodynamic quantities such as the incidence and drag coefficient are known in terms of C_L. If in (2.4) we use the maximum lift coefficient we obtain the 'stalling speed' in level flight as

$$V_{s_1} = \sqrt{\frac{2mg}{\rho S C_{L\max}}} \tag{2.5}$$

This is the lowest speed that an aircraft can fly at steadily, assuming that the pilot can maintain control.

Another variable often used in place of the speed is the 'equivalent airspeed', V_E, normally abbreviated to eas. This is defined such that

$$\tfrac{1}{2}\rho_0 V_E^2 = \tfrac{1}{2}\rho V^2 \tag{2.6}$$

where ρ_0 is the sea-level value of the air density. We can rewrite (2.6) in the form

$$V_E = V\sqrt{\sigma} \tag{2.7}$$

where $\sigma (= \rho/\rho_0)$ is the relative density. To emphasize the difference between these speeds we may refer to V as the 'true airspeed' or tas, but it is only true in the sense that it is the speed relative to the air; the speed relative to the ground is also a function of the wind speed. The main advantage of using eas is that if the combination ρV^2 is replaced by $\rho_0 V_E^2$ then one variable quantity (V_E) has replaced two (ρ and V), since the sea-level density is a constant. It also follows that a speed which is determined by a constant value of the lift coefficient is constant with altitude when expressed as an equivalent airspeed, but most usefully it means that the stalling speed is constant as an eas. A secondary advantage is that the eas is approximately the speed indicated to the pilot by his airspeed indicator. This operates on the static and pitot pressures fed to it from a static vent and a pitot tube and is calibrated on the assumption of sea-level density. Errors in sensing the correct pressures due to the presence of the aircraft, compressibility effects and errors within the instrument account for the difference between the airspeed indicated and the eas.

2.3 Minimum drag and power in level flight

In Chapter 1 we saw that the drag coefficient is a function of C_L, Reynolds and Mach numbers; however, for the present we will concentrate on the variation with C_L and write $C_D = f(C_L)$, or adopt the approximate form $C_D = a + bC_L^2$ from (1.10). In this last case using the definitions of C_L and C_D and (2.6) we find

$$\begin{aligned} D &= \tfrac{1}{2}\rho_0 V_E^2 S\left(a + bC_L^2\right) \\ &= \tfrac{1}{2}\rho_0 V_E^2 S a + \frac{bL^2}{\tfrac{1}{2}\rho_0 V_E^2 S} \end{aligned}$$

and using (2.3)

$$D = \tfrac{1}{2}\rho_0 V_E^2 Sa + \frac{b(mg)^2}{\tfrac{1}{2}\rho_0 V_E^2 S} \tag{2.8}$$

The first term in this equation is the no-lift drag and can be seen to be proportional to the square of the eas, whilst the second term is the lift-dependent drag[2] and is inversely proportional to (eas)^2. The variation of these drags with eas is sketched in figure 2.2(a), as is the total drag.

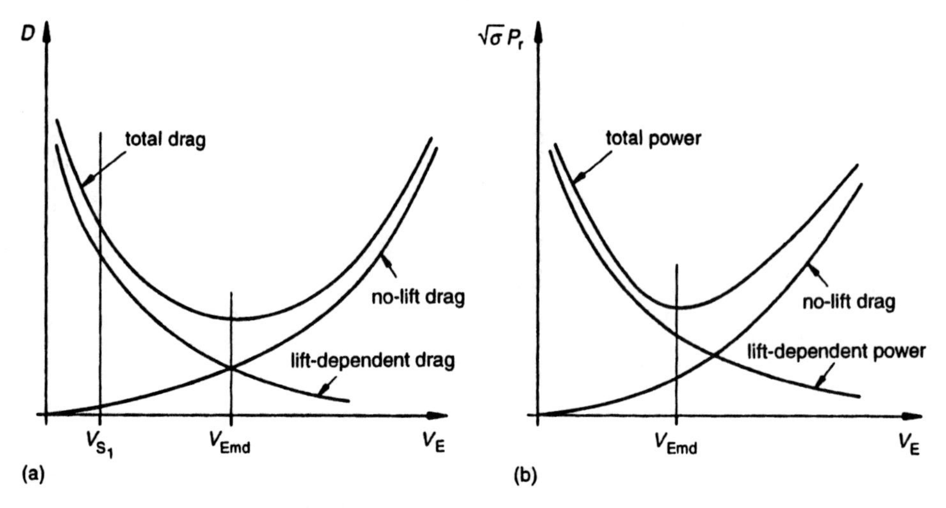

Fig. 2.2 (a) Variation of no-lift, lift dependent and total drags with equivalent airspeed. (b) Variation of no-lift, lift dependent and total powers with equivalent airspeed

It is immediately clear that the drag has a minimum at a certain value of speed. This is also true for the more general case of $C_D = f(C_L)$, as the effects of speed dominate over variations in no-lift drag coefficient. Provided that a and b in the drag equation are constant the curve is independent of height.

The power, P_r, absorbed in overcoming the drag, is obtained from (2.8) by multiplying through by V and using (2.7) to obtain

$$\sqrt{\sigma}P_r = \tfrac{1}{2}\rho_0 V_E^3 Sa + \frac{b(mg)^2}{\tfrac{1}{2}\rho_0 V_E S} \tag{2.9}$$

The variation of the no-lift, lift-dependent and total powers with eas is shown in figure 2.2(b), and again there is a minimum. Aircraft are somewhat unusual amongst vehicles in general having these minima; more usually the resistance to motion rises continuously with speed. We shall see that this behaviour leads to equivalent airspeeds and corresponding lift coefficients which are optimum for some criterion.

To find these minima we choose to use C_L in place of speed as the independent variable and start by considering the drag. We are interested only in conditions in which the aircraft weight is exactly supported by the lift and so must use $L = mg$ from (2.10) and write

$$D = \frac{D}{L}\cdot mg = \frac{C_D}{C_L}\cdot mg \tag{2.10}$$

on cancelling the factors common to C_L and C_D. The condition therefore for minimum drag is then

$$\boxed{(C_D / C_L) \text{ is a minimum}} \tag{2.11}$$

If we multiply (2.10) by the speed V and use (2.4), the power absorbed becomes

$$\begin{aligned} P_r = DV &= \frac{C_D}{C_L} \cdot mg\sqrt{\frac{2mg}{\rho S C_L}} \\ &= \sqrt{\frac{2}{\rho S}}(mg)^{3/2}\left(C_D / C_L^{3/2}\right) \end{aligned} \tag{2.12}$$

The condition therefore for minimum power absorbed is then

$$\boxed{\left(C_D / C_L^{3/2}\right) \text{ is a minimum}} \tag{2.13}$$

Determining the speed or C_L values for the minima now depends on the form in which the aerodynamic data are given. If we assume the form $C_D = a + bC_L^2$ is valid then we can write the condition for minimum drag from (2.11)

$$\frac{d}{dC_L}(C_D / C_L) = 0$$

as

$$\frac{d}{dC_L}\left(\frac{a + bC_L^2}{C_L}\right) = \frac{d}{dC_L}\left(\frac{a}{C_L} + bC_L\right) = -aC_L^{-2} + b = 0$$

Then solving for C_L we find the lift coefficient for minimum drag as

$$C_{\mathrm{Lmd}} = \sqrt{\frac{a}{b}} \tag{2.14}$$

We can now find the equivalent airspeed for minimum drag using (2.4) and (2.6) as

$$V_{\mathrm{Emd}} = \sqrt{\frac{2mg}{\rho_0 S}}\left(\frac{b}{a}\right)^{\frac{1}{4}} \tag{2.15}$$

Alternatively this result could have been obtained directly from (2.8). We now find the no-lift and lift-dependent drags, respectively, to be

$$\begin{aligned} \tfrac{1}{2}\rho_0 V_E^2 Sa \text{ and } \tfrac{1}{2}\rho_0 V_E^2 S\left(bC_L^2\right) \\ = \tfrac{1}{2}\rho_0 V_E^2 Sa, \text{ on using (2.14)} \end{aligned}$$

Hence we see that the no-lift and lift-dependent drags are equal in the minimum drag condition.

In a similar manner we apply the condition for minimum power absorbed, which from (2.13) is

$$\frac{\mathrm{d}}{\mathrm{d}C_L}\left(C_D / C_L^{3/2}\right) = 0$$

giving

$$\frac{\mathrm{d}}{\mathrm{d}C_L}\left(\frac{a + bC_L^2}{C_L^{3/2}}\right) = \frac{\mathrm{d}}{\mathrm{d}C_L}\left(\frac{a}{C_L^{3/2}} + bC_L^{1/2}\right) = -\frac{3a}{2}C_L^{-5/2} + \frac{b}{2}C_L^{-1/2} = 0$$

Then solving for C_L we find the lift coefficient for minimum power absorbed as

$$C_{Lmp} = \sqrt{\frac{3a}{b}} \tag{2.16}$$

We can now find the equivalent airspeed for minimum power absorbed as

$$V_{Emp} = \sqrt{\frac{2mg}{\rho_0 S}}\left(\frac{b}{3a}\right)^{\frac{1}{4}} \tag{2.17}$$

Alternatively this result could have been obtained directly from (2.9). We note from (2.14)–(2.17) that $C_{Lmp} = \sqrt{3}\ C_{Lmd}$ and $V_{Emp} = V_{Emd}/\sqrt[4]{3}$. We now find the no-lift and lift-dependent absorbed powers, respectively, to be

$$\tfrac{1}{2}\rho_0 V_E^3 Sa \text{ and } \tfrac{1}{2}\rho_0 V_E^3 S\left(bC_L^2\right)$$
$$= \tfrac{1}{2}\rho_0 V_E^3 S.3a, \text{ using (2.16)}$$

Hence we see that the no-lift and lift-dependent powers are equal to one-quarter and three-quarters, respectively, of the total power in the minimum power condition.

If the form $C_D = a + bC_L^2$ is not valid and the data are given in tabular form, then graphical or equivalent computer methods must be used.

2.4 Shaft and equivalent powers for turboprop engines

In Section 1.4 we discussed the characteristics of turboprop engines in terms of the shaft power produced and that of jet engines in terms of the thrust and noted that in the case of turboprop engines there is also a small thrust. Considering turboprop engines, if the power transmitted by the output shaft of the engine to the propeller is P we have, using the definition of propeller efficiency,

$$\eta = (\text{useful power out})/(\text{power input}) = TV/P \tag{2.18}$$

Transposing to find the thrust T and adding the direct thrust from the engine, T_j, we find

$$T = \eta P/V + T_j \tag{2.19}$$

which we can take as a general form for the thrust for any type of engine. In the case of the turboprop engine which has only a small thrust from the turbine exhaust we define an equivalent shaft power P_e such that

$$T = \eta P_e/V \tag{2.20}$$

Equating (2.19) and (2.20) we find that the equivalent shaft power is

$$P_e = P + T_j V/\eta \tag{2.21}$$

Using P_e we can treat aircraft with turboprop engines on the same basis as those with piston engines. Pure turbojet and turbofan engines have similar characteristics so that we can divide aircraft into two broad groups, propeller-driven and jet-driven.

2.5 Maximum speed and level acceleration

To discuss these we superpose the aircraft drag curve onto the engine thrust curve at full throttle at some height, as shown in figure 2.3.

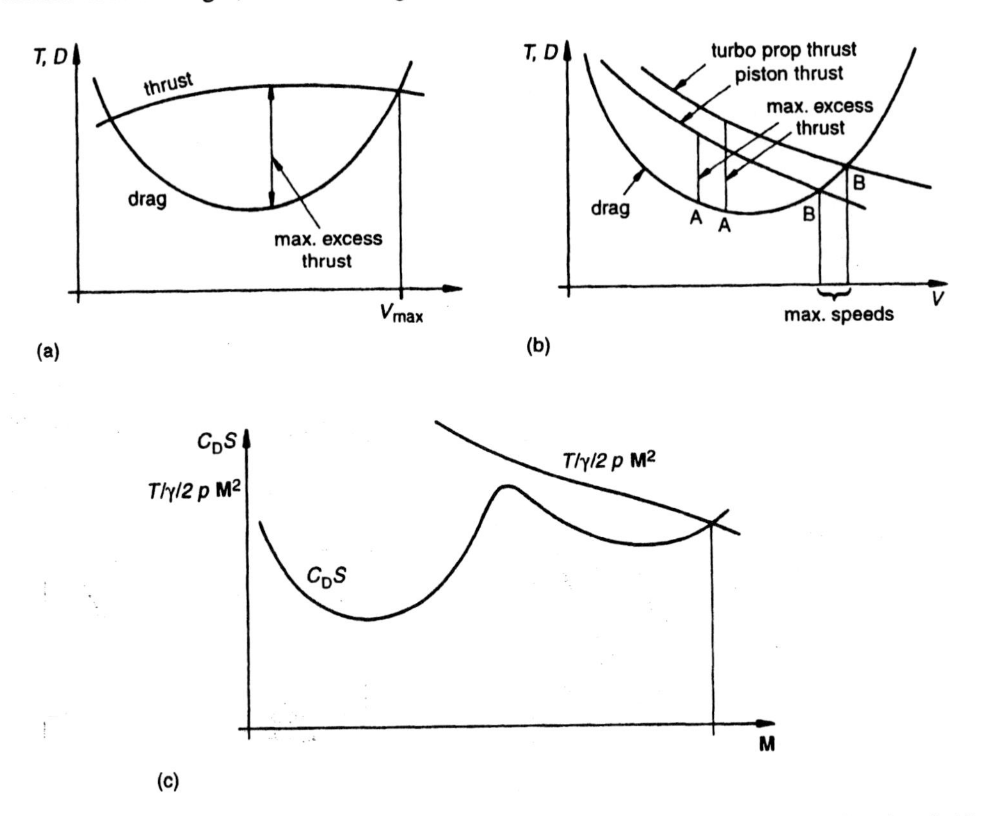

Fig. 2.3 Determination of maximum speed using excess thrust: (a) jet aircraft, (b) propeller aircraft, (c) high speed aircraft

Figure 2.3(a) shows the curves for a jet-driven aircraft and figure 2.3(b) those for propeller-driven. From (2.1) the acceleration is proportional to $(T - D)$, i.e. the 'excess thrust', and is a maximum where the latter is a maximum, that is at points A. The maximum speed occurs where thrust and drag are equal at the points B. For high-speed aircraft it is more convenient to plot the characteristics against Mach number as in figure 2.3(c). In this connection we note that

$$V = c_s\mathbf{M}$$

where $c_s = \sqrt{\gamma p/\rho}$, in which c_s is the speed of sound, γ is the ratio of specific heats, p is the atmospheric pressure and $\mathbf{M}$ is the Mach number. The dynamic pressure can now be written

$$\tfrac{1}{2}\rho V^2 = \tfrac{1}{2}\rho\left(\mathbf{M}\sqrt{\gamma p/\rho}\right)^2 = \frac{\gamma}{2}p\mathbf{M}^2 \tag{2.22}$$

so that all quantities involved can be expressed in terms of Mach number.

As an alternative, to find the maximum speed we superpose the power required curve onto the engine plus propeller characteristics, as shown in figure 2.4(a). The intersection point of the curves again gives the maximum speed. Figure 2.4(b) shows the method for jet-driven aircraft. In both figures 2.3 and 2.4 there are in fact two intersections of the curves, one corresponding to the maximum speed and one to the minimum speed; the latter may be below the stalling speed, and is in any case of little interest.

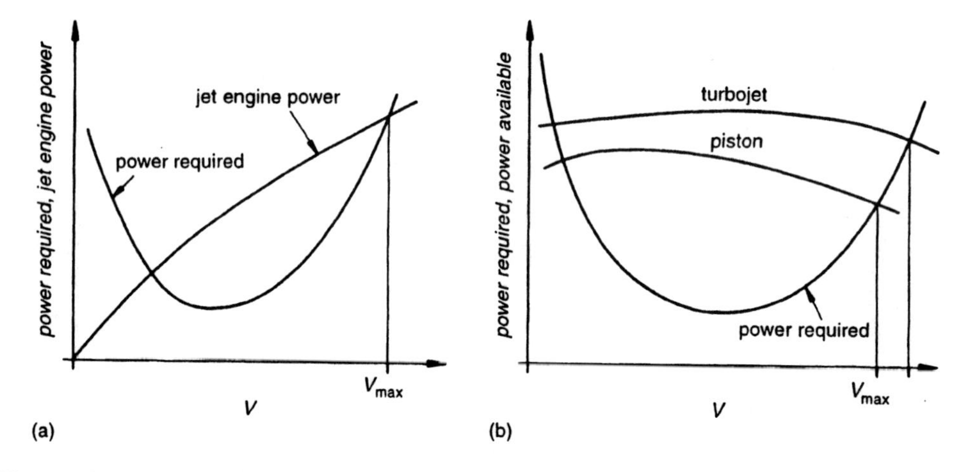

Fig. 2.4 Determination of maximum speed using excess power: (a) jet aircraft, (b) propeller aircraft

These graphical methods, or computerized equivalents, find the exact maximum speeds but give little information on the factors on which they depend. We can sacrifice accuracy to obtain this information by assuming that the thrust for a jet-driven aircraft is constant with speed or that the power available (the product of equivalent shaft power and propeller efficiency) for a propeller-driven aircraft is constant.

Worked example 2.1

An aircraft of mass 30 000 kg and a wing area of 95 m² has turboprop engines giving a maximum power of 3500 kW which can be assumed not to vary with speed. Find the maximum speed at sea-level if the drag coefficient is given by $C_D = 0.035 + 0.042C_L^2$ and the propeller efficiency is 0.82.

Solution

Using (2.9) and equating it to the available power, ηP, we find

$$\eta P = \tfrac{1}{2}\rho_0 V_E^3 Sa + \frac{b(mg)^2}{\tfrac{1}{2}\rho_0 V_E S}$$

or, since at sea-level $\rho = \rho_0 = 1.225$ and $V_E = V$, we have

$$3500 \times 10^3 \times 0.82 = 0.5 \times 1.225 \times 95 \times 0.035V^3 + \frac{(30\,000 \times 9.81)^2 \times 0.042}{0.5 \times 1.225 \times 95 \times V}$$

or $2.87 \times 10^6 = 2.036 \times V^3 + 6.2517 \times 10^7 \times V^{-1}$. We anticipate that the no-lift drag will be much larger than the lift-dependent drag, so neglecting the second term on the right above we obtain a first approximation of

$$V = \sqrt[3]{\frac{2.87 \times 10^6}{2.036}} = 112.1 \text{ m s}^{-1}$$

We can now improve this result by subtracting the lift-dependent power given by the neglected term $6.2517 \times 10^7 \times V^{-1} = 6.2517 \times 10^7/112.1 = 5.577 \times 10^5$. Subtracting this from the nett engine power leaves 2.312×10^6 and hence an improved estimate is

$$V = \sqrt[3]{\frac{2.312 \times 10^6}{2.036}} = 104.38 \text{ m s}^{-1}$$

Repeating this process gives 103.7 m s^{-1}. This result is clearly within 1 per cent of the correct result and there is little point in further improvement.

This method would also work if the power is not taken as constant with speed as the power available could be adjusted at each iteration. Similarly the assumption that the thrust is constant can be made for jet-driven aircraft which leads to the solution of a quadratic equation in the square of the maximum speed.

2.6 Range and endurance

The distance that an aircraft can fly (or possibly the endurance) is one of the most important items in the specification of an aircraft, and needs to be defined accurately. For civil aircraft operations extra fuel must be carried to cope with various contingencies during a flight. It must fly the distance required allowing for the forecast headwind. On arrival near to the destination airport it may not be possible to land and the aircraft may need fuel to fly to a diversion airport. At either the destination airport or the diversion it may have to wait for an opportunity to land and have to 'join the stack'. The stack is a reserved area in which aircraft spiral downwards keeping a safe distance apart. Fuel is obviously required also for take-off, climb, landing and manouevring on the ground; there is also a small amount of fuel which cannot be extracted from the fuel tanks and piping, known as 'unusable fuel'. We will only discuss two definitions of range although others have been used.

- **Safe operating range.** This is the maximum distance between two airfields which the aircraft can fly with full allowance for headwind, diversion and stacking. The distance flown in the climb is included, as is the fuel for take-off, climb and landing. Values for the diversion distance and the stacking time are standardized according to the type of operation.
- **Gross still air range.** This assumes that the aircraft starts the flight at the initial cruising altitude above the first airfield and runs out of fuel above the destination airfield.

The concept of gross still air range is quite artificial, but it can be related to the safe operating range. Fuel for diversion, stacking and landing can be estimated and then regarded as unusable fuel. The fuel for take-off and climb can be subtracted from the initial fuel load and the aircraft placed initially along the flight by the climb distance. Given the cruise speed and the endurance, the distance that the aircraft is 'blown backwards' by the headwind can be estimated. From now on we will concentrate on the calculation of gross still air ranges and will use the term 'range' only in this sense.

2.6.1 General equations for range and endurance

This is the one instance when we must drop the assumption made in Section 1.2 that the aircraft mass is constant and define carefully the various masses involved. Figure 2.5 shows the variation of total aircraft mass with distance flown in a cruise. In this figure m_0 = initial mass, m_1 = final mass and m = mass at time t or distance x. The initial fuel mass is then $m_0 - m_1$.

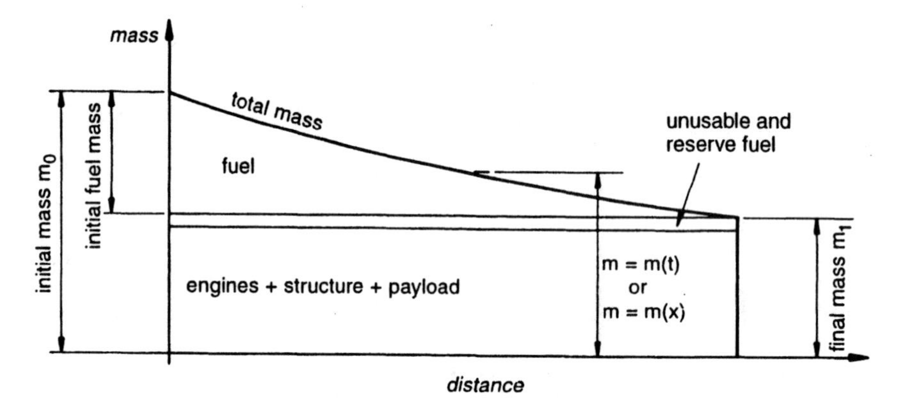

Fig. 2.5 Definition of terms and symbols used in determination of range

In discussing range it is convenient to introduce the 'specific air range' r, defined as

$$r = (\text{distance gone in a short time})/(\text{fuel used in the same time})$$

If we write the fuel mass flow rate as $\dot{m}_f$ then the fuel used in a time δt is $\dot{m}_f\delta t$ and the distance flown is $V\delta t$. The specific air range then becomes

$$r = V/\dot{m}_f \tag{2.23}$$

The range is

$$R = \int V\mathrm{d}t = \int \frac{V\mathrm{d}m}{\mathrm{d}m/\mathrm{d}t} \tag{2.24}$$

on changing the variable of integration to mass for convenience. Now as fuel is burnt the aircraft mass decreases so that

$$dm/dt = -\dot{m}_f \tag{2.25}$$

Using this in (2.24) and integrating from the initial mass to the final mass, we find

$$R = -\int_{m_0}^{m_1} \frac{V}{\dot{m}_f} dm = \int_{m_1}^{m_0} r dm \tag{2.26}$$

We deal with the endurance in a similar manner, defining the 'specific endurance' as

$$e = \text{(time for a short distance)/(mass of fuel used)} = 1/\dot{m}_f \tag{2.27}$$

The endurance E is then

$$E = \int dt = \int \frac{dm}{dm/dt}$$

and then using (2.25) we find

$$E = -\int_{m_0}^{m_1} \frac{dm}{\dot{m}_f} = \int_{m_1}^{m_0} e dm \tag{2.28}$$

It remains to determine the rate of using fuel.

The fuel rate is determined from the specific fuel consumption, expressed either as the fuel mass flow rate for unit equivalent shaft power, or for unit thrust as discussed in Section 1.4. For a propeller-driven aircraft we can write, using (2.20),

$$\dot{m}_f = fP_e = fTV/\eta = fDV/\eta \tag{2.29}$$

where f is the specific fuel consumption (sfc). In a similar manner for a jet-driven aircraft we have

$$\dot{m}_f = f_j T = f_j D \tag{2.30}$$

where f_j is the sfc for this case. Using these results we can write general expressions for the specific air range, specific endurance, range and endurance. From (2.23) and (2.26)–(2.29) we find, for propeller-driven aircraft,

$$r = \eta/fD \tag{2.31}$$

$$e = \eta/fDV \tag{2.32}$$

$$R = \int_{m_1}^{m_0} \frac{\eta dm}{fD} = \int_{m_1}^{m_0} \frac{\eta}{f}\left(\frac{C_L}{C_D}\right)\frac{dm}{mg} \tag{2.33}$$

and

$$E = \int_{m_1}^{m_0} \frac{\eta dm}{fDV} = \int_{m_1}^{m_0} \frac{\eta}{fV}\left(\frac{C_L}{C_D}\right)\frac{dm}{mg} \tag{2.34}$$

From (2.23), (2.26)–(2.28) and (2.30) we find, for jet-driven aircraft,

$$r = V/f_j D \tag{2.35}$$

$$e = 1/f_j D \tag{2.36}$$

$$R = \int_{m_1}^{m_0} \frac{V dm}{f_j D} = \int_{m_1}^{m_0} \frac{V}{f_j}\left(\frac{C_L}{C_D}\right)\frac{dm}{mg} \tag{2.37}$$

and

$$E = \int_{m_1}^{m_0} \frac{dm}{f_j D} = \int_{m_1}^{m_0} \frac{1}{f_j}\left(\frac{C_L}{C_D}\right)\frac{dm}{mg} \tag{2.38}$$

In the last parts of (2.33)–(2.36) we have used $L = mg$ and, if it is appropriate, we can use (2.4) to substitute for V. To evaluate the integrals we need the quantities involved to be constant or a known function of the mass. To optimize the range or endurance the specific air range or specific endurance must be maximized at all stages of the cruise, subject to satisfying the condition that the lift and weight are equal.

2.6.1.1 *Application of general equations*

The best way to describe the application of the equations is to work through a few examples.

Worked example 2.2

At the start of a long distance cruise a turboprop aircraft has a total mass of 67 000 kg of which 13 000 kg is fuel. The drag coefficient is given by $C_D = 0.021 + 0.052C_L^2$, the propeller efficiency is 0.84 and the sfc is 1.0×10^{-7} kg J^{-1}. Determine the maximum range. If the speed is kept constant during the cruise determine the ratio of the final to the initial air densities.

Solution

Since we are dealing with the range of a turboprop we repeat part of (2.33) here:

$$R = \int_{m_1}^{m_0} \frac{\eta}{f}\left(\frac{C_L}{C_D}\right)\frac{dm}{mg}$$

The maximum range will occur when the term (C_L/C_D) is at a maximum and this will be a constant value. Taking this and the other constant terms outside the integral gives

$$R_{max} = \frac{\eta}{f}\left(\frac{C_L}{C_D}\right)_{max} \cdot \int_{m_1}^{m_0} \frac{dm}{mg}$$

and integrating gives

$$R_{max} = \frac{\eta}{fg}\left(\frac{C_L}{C_D}\right)_{max} \ln\left(\frac{m_0}{m_1}\right)$$

This expression is known as the 'Breguet Equation' (for propeller aircraft). Note that if the flight conditions are constrained in some way, for example flight at constant eas, then the resulting expression for the maximum range will be different.

The condition for (C_L/C_D) to be a maximum is that for (C_D/C_L) to be a minimum and therefore we use (2.14), that is

$$C_{Lmd} = \sqrt{a/b} = \sqrt{0.021/0.052} = 0.6355$$

The drag coefficient is then $C_D = 0.021 + 0.052 \times 0.6355^2 = 0.042$ and $(C_L/C_D) = 15.13$. Then substituting values in to find the range we have

$$R_{max} = \frac{0.84 \times 15.13}{1 \times 10^{-7} \times 9.81} \ln\left(\frac{67\,000}{67\,000 - 13\,000}\right) = 2.7946 \times 10^6$$

Since we are using SI units this is expressed in metres and the final result is 2795 km (Ans).

To find the ratio of densities we note that the lift coefficient is constant during the cruise so we have

$$C_L = \frac{m_0}{\frac{1}{2}\rho_0 V^2 S} = \frac{m_1}{\frac{1}{2}\rho_1 V^2 S}$$

where, just in this case, ρ_0 and ρ_1 are the initial and final densities. Solving for the density ratio we have

$$\rho_1/\rho_0 = m_1/m_0 = 54\,000/67\,000 = 0.806 \text{ (Ans)}$$

This shows that the optimum climb technique is for the aircraft to climb during the cruise keeping C_L constant. This 'cruise-climb' is not normally practical for airliners owing to air traffic control procedures, and a stepped climb technique is used. Note that in the Breguet Equation we have terms relating to the propeller efficiency, η, and the engine efficiency, $1/f$, while the term (C_L/C_D) can be regarded as the aerodynamic efficiency of the aircraft.

Worked example 2.3

A turbojet aircraft having a wing area of 75 m^2 starts a cruise at mass of 18 000 kg which includes 3500 kg of usable fuel. The drag coefficient is given by $C_D = 0.025 + 0.065C_L^2$ and the sfc is 2.8×10^{-5} kg N^{-1} s^{-1}. Determine the maximum range for flight at constant speed with the cruise starting at the altitude where $\sigma = 0.53$. Also find the fuel remaining at half the distance.

Solution

Since we are dealing with the range of a turbojet we repeat part of (2.37) here:

$$R = \int_{m_1}^{m_0} \frac{V}{f_j}\left(\frac{C_L}{C_D}\right)\frac{dm}{mg}$$

Again the maximum range will occur when the the term (C_L/C_D) is at a maximum and this will be a constant value. Taking this and the other constant terms outside the integral gives

$$R_{max} = \frac{V}{f_j}\left(\frac{C_L}{C_D}\right)_{max} \int_{m_1}^{m_0} \frac{dm}{mg}$$

and integrating gives

$$R_{max} = \frac{V}{f_j g}\left(\frac{C_L}{C_D}\right)_{max} \ln\left(\frac{m_0}{m_1}\right)$$

This is also known as the Breguet Equation (for jet aircraft). Note that flight was constrained to be at constant speed; if flight is constrained in some other way, for instance flight at constant altitude, then the resulting expression for the maximum range will be different.

Again using (2.14) to find $(C_L/C_D)_{max}$ we have

$$C_{Lmd} = \sqrt{a/b} = \sqrt{0.025/0.065} = 0.6202$$

The drag coefficient is then $C_D = 0.05$ and $(C_L/C_D)_{max} = 12.4$. We use C_L to find the speed,

$$V = \sqrt{\frac{2m_0 g}{\rho S C_L}} = \sqrt{\frac{2 \times 18\,000 \times 9.81}{1.225 \times 0.53 \times 75 \times 0.6202}} = 108.1 \text{ m s}^{-1}$$

The range is then

$$R_{max} = \frac{108.1 \times 12.4}{2.8 \times 10^{-5} \times 9.81} \ln\left(\frac{18\,000}{18\,000 - 3500}\right) = 1.055 \times 10^6 \text{ m} = 1055 \text{ km (Ans)}$$

To find the fuel at half way, let $m_{0.5}$ be the corresponding aircraft mass; then we must have

$$R_{0.5} = \frac{V}{f_j g}\left(\frac{C_L}{C_D}\right)_{max} \ln\left(\frac{m_0}{m_{0.5}}\right)$$

Then dividing,

$$R_{0.5} / R_{max} = \frac{\ln(m_0 / m_{0.5})}{\ln(m_0 / m_1)} = \tfrac{1}{2}$$

and taking exponentials gives $(m_0/m_{0.5}) = (m_0/m_1)^{0.5} = \sqrt{18\,000/14\,500} = 1.114$. The mass at half distance is then 16 158 kg and the fuel remaining 1658 kg.

Throughout these examples we have assumed that the sfc is constant; this is some way from being correct. The variation in sfc during a cruise due to change in height or speed is small and no great error is involved if an average value is used. The determination of the optimum conditions for a cruise are, however, dependent on the variation of sfc with Mach number, altitude and throttle setting. A more accurate expression for the fuel flow rate than (2.31) has been given by Bore, reference (2.1).

2.6.2 Cruise in the stratosphere

We have seen that a jet-driven aircraft flies further the higher and faster it flies; however, compressibility sets a limit to increases in speed. For optimum range in many cases the cruise takes place in the stratosphere where the temperature and therefore the speed of sound are constant. In view of this we recast (2.37) in terms of Mach number, **M**, thus:

$$R = \frac{c_s}{f_j g} \int_{m_1}^{m_0} \left(\frac{\mathbf{M} C_L}{C_D} \right) dm \tag{2.39}$$

From (2.39) we see that $\mathbf{M}C_L/C_D$ must be maximized for maximum range. If we consider the typical plot of no-lift drag coefficient with Mach number shown in figure 1.13, then $\mathbf{M}C_L/C_D$ increases in proportion to Mach number up to the point where the no-lift drag starts to rise. It will continue to rise for a further small increase in **M** as this increase will offset the increase in the drag. The optimum cruise Mach number is therefore just beyond the point where the drag starts to rise.

In practice the optimum has to be found by plotting $\mathbf{M}C_L/C_D$ against Mach number, or a computerized equivalent, but examples can be constructed to show the principles involved.

Worked example 2.4

A long range turbojet aircraft has a drag coefficient which can be expressed as follows:

$$C_D = (a + bC_L^2) f(\mathbf{M})$$

where

$$f(\mathbf{M}) \begin{cases} = 1 \text{ for } \mathbf{M} \leq \mathbf{M}_c \\ = \exp\left[c(\mathbf{M} - \mathbf{M}_c)^2\right] \textit{for } \mathbf{M} > \mathbf{M}_c \end{cases}$$

where $a = 0.019$, $b = 0.059$, $c = 25$ and $\mathbf{M}_c = 0.84$. Determine the maximum range in the stratosphere given initial mass = 160 000 kg, final mass = 110 000 kg, $c_s = 295$ m s^{-1} and sfc = 1.5×10^{-5} kg N^{-1} s^{-1}.

Solution

We first consider maximizing $\mathbf{M}C_L/C_D$ or, what is equivalent, minimizing $C_D/\mathbf{M}C_L$:

$$\frac{C_D}{\mathbf{M}C_L} = \left(\frac{a}{C_L} + bC_L \right) \frac{\exp\left[c(\mathbf{M} - \mathbf{M}_c)^2\right]}{\mathbf{M}}$$

Hence

$$\frac{\partial}{\partial C_L}\left(\frac{C_D}{\mathbf{M}C_L} \right) = \left(-\frac{a}{C_L^2} + bC_L \right) \frac{\exp\left[c(\mathbf{M} - \mathbf{M}_c)^2\right]}{\mathbf{M}} \tag{i}$$

and

$$\frac{\partial}{\partial \mathbf{M}}\left(\frac{C_D}{\mathbf{M}C_L} \right) = \left(\frac{a}{C_L} + bC_L \right) \left(\frac{2c\mathbf{M}(\mathbf{M} - \mathbf{M}_c)\exp\left[c(\mathbf{M} - \mathbf{M}_c)^2\right] - \exp\left[c(\mathbf{M} - \mathbf{M}_c)^2\right]}{\mathbf{M}^2} \right) \tag{ii}$$

and equating these to zero gives, from (i) $C_{\text{Lopt}} = \sqrt{a/b}$ and from (ii)

$$2c\mathbf{M}(\mathbf{M} - \mathbf{M}_c) - 1 = 0 = 2c\mathbf{M}^2 - 2c\mathbf{M}_c\mathbf{M} - 1$$

Hence solving the quadratic, the roots are

$$\mathbf{M}_{\text{opt}} = \frac{2c\mathbf{M}_c \pm \sqrt{(2c\mathbf{M}_c)^2 + 4.2c}}{2.2c}$$

Substituting numbers, $C_{\text{Lopt}} = \sqrt{0.019/0.059} = 0.5675$ and

$$\mathbf{M}_{\text{opt}} = \frac{2 \times 25 \times 0.84 \pm \sqrt{(2 \times 25 \times 0.84)^2 + 8 \times 25}}{2 \times 2 \times 25} = 0.864$$

Then $C_D = 2 \times 0.019 \times \exp\left[25(0.863 - 0.84)^2\right] = 0.0385$ and $\left(\mathbf{M}C_L / C_D\right)_{\text{opt}} = 12.72.$

Then using (2.39) and integrating as before,

$$R_{\text{opt}} = \frac{c_s}{f_j g}\left(\frac{\mathbf{M}C_L}{C_D}\right)_{\text{opt}} \ln\left(\frac{m_0}{m_1}\right) = \frac{295 \times 12.72}{1.5 \times 10^{-5} \times 9.81} \ln\left(\frac{160}{110}\right) = 9555 \text{ km (Ans)}$$

2.6.3 Range–payload curves

Aircraft manufacturers use various plots to illustrate the capabilities of their airliners; one of the commonest is the range–payload curve as shown in figure 2.6(a).

Suppose an airliner is loaded to its full capacity with passengers, their baggage and freight and a small quantity of fuel; it can then take off but has to land immediately. This corresponds to the point A in figure 2.6(a). Adding more and more fuel then enables the aircraft to fly further and further up to the point B where the aircraft has reached its maximum take-off weight, which is fixed by the designer, bearing in mind such constraints as take-off requirements and stressing. Further range can be obtained only by leaving out some of the payload and adding fuel, keeping the weight constant; this corresponds to BC in the diagram. Ultimately the fuel tanks are full and the only way to increase range is to further reduce the payload; this is section CD of the diagram. It is conventional to calculate the position of the points B, C and D and then complete the diagram by joining them with straight lines rather than the gentle curves they should be. Figure 2.6(b) shows the increase of fuel required as the range increases. Further range could be obtained by removing unnecessary crew or equipment as might be the case if the operator required to ferry the aircraft to another airport, giving the 'ferry range'. A number of range–payload curves have to be drawn for a given airliner as the maximum take-off weight will depend on the altitude and air temperature at the take-off airport, and the weight of the passengers will depend on the split between numbers in first, second and tourist classes.

2.7 Incremental performance

Once an aircraft type enters into service it is likely to be subject to continual improvements; improved engines or other equipment become available or it becomes apparent that there is a

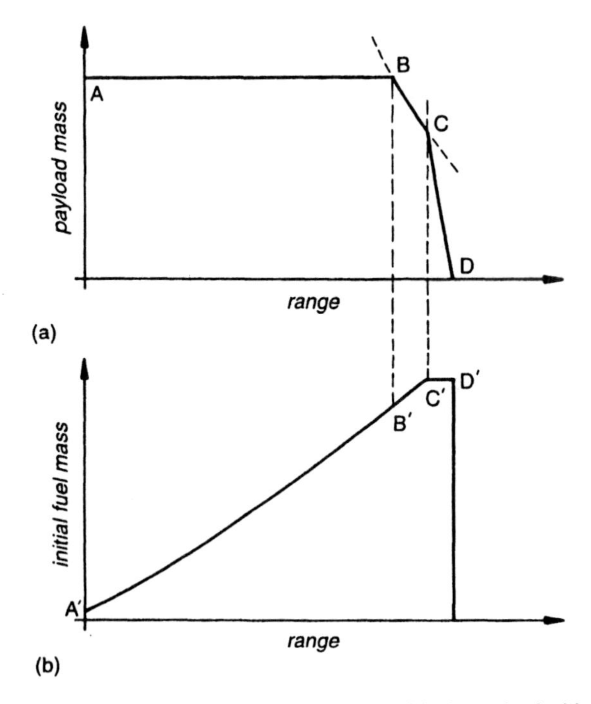

Fig. 2.6 (a) Range–payload curve. (b) Variation of fuel required with range

market for a variant with a 'stretched' fuselage. Changes may also occur which reduce the performance, such as the addition of external stores. There then arises the need to estimate the effects of these changes and this is more accurately done as an increment applied to the established performance of the basic type than *ab initio*. As an example of the technique we find the effect of a small increase in the thrust of the engine on the maximum speed of a jet-driven aircraft.

We first rewrite (2.8) in the form

$$D = \mathscr{A}V^2 + \mathscr{B}V^{-2} \tag{2.40}$$

where $\mathscr{A} = \frac{1}{2}\rho Sa$ and $\mathscr{B} = b(mg)^2/\frac{1}{2}\rho S$ and use has been made of (2.6). We next differentiate with respect to V to find the speed for minimum drag:

$$\frac{dD}{dV} = 2\mathscr{A}V - 2\mathscr{B}V^{-3} = 0$$

Solving gives

$$V_{md}^{\ 2} = \sqrt{\mathscr{B}/\mathscr{A}} \tag{2.41}$$

and then the minimum drag is

$$D_{min} = 2\sqrt{\mathscr{A}\mathscr{B}}$$

We then write the thrust required for steady level flight in the form

$$\frac{T}{D_{\text{min}}} = \frac{D}{D_{\text{min}}} = \frac{1}{2\sqrt{\mathscr{A}\mathscr{B}}}\left(\mathscr{A}\left(\frac{V}{V_{\text{md}}}\right)^2\sqrt{\mathscr{B}/\mathscr{A}} + \mathscr{B}\left(\frac{V_{\text{md}}}{V}\right)^2\sqrt{\mathscr{A}/\mathscr{B}}\right)$$

using (2.40) where we have also introduced V_{md} from (2.41). Then writing $y = V/V_{\text{md}}$ the above expression reduces to

$$\frac{T}{D_{\text{min}}} = \tfrac{1}{2}\left(y^2 + y^{-2}\right) \tag{2.42}$$

Differentiating this implicitly, assuming that thrust is independent of speed, gives

$$\frac{\text{d}T}{D_{\text{min}}} = \tfrac{1}{2}\left(2y - 2y^{-3}\right)\text{d}y$$

Then approximating differentials by small differences and dividing through by (2.42) we find

$$\frac{\delta T}{T} = 2\left(\frac{y - y^{-3}}{y + y^{-3}}\right)\frac{\delta y}{y}$$

and rearranging gives

$$\frac{\delta V}{V} = \tfrac{1}{2}\left(\frac{y^4 + 1}{y^4 - 1}\right)\frac{\delta T}{T} \tag{2.43}$$

For values of y greater than about 3 we can approximate further and find that the percentage change in speed is half the percentage change in thrust. The concept of incremental performance can be applied to most aspects of the performance.

Student problems

Take the density of air at sea-level as 1.225 kg m^{-3} and $g = 9.81$ m s^{-1}.

2.1 An aircraft has mass of 50 000 kg and a wing area of 210 m^2. With the engines giving a thrust of 23.4 kN it has a speed of 100 m s^{-1} at an altitude where the relative density is 0.75. Find the lift and drag coefficients. If the maximum lift coefficient is 1.42, find the stalling speed at this height. (A)

2.2 An aircraft has a mass of 40 000 kg and a wing area of 180 m^2. If the drag coefficient is given by $C_D = 0.015\,75 + 0.033\,34C_L^2$ and $\sigma = 0.75$, find:
   (a) the C_L for minimum power required in level flight, the corresponding eas and power;
   (b) the C_L for the minimum thrust required in level flight, the corresponding eas and thrust. (A)

2.3 If the aircraft of problem (2.2) has jet engines with a maximum thrust of 30 kN find the maximum steady speed at the same height assuming that the thrust is independent of speed. (A)

2.4 An aircraft of mass 25 000 kg and a wing area 110 m^2 has turboprop engines giving a maximum power of 3200 kW which can be assumed not to vary with speed. Find the maximum speed at a height at which the relative density is 0.69 if the drag coefficient is given by $C_D = 0.021 + 0.032C_L^2$ and the propeller efficiency is 0.88. (A)

2.5 An aircraft has the following characteristics as a function of Mach number at a certain altitude:

Mach number	1.7	1.9	2.1	2.3
No-lift drag coefficient	0.01	0.00875	0.00775	0.007
(Lift-dependent coeff.)/C_L^2	0.41	0.46	0.56	0.63
Thrust (kN)	128	131	126	116

The aircraft has a mass of 90 000 kg and the wing area is 200 m^2. Find the maximum level flight Mach number if the atmospheric pressure at this altitude is 9.6 kN m^{-2}. (A)

2.6 A jet aircraft has a mass of 5800 kg and a wing area of 21.5 m^2. It is flying at a speed of 240 m s^{-1} at a height where $\sigma = 0.84$, when the throttle lever is suddenly moved to the flight-idle position at which it may be assumed that the engine gives zero nett thrust. Determine the time taken for the speed to fall to 160 m s^{-1}, while maintaining level flight. The drag coefficient is given by $C_D = 0.0276 + 0.075C_L^2$; integrate analytically or use Simpson's rule. (A)

2.7 A turboprop aircraft has a mass of 48 000 kg and a wing area of 260 m^2 and is flying at sea level at a speed of 70 m s^{-1}. The throttles are suddenly opened to give a power of 5500 kW. Find the distance travelled in accelerating to a speed of 100 m s^{-1}, while maintaining level flight. The drag coefficient is given by $C_D = 0.022 + 0.055C_L^2$, the propeller efficiency is 0.87 and it can be assumed that the engine power is independent of speed. (A)

2.8 At the start of a long distance cruise a turboprop aircraft has a total mass of 70 000 kg of which 12 000 kg is fuel, and the initial wing loading is 3.2 kN m^{-2}. The drag coefficient is given by $C_D = 0.014 + 0.05C_L^2$, the propeller efficiency is 0.82 and the sfc is 1.0×10^{-7} kg J^{-1}. Find:

(a) the maximum range;

(b) an expression for the range for an aircraft constrained to fly at constant eas and use this to find the range for this aircraft when it flies at a constant eas 40 per cent above that for minimum drag at the initial mass;

(c) an expression for the speed for maximum range of an aircraft constrained to fly at constant eas and use this to find the speed for maximum range and the range under these conditions. (A)

2.9 A turbojet aircraft has a wing area of 80 m^2 and at the start of a cruise the total mass is 19 000 kg which includes 4000 kg of usable fuel. The drag coefficient is given by $C_D = 0.015 + 0.075C_L^2$, and the sfc is 3.0×10^{-5} kg N^{-1} s^{-1}. Find:

(a) the maximum range for flight at a constant altitude where $\sigma = 0.53$ (hint: see comments in solution to Worked example 2.3);

(b) the tas for minimum drag at this altitude at the initial mass and determine the range for flight constrained to this speed; also find the value of the relative density at the end of the flight. (A)

2.10 A turboprop aircraft of mass 70 000 kg including a fuel mass of 12 000 kg starts a long distance flight. After 2400 km it receives another 12 000 kg of fuel by flight refuelling. Neglecting the fuel required for take-off and climb, but allowing for 2200 kg of fuel reserves, find the maximum distance it can fly. The drag is given by $C_D = 0.014 + 0.05C_L^2$, the sfc is 0.9×10^{-7} kg J^{-1} and the propeller efficiency is 0.82. (A)

2.11 Find the maximum endurance of a turbojet aircraft having a drag given by $C_D = 0.02 + 0.055C_L^2$ and a sfc of 2.5×10^{-5} kg N^{-1} s^{-1}. At the start of the cruise the usable fuel mass is 30 per cent of the total mass. (A)

2.12 The series for $\ln(1 + x)$ truncated to two terms is $x - x^2/2$ valid for $-1 \le x \le 1$. Use this to show that the logarithmic term in the Breguet Equation can be approximated by (fuel mass)/(average mass).

2.13 A long range turbojet aircraft has a drag coefficient which can be expressed as follows:

$$C_D = (a + bC_L^2)f(\mathbf{M})$$

where

$$f(\mathbf{M})\begin{cases} = 1 \text{ for } \mathbf{M} \le \mathbf{M}_c \\ = 1 + 0.5 \times 10^3 \times \mathbf{M} \times (\mathbf{M} - \mathbf{M}_c)^2 \text{ for } \mathbf{M} > \mathbf{M}_c \end{cases}$$

where $a = 0.018$, $b = 0.060$ and $\mathbf{M}_c = 0.85$. Determine the maximum range in the stratosphere given initial mass = 200 000 kg, fuel mass = 60 000 kg, $c_s = 295$ m s^{-1} and sfc = 1.5×10^{-5} kg N^{-1} s^{-1}.

2.14 Show that, if C_D is given in the form $C_D = a + bC_L^2$, the power P at any speed V is related to the minimum power P_{min} which occurs at speed V_{mp} by

$$\frac{P}{P_{min}} = \frac{1}{4}\left(z^3 + 3/z\right)$$

where $z = V/V_{mp}$. Repeat the analysis leading to (2.43) for the case of a change in engine power.

Notes

1. The reader is reminded that it is common practice to quote weights in kilograms, whereas strictly speaking the kg is the unit of *mass* in SI. In the former aeronautical version of Imperial units mass was measured in slugs, one slug being 'g' pounds (mass).
2. From here on we use the term lift-dependent drag somewhat loosely; it was pointed out in Section 1.3.4.2 that it may include small drag contributions not directly associated with the production of lift.

3
Performance – other flight manoeuvres

3.1 Introduction

In this chapter we consider the performance of the aircraft in other manoeuvres. We need to know such things as the minimum rate of descent in a glide, the time to reach a given height in a climb, the radius of turn in given conditions and the distances required for take-off and landing. Some of the manoeuvres currently practised by combat aircraft, such as the 'cobra', involve stalling the wing and are beyond the scope of this book. We also exclude manoeuvres which involve varying the direction of the thrust relative to the aircraft centre-line, known as 'thrust-vectoring'. Flight in a vertical circle is briefly dealt with in Section 5.5.1.

3.2 Steady gliding flight

Consider an aircraft in a steady glide at an angle γ to the horizontal. The forces acting on it are the lift, the drag and the weight as shown in figure 3.1; the forces are in equilibrium and a triangle of forces can be drawn as shown.

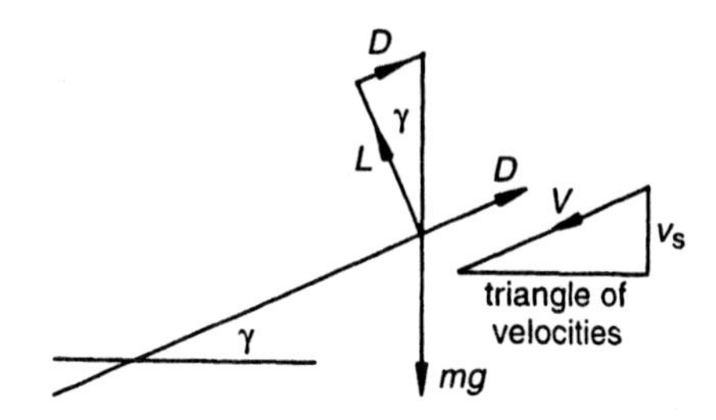

Fig. 3.1 Forces acting on an aircraft in a glide

We resolve the forces along the direction of flight to give

$$D + mg \sin \gamma = 0 \tag{3.1}$$

and resolving normal to the direction of flight

$$L - mg \cos \gamma = 0 \tag{3.2}$$

Solving these to find $\sin \gamma = D/mg$ and $\cos \gamma = L/mg$ and dividing gives

$$\tan \gamma = \frac{D}{L} = \frac{C_D}{C_L} \tag{3.3}$$

The minimum gliding angle is therefore obtained when C_D/C_L is a minimum which is the condition for the minimum drag considered in Section 2.2. It follows that to achieve the maximum distance travelled, starting from a given initial height, the aircraft must fly at the minimum drag speed.

To find the condition for the aircraft to remain in the air for the maximum time we need to find an expression for the minimum rate of descent or 'sinking speed'. From the triangle of velocities the sinking speed v_s is

$$v_s = V \sin \gamma \tag{3.4}$$

Now from the triangle of forces

$$\sin \gamma = \frac{D}{\sqrt{L^2 + D^2}} = \frac{C_D}{\sqrt{C_L^2 + C_D^2}} \tag{3.5}$$

and from (3.2), after writing $L = \frac{1}{2}\rho SV^2 C_L$ and solving,

$$V = \sqrt{\frac{2mg \cos \gamma}{\rho S C_L}} \tag{3.6}$$

where

$$\cos \gamma = \frac{C_L}{\sqrt{C_L^2 + C_D^2}} \tag{3.7}$$

Then on combining (3.4)–(3.7) we have

$$v_s = \sqrt{\frac{2mg}{\rho S}} \frac{C_D}{\left(C_L^2 + C_D^2\right)^{3/4}} \tag{3.8}$$

In deriving this equation we have made no assumptions as to the magnitude of the glide angle or the relative sizes of C_L and C_D, and (3.8) applies equally well to bluff bodies for which C_L and C_D may be of similar sizes. However, aircraft are designed such that $C_D \ll C_L$ for most of the useful speed range, so that on neglecting C_D^2 compared with C_L^2 we find

$$v_s = \sqrt{\frac{2mg}{\rho S}} \left(C_D / C_L^{3/2}\right) \tag{3.9}$$

The condition for the minimum rate of sink is therefore that for the minimum power considered in Section 2.2. It follows that to achieve the maximum duration of flight starting from a given initial height the aircraft must fly at the minimum power speed. In addition, from (3.7) we see in this case that $\cos \gamma \simeq 1$ and from (3.6) we again have $V \propto C_L^{-1/2}$, so if we wish we can use C_L again as an alternative to V as an independent variable with its consequent advantages.

3.3 Climbing flight, the 'Performance Equation'

Consider now an aircraft in a climb at a constant angle Θ to the horizontal.[1] The forces acting on it are the lift, the drag, the thrust and the weight as shown in figure 3.2.

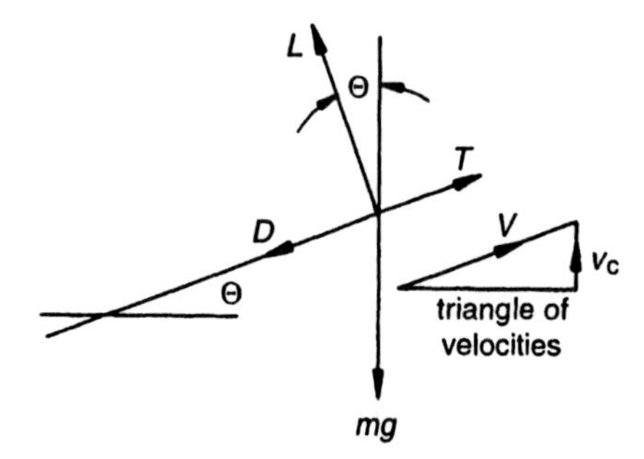

Fig. 3.2 Forces acting on an aircraft in a climb

We again resolve the forces along and normal to the direction of flight giving

$$T - D - mg\sin\Theta = m\frac{\mathrm{d}V}{\mathrm{d}t} \tag{3.10}$$

allowing for a forward acceleration, and

$$\mathrm{L} - mg\sin\Theta = 0 \tag{3.11}$$

As in the case of the glide we can rewrite this to give the velocity in terms of the lift coefficient and other parameters:

$$V = \sqrt{\frac{2mg\cos\Theta}{\rho S C_{\mathrm{L}}}} \tag{3.12}$$

If in (3.10) we now substitute for the thrust from (2.19) and multiply through by the velocity we can write it in the form

$$\eta P + T_{\mathrm{j}}V = DV + mV\frac{\mathrm{d}V}{\mathrm{d}t} + mgV\sin\Theta \tag{3.13}$$

We can write this equation in various other forms; first, noting that

$$\frac{\mathrm{d}}{\mathrm{d}t}\left(\tfrac{1}{2}V^2\right) = V\frac{\mathrm{d}V}{\mathrm{d}t}$$

and that the rate of climb, v_{c}, is

$$v_{\mathrm{c}} = V\sin\Theta \tag{3.14}$$

then (3.13) becomes

$$\eta P + T_{\mathrm{j}}V = DV + m\frac{\mathrm{d}}{\mathrm{d}t}\left(\tfrac{1}{2}V^2\right) + mgv_{\mathrm{c}} \tag{3.15}$$

This is in fact an energy equation and the various terms can be identified as such. On the left-hand side we have the useful power input from the propeller and the rate of work done by the thrust. On the right-hand side the first term is the rate of doing work against the drag, the next is the rate of gain of the kinetic energy and the last is the rate of gain of potential energy. We can further write the rate of climb as dh/dt; then

$$m\frac{d}{dt}\left(\tfrac{1}{2}V^2\right) + mgv_c = m\frac{d}{dt}\left(\tfrac{1}{2}V^2 + gh\right)$$

If we now write $h_{en} = h + \dfrac{V^2}{2g}$ = 'energy height', we can write (3.15) as

$$\eta P + \left(T_j - D\right)V = mg\frac{dh_{en}}{dt} \tag{3.16}$$

The energy height can be described as the total mechanical energy per unit mass or as the height to which the aircraft would climb if all the kinetic energy were converted to potential energy. The variation of energy height with height and speed is illustrated in figure 3.3.

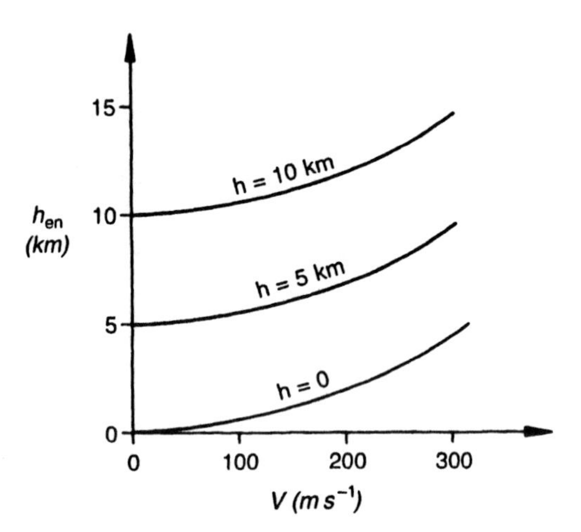

Fig. 3.3 Variation of energy height with speed and height

Methods of performance calculation based on the concept of energy height are well suited to high speed aircraft and will be dealt with in Section 3.3.5.

A further form of (3.13) can be found by multiplying through by $\sqrt{\sigma}$ and replacing the airspeed with the equivalent airspeed, giving

$$\sqrt{\sigma}\eta P + T_j V_E = DV_E + mV_E\frac{dV}{dt} + \sqrt{\sigma}mgv_c \tag{3.17}$$

The various forms (3.13), (3.15), (3.16) and (3.17) are known as the 'Performance Equation'.

In the last chapter we divided engines into two broad groups, propeller-driven and jet-driven, and it is useful (but not essential) to have performance equations for each type. Introducing the equivalent shaft power, P_e, from (2.21) we can rewrite (3.13) as

$$\eta P_e = DV + mV\frac{dV}{dt} + mgv_c \tag{3.18}$$

or, alternatively, from (3.17) we have

$$\sqrt{\sigma}\eta P_e = DV_E + mV_E\frac{dV}{dt} + \sqrt{\sigma}mgv_c \tag{3.19}$$

For jet-driven aircraft it is sufficient to rearrange (3.10) to give

$$T = D + m\frac{dV}{dt} + mg\sin\Theta \tag{3.20}$$

where T is now understood to be the jet thrust.

3.3.1 Climb at constant speed

We consider first the case of climb at constant speed, determining the angle and rate of climb and the speeds for which these are a maximum. The rate of climb is required in order to find the time to a given height, whilst the angle is required in order to determine if an aircraft can climb sufficiently to clear an obstacle in its path. In Section 3.3.2 we determine how the effect of acceleration modifies the results.

3.3.1.1 Propeller-driven aircraft

Putting the acceleration to zero we find directly from (3.18)

$$v_{co} = \frac{\eta P_e - DV}{mg} \tag{3.21}$$

where we have written v_{co} for the rate of climb at constant speed. If we substitute for mg from (3.11) we find

$$v_{co} = \frac{\eta P_e}{mg} - \frac{DV}{L}\cos\Theta \tag{3.22}$$

We now substitute for V from (3.12) for L and D in terms of C_L and C_D and find

$$v_{co} = \frac{\eta P_e}{mg} - (\cos\Theta)^{3/2}\sqrt{\frac{2mg}{\rho S}}\left(C_D / C_L^{3/2}\right) \tag{3.23}$$

In this expression the climb angle Θ is unknown, but for most aircraft it will be less than about 10°. The cosine of this angle is 0.9848 and for this value the error in the second term incurred by putting $(\cos\Theta)^{3/2} = 1$ is just over 2 per cent. We then simplify (3.23) to

$$v_{co} = \frac{\eta P_e}{mg} - \sqrt{\frac{2mg}{\rho S}}\left(C_D / C_L^{3/2}\right) \tag{3.24}$$

If the final error is unacceptable we can still use (3.24) to obtain a first approximation, calculate the climb angle using (3.14) and then use (3.23) to find an improved result. However, from this point on we will assume that the climb angle is small enough for (3.24) or similar expressions to be valid.

An alternative expression for the rate of climb can be found from (3.19):

$$\sqrt{\sigma}v_{co} = \frac{\sqrt{\sigma}\eta P_e - DV_E}{mg} \tag{3.25}$$

We noted in Section 2.3 that the curve of $\sqrt{\sigma}P_r = DV_E$ plotted against V_E is independent of altitude. We now superimpose curves of $\sqrt{\sigma}\eta P_e$ against V_E for various altitudes as shown in figure 3.4(a).

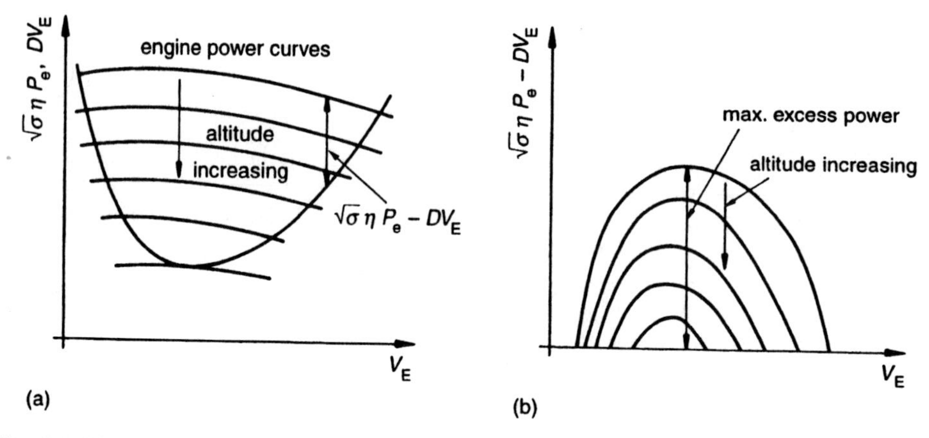

Fig. 3.4 (a) Engine power curves superimposed onto drag power curve. (b) Curves of excess power for various heights

We can now determine the values of $\sqrt{\sigma}\eta P_E - DV_E$ and plot these against V_E as shown in figure 3.4(b) and hence determine the maximum rate of climb at each altitude and the speed at which it occurs. It can be seen that if the power curves are fairly flat then the speed for maximum rate of climb is close to the speed for minimum power V_{mp} discussed in Section 2.3. In some cases it may be acceptable to assume that the power ηP_e is constant with speed; the speed for maximum rate of climb is then V_{mp} as is evident from (3.24) and (2.13). It should be noted that in this case the optimum speed is a constant eas and hence the true airspeed increases with height.

To determine the angle of climb we use (3.10), (3.22) and (2.20) and assume that the angle is small, so that cos Θ can be taken as unity, giving

$$\sin\Theta = \frac{T - D}{mg} = \frac{\eta P_e}{mVg} - \frac{D}{L} \tag{3.26}$$

then using (2.10)

$$\sin\Theta = \eta P_e\sqrt{\frac{\rho S}{2}}(mg)^{-3/2}\cdot C_L^{1/2} - \frac{C_D}{C_L} \tag{3.27}$$

To find the speed for maximum climb angle, graphical methods or their computer equivalent are necessary. In cases where it is reasonable to assume that the drag can be expressed in the form $C_D = a + bC_L^2$ and the power is constant a quartic equation has to be solved. Figure 3.5 shows a typical plot of thrust and drag against speed for propeller-driven aircraft.

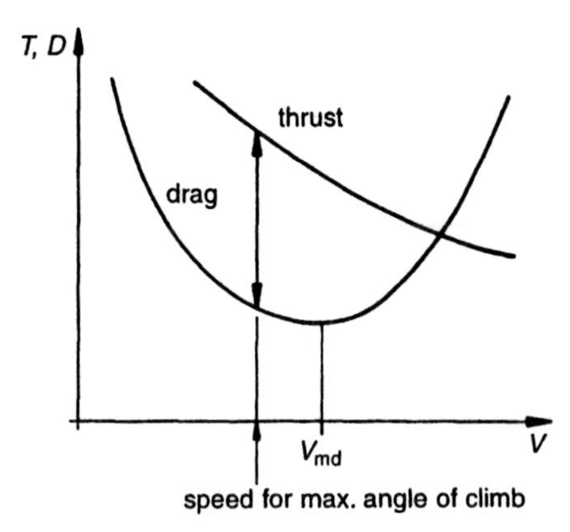

Fig. 3.5 Determination of maximum angle of climb for a propeller-driven aircraft

From the first part of (3.26) the speed for the maximum angle occurs where $(T - D)$ is at a maximum and from figure 3.5 we see that it is less than that for minimum drag.

Worked example 3.1

An aircraft having a mass of 42 000 kg and a wing area of 100 m^2 has turboprop engines of total equivalent power 8000 kW, assumed not to vary with speed, and propellers of 83 per cent efficiency. The drag coefficient is given by $C_D = 0.015 + 0.055C_L^2$. Find the maximum steady rate of climb at a height where the relative density, σ, is 0.766.

Solution

The maximum rate of climb occurs at the speed for minimum power, so from (2.16) we have $C_{Lmp} = \sqrt{3a/b} = \sqrt{3 \times 0.015/0.055} = 0.9045$ and $C_D = 0.06$. We then use (3.24) to find the rate of climb:

$$v_{co} = \frac{0.83 \times 8000 \times 10^3}{42\,000 \times 9.81} - \sqrt{\frac{2 \times 42\,000 \times 9.81}{1.225 \times 0.766 \times 100}}\left(0.06/0.9045^{3/2}\right) = 9.58 \text{ m s}^{-1} \text{ (Ans)}$$

3.3.1.2 *Jet-driven aircraft*

In this case we start by discussing the angle of climb; the first part of (3.26) is applicable in this case also. Figure 3.6 shows a typical plot of thrust and drag for a jet-driven aircraft and it can be seen that the speed for the maximum angle occurs near to that for minimum drag.

We can also write the climb angle in the form

$$\sin \Theta = \frac{T - D}{mg} = \frac{T}{mg} - \frac{C_D}{C_L} \tag{3.28}$$

where we have again assumed that the climb angle is small, so that cos Θ can be taken as unity. If the thrust can be taken as independent of the speed then (3.28) and (2.11) show that the maximum angle of climb occurs at the speed for minimum drag.

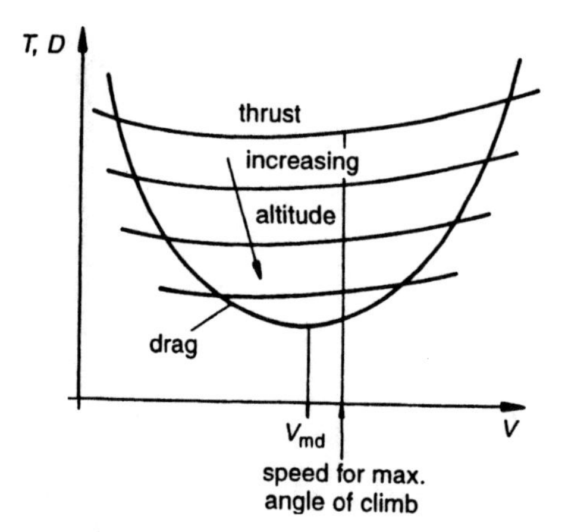

Fig. 3.6 Determination of maximum angle of climb for a jet aircraft

To find the rate of climb we use (3.14) and (3.28) giving

$$v_{co} = V \sin \Theta = \left(\frac{T}{mg} - \frac{C_D}{C_L} \right) V \tag{3.29}$$

and then using (3.12)

$$v_{co} = \left(\frac{T}{mg} - \frac{C_D}{C_L} \right) \sqrt{\frac{2mg}{\rho S C_L}}$$

or

$$v_{co} = \left(\tau C_L^{-1/2} - C_D / C_L^{3/2} \right) \sqrt{\frac{2mg}{\rho S}} \tag{3.30}$$

where we have written $T/mg = \tau$, which is the (thrust/weight) ratio. For the maximum rate of climb the quantity in the square brackets in (3.30) has to be maximized, which will usually need a graphical method. If the thrust can be taken as constant with speed and we can write the drag in the form $C_D = a + bC_L^2$ we can proceed as follows. Substituting for C_D, the condition for maximum rate of climb can be written

$$\frac{d}{dC_L}\left[\tau C_L^{-1/2} - aC_L^{-3/2} - bC_L^{1/2}\right] = 0$$

and performing the differentiation

$$-\tfrac{1}{2}\tau C_L^{-3/2} + \tfrac{3}{2}aC_L^{-5/2} - \tfrac{1}{2}bC_L^{-1/2} = 0$$

Because $V^2 \propto C_L^{-1}$ we choose to convert this expression into a quadratic in the latter quantity by multiplying through by $2C_L^{1/2}$, giving

$$3aC_L^{-1/2} - \tau C_L^{-1} - b = 0$$

and on using the standard solution for a quadratic we obtain

$$C_L^{-1} = \frac{\tau \pm \sqrt{\tau^2 + 12ab}}{6a}$$

We expect the lift coefficient to be positive and so take the upper sign of the pair to give the result

$$C_L^{-1} = \frac{\tau + \sqrt{\tau^2 + 12ab}}{6a} \tag{3.31}$$

We now find the speed for maximum rate of climb as

$$V_{mc}^2 = \frac{2mg}{\rho S C_L} = \frac{mg}{3\rho a S}\left(\tau + \sqrt{\tau^2 + 12ab}\right) \tag{3.32}$$

or taking mg inside the bracket

$$V_{mc}^2 = \frac{1}{3\rho a S}\left(T + \sqrt{T^2 + 12ab(mg)^2}\right) \tag{3.33}$$

Since the thrust normally falls off with height the eas for optimum climb will also fall with height, in contrast to the case of the propeller-driven aircraft.

Worked example 3.2

An aircraft has a mass of 40 000 kg, a wing area of 120 m^2 and jet engines giving a thrust of 80 kN, assumed not to vary with speed. If the drag coefficient is given by $C_D = 0.015\,75 + 0.033\,35 C_L^2$ find the maximum steady rate of climb at sea-level.

Solution

The maximum rate occurs at the lift coefficient from (3.31) given by

$$C_L^{-1} = \frac{\tau + \sqrt{\tau^2 + 12ab}}{6a}$$

Now $\tau = T/mg = 80 \times 10^3/40\,000 \times 9.81 = 0.2039$. Then

$$C_L^{-1} = \frac{0.2039 + \sqrt{0.2039^2 + 12 \times 0.015\,75 \times 0.033\,35}}{6 \times 0.015\,75} = 4.473$$

The drag coefficient is then $0.015\,75 + 0.033\,35/4.473^2 = 0.017\,42$ and the speed is

$$V = \sqrt{\frac{2mg}{\rho S C_L}} = \sqrt{\frac{2 \times 40\,000 \times 9.81 \times 4.473}{1.225 \times 120}} = 154.5 \text{ m s}^{-1}$$

The rate of climb is then found from (3.29):

$$v_{co} = (0.2039 - 0.017\,42 \times 4.473) \times 154.5 = 19.45 \text{ m s}^{-1} \text{ (Ans)}$$

As a check we calculate the corresponding angle of climb as $\sin \Theta = v_{co}/V = 19.45/154.5$, or $\Theta = 7.23°$, which is within our limit of 10°.

3.3.2 Climb with acceleration

We have seen that the optimum climb procedure indicates that the true airspeed must vary with height for both propeller and jet-driven aircraft, so to maintain optimum conditions some acceleration must occur. Using (3.13) and (3.14) we solve for the rate of climb giving

$$v_c = \frac{\eta P + T_j V - DV}{mg} - \frac{V}{g} \cdot \frac{dV}{dt}$$

The first term on the right is the rate of climb with zero acceleration v_{co} – see (3.21) and (2.21) – and we can write v_c as

$$v_c = v_{co} - \frac{V}{g} \cdot \frac{dV}{dt} \tag{3.34}$$

We now assume that airspeed is some defined function of the height, that is $V = V(h)$; then we have

$$\frac{dV}{dt} = \frac{dV}{dh} \cdot \frac{dh}{dt} = \frac{dV}{dh} \cdot v_c$$

Substituting into (3.34) gives

$$v_c = v_{co} - \frac{V}{g} \cdot \frac{dV}{dh} \cdot v_c$$

and on solving we find

$$v_c = \frac{v_{co}}{1 + V/g \cdot dV/dh} \tag{3.35}$$

The reciprocal of the denominator on the right-hand side of the above expression is known as the 'kinetic energy factor' and is independent of the engine type. Since it contains the airspeed, the earlier determinations of the optimum speeds for climbing are very slightly in error. For this reason the use of the energy height methods to be described in Section 3.3.5 is to be preferred for aircraft capable of high speeds and rates of climb.

We will obtain explicit relations for the denominator of (3.35) in two cases: climb at constant eas and at constant Mach number. Dealing with the first of these, we have from (2.7)

$$V = V_E/\sqrt{\sigma}$$

then on differentiating

$$\frac{dV}{dh} = -\frac{V_E}{2}\sigma^{-3/2}\frac{d\sigma}{dh}$$

The kinetic energy factor then becomes

$$\left(1 - \frac{V_E^2}{2g\sigma^2}\frac{d\sigma}{dh}\right)^{-1} \tag{3.36}$$

The quantity $d\sigma/dh$ can be obtained by differentiating the expressions for σ given in Section 1.4 (or see Student problem 1.4). It is a negative quantity and hence the rate of climb is reduced as would be expected since the true airspeed will increase with height.

Turning now to climb at constant Mach number, we have $V = c_s\mathbf{M}$, where c_s = speed of sound = $\sqrt{\gamma R\vartheta}$ and $\mathbf{M}$ is the Mach number. The temperature ϑ in the troposphere is given by (1.12) as $\vartheta = \vartheta_0 - \lambda h$. We then have

$$V = \sqrt{\gamma R(\vartheta_0 - \lambda h)}\mathbf{M}$$

and on differentiating we find

$$\frac{dV}{dh} = -\frac{\lambda\gamma R\mathbf{M}}{2\sqrt{\gamma R(\vartheta_0 - \lambda h)}} = -\frac{\lambda\gamma R\mathbf{M}}{2c_s}$$

The kinetic energy factor then becomes

$$\left(1 - \frac{1}{2g}\lambda\gamma R\mathbf{M}^2\right)^{-1} \tag{3.37}$$

In this case the rate of climb is increased, as would be expected as the temperature decreases with height so that the speed of sound also decreases and the true airspeed decreases to keep $\mathbf{M}$ constant.

3.3.3 Ceiling

The ceiling is defined as the maximum height to which the aircraft can climb, both the angle and rate of climb having fallen to zero. In the case of propeller-driven aircraft, (3.21) shows that the excess power ($\eta P_e - DV$) is zero and the engine power curve must be tangent to the power required curve as shown for the lowest engine power curve in figure 3.4(a).[2] In the case of jet-driven aircraft, (3.28) shows that the nett forward force ($T - D$) is zero and the thrust curve must be tangent to the drag curve as shown for the lowest curve in figure 3.6. If the power in the first case is constant with speed the speed will be that for minimum power; similarly if thrust is constant the speed is that for minimum drag. At the ceiling the pilot has no freedom to manoeuvre without losing height as there is no excess thrust. The ceiling defined above is often known as the 'absolute ceiling'; other more practical ceilings have been defined. The 'service ceiling' is defined as the height at which the rate of climb has fallen to 100 ft min^{-1} (0.51 m s^{-1}) and ceilings corresponding to rates of climb of 500 ft min^{-1} and

1.5 m s^{-1} have also been proposed. All these ceilings can be determined by plotting rate of climb against height.

3.3.4 Time to height

The rate of climb can be written in the form $v_c = \mathrm{d}h/\mathrm{d}t$, so that the time taken to climb from height h_1 to h_2 is

$$t_{12} = \int_{h_1}^{h_2} \frac{\mathrm{d}h}{v_c} \tag{3.38}$$

The time taken is therefore equal to the area under a plot of $1/v_c$ against h and graphical or numerical methods of integration are indicated.

3.3.5 Energy height methods

In previous sections we have considered the estimation of the rate of climb at constant speed and the correction required to allow for acceleration, and in the previous chapter we considered level acceleration. These topics can all be treated in a unified manner using the concept of energy height introduced in Section 3.3. The effects of changing flight speed on climb performance are most important for supersonic aircraft for which the transonic rise in drag coefficient followed by the fall at supersonic speed (see figure 1.13) means that there may be two ranges of speeds for best climbing separated by a poor region.

Suppose an aircraft is required to achieve a certain speed and height starting from another pair of values of V and h while minimizing the time taken, the fuel used or some other quantity. The problem is then to find the 'climb schedule', that is the function $V = V(h)$, which achieves this. We make one special assumption which is that exchange of V and h at a given h_{en} can be carried out instantaneously and without loss. The effect of this is easily corrected subsequently. We also assume that the climb angle is small, so that $\cos\Theta$ can be taken as unity; the method can, however, be developed without this. This technique is really only relevant to jet-driven aircraft so that we write (3.16) as

$$mg\frac{\mathrm{d}h_{en}}{\mathrm{d}t} = (T - D)V \tag{3.39}$$

where

$$h_{en} = h + \frac{V^2}{2g} \tag{3.40}$$

If we now write

$$w = \frac{\mathrm{d}h_{en}}{\mathrm{d}t} \tag{3.41}$$

then the time taken for an aircraft to reach h_{en2} from h_{en1} is

$$t_{12} = \int_{h_{en1}}^{h_{en2}} \frac{\mathrm{d}h_{en}}{w} \tag{3.42}$$

where

$$w = \frac{T - D}{mg} V \tag{3.43}$$

The thrust and drag are functions of speed and height and therefore w is also. Thus to minimize t_{12} we have to make the integrand as small as possible at each value of h_{en} in the range of integration by varying V and h at constant h_{en}. That is, we must maximize w for each point of the climb schedule. Graphical methods or their computer equivalent are essential and we will describe two graphical methods. The first, which has the advantage of illustrating the climb schedule of the aircraft, is shown in figure 3.7.

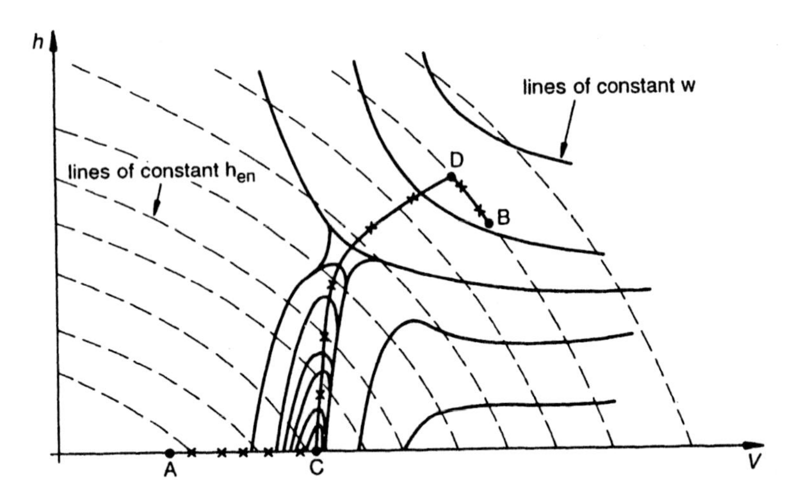

Fig. 3.7 Determination of optimum climb programme

The axes of the graph are height and speed and three kinds of curve are shown. The dashed curves are simply curves of constant energy height and the continuous lines are curves of constant $w = dh_{en}/dt$. The aircraft in this problem is required to achieve the conditions represented by the point B starting from point A, which it does via the climb schedule ACDB, marked by crosses. From point C to D the curve is chosen such that at every point w is maximized with respect to the curves of constant energy height. At D the energy height value of the final point B is achieved and the curve from D to B to the aircraft follows a path at constant energy height and is in fact diving. From A to D the aircraft is simply accelerating. In some cases it has been found that the optimum path may involve the aircraft diving through the transonic speed range as the aircraft has insufficient thrust for level flight there. In determining the optimum path no account is taken of possible limitations on the aircraft such as buffet, or on the engine such as height limitations.

Figure 3.8 shows the second method. The axes in this case are $1/w$ and h. The dashed curves are plots of $1/w$ at constant height for a number of heights; the minima of these curves represent optimum conditions and are joined to produce the curve CD, corresponding to the curve CD in the previous figure. Since $1/w$ is the integrand in (3.42) the area below the curve is proportional to the time taken. In addition the time taken for the sea-level acceleration phase is proportional to the area below the curve AC. The time taken from D to B has to be obtained by other means as these points are at the same value of energy height.

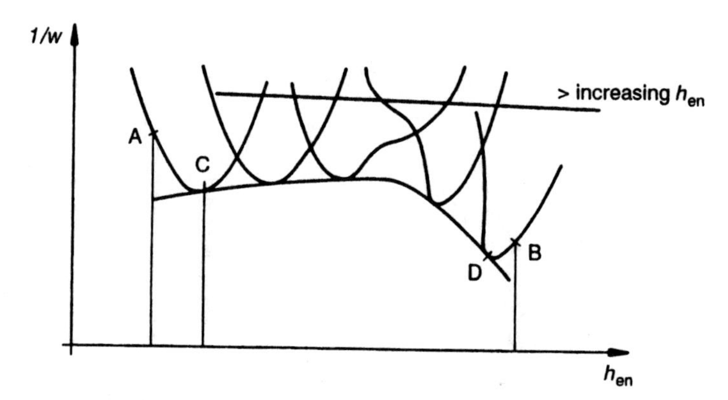

Fig. 3.8 Determination of optimum climb programme, alternative method

Very similar techniques can be used to minimize the fuel used. From (2.30) we have the fuel mass flow rate as

$$\dot{m}_f = f_j T$$

(in this case $T \neq D$). The total fuel used is then

$$m_F = \int \dot{m}_f \mathrm{d}t = \int f_j T \mathrm{d}t$$

or

$$m_F = \int \frac{f_j T}{\mathrm{d}h_{en} / \mathrm{d}t} \mathrm{d}h_{en}$$

and then using (3.39) we find

$$m_F = \int_{h_{en1}}^{h_{en2}} \frac{\mathrm{d}h_{en}}{w_f} \tag{3.44}$$

where

$$w_f = \frac{V(T - D)}{mgf_j T} \tag{3.45}$$

Thus to minimize the fuel used we have to make the integrand as small as possible at each value of h_{en} in the range of integration by varying V and h at constant h_{en}. That is we must maximize w_f for each point of the climb schedule.

3.3.6 Standardized performance

A problem arises in the flight testing of aircraft because we cannot control the atmospheric test conditions closely enough. We therefore seek help by means of dimensional analysis. Taking an aircraft powered by a turbofan engine as an example we look first at the variables that the engine thrust and specific fuel consumption depend on. The relevant parameters are the airspeed V, air pressure p, air temperature ϑ, engine diameter D and rotational speed N. If there is more than one shaft it is usual to take the high-speed compressor shaft speed; the steady speed of the other shafts is a unique function of N. The introduction of the viscosity of the air would simply result in the appearance of the Reynolds number which we can take for granted; also we can omit the air density as it is a function of p and ϑ. We assume that the relation for the thrust is of the form

$$T = \text{function}(V, p, \vartheta, D, N) \tag{3.46}$$

The dimensions of these quantities are as follows: T: $ML^{-1}T^{-2}$; V: LT^{-1}; p: $ML^{-1}T^{-2}$; ϑ: L^2T^{-2}; D: L and N: T^{-1}. Then using Buckingham's π-theorem, which is the result of the necessity of dimensional homogeneity in an equation, we have six quantities with three fundamental dimensions; the result we require must then be one involving three nondimensional groups. We choose the form

$$\frac{T}{pD^2} = \text{function}\left[\left(\frac{V}{\sqrt{\vartheta}}\right),\left(\frac{ND}{\sqrt{\vartheta}}\right)\right] \tag{3.47}$$

If we are considering a particular engine the diameter is constant and can be omitted. We can also write the air pressure in terms of the standard sea-level pressure as $p = \delta p_0$ and omit the constant p_0. Bearing in mind that we are going to apply this to aircraft performance we replace $\sqrt{\vartheta}$ in the first group on the right-hand side of (3.47) by the speed of sound, to which it is proportional. This first group then becomes the Mach number $\mathbf{M}$ assuming that γ and the gas constant, R, are constants. We then write this relation in the form

$$\frac{T}{\delta} = f_1\left(\mathbf{M}, N/\sqrt{\vartheta}\right) \tag{3.48}$$

The variables T/δ, $V/\sqrt{\vartheta}$ and $N/\sqrt{\vartheta}$ are known as 'standardized variables'. The relative temperature could also be substituted for the actual temperature. A similar analysis for the specific fuel consumption gives

$$f_j/\sqrt{\vartheta} = f_2\left(\mathbf{M}, N/\sqrt{\vartheta}\right) \tag{3.49}$$

We turn now to aircraft performance, specifically to the lift coefficient. Assuming that the climb angle is small, using (2.22) we have

$$C_L = \frac{mg}{\frac{1}{2}\rho V^2 S} = \frac{mg}{\frac{\gamma}{2} p\mathbf{M}^2 S}$$

We can then write

$$C_L = f_3(\mathbf{M}, m/\delta) \tag{3.50}$$

taking γ, g and S as constants. Consider now the drag coefficient which we write in the form $C_D = a + bC_L^2$. In the most general case the coefficient a is a function of $\mathbf{M}$ and C_L and the coefficient b is function of $\mathbf{M}$. Using (2.22) again, the drag is given by

$$D = \frac{\gamma}{2} p\mathbf{M}^2 S C_D = \frac{\gamma}{2} p\mathbf{M}^2 S\left(a + bC_L^2\right)$$

or, using (3.50),

$$\frac{D}{\delta} = f_4(\mathbf{M},\ m/\delta) \tag{3.51}$$

Let us now consider the rate of change of energy height. From (3.39) we have

$$mg\frac{\mathrm{d}h_{en}}{\mathrm{d}t} = (T - D)V = \delta\left(\frac{T}{\delta} - \frac{D}{\delta}\right)\frac{V}{\sqrt{\vartheta}}\sqrt{\vartheta}$$

then using (3.48) and (3.51) we find

$$\frac{1}{\sqrt{\vartheta}}\cdot\frac{\mathrm{d}h_{en}}{\mathrm{d}t} = f_5\left(\mathbf{M},\ m/\delta,\ N/\sqrt{\vartheta}\right) \tag{3.52}$$

In the case of level flight at constant speed, where we have $T = D$, or from (3.48) and (3.51),

$$\mathbf{M} = f_6\left(m/\delta,\ N/\sqrt{\vartheta}\right) \tag{3.53}$$

This analysis can be extended to other performance parameters and other types of engine. The use of these variables enables the aircraft performance to be presented in a much more concise form than would otherwise be possible.

3.4 Correctly banked level turns

Any body following a curved path must have a centripetal force applied to it. In the case of an aircraft the centripetal force, which must be normal to the direction of flight and in the horizontal plane, can be generated in two ways. The first is for the pilot to roll the aircraft around the forward axis, as in figure 3.9, so that the lift has a component towards the centre of the turn; this is known as 'banking' the aircraft.

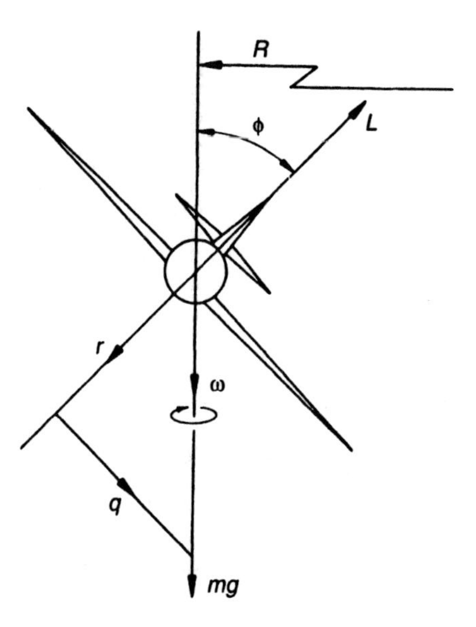

Fig. 3.9 Aircraft in a correctly banked turn

The second is for the pilot to rotate the aircraft about a vertical axis thus giving the fin and fuselage an incidence and thus producing a sideways force; this is known as 'sideslipping' the aircraft and is illustrated in figure 6.2(a). The first method is much the more effective way to produce the necessary centripetal force; the control actions required from the pilot are discussed in Section 6.1.2. A correctly banked turn is one in which the centripetal force is provided solely by the inward component of lift due to banking the aircraft.

We now consider the forces on the aircraft shown in figure 3.9. Resolving vertically we have

$$L \cos \phi - mg = 0 \tag{3.54}$$

where ϕ is the bank angle. The horizontal component of the lift is $L \sin \phi$ and equating this to the product of mass and centripetal acceleration we find

$$L \sin \phi = m\frac{V^2}{R} \tag{3.55}$$

where R is the radius of the turn. Solving (3.54) and (3.55) for $\cos \phi$ and $\sin \phi$ respectively and dividing we find

$$\tan \phi = \frac{V^2}{gR} \tag{3.56}$$

which shows that the bank angle increases rapidly with speed for a given radius of turn. An important parameter in the stressing of an aircraft is the 'load factor', N, which is defined by

$$N = (\text{lift in turn})/(\text{weight}) = L/mg \tag{3.57}$$

Using (3.54) we find

$$N = \sec \phi \tag{3.58}$$

On any given aircraft there is a maximum value of the load factor which the aircraft is designed to withstand. Typical figures are slightly above 2.5 for airliners and 6 to 8 or more for highly manoeuvrable aircraft. This makes load factor a significant parameter to use in equations for the turning performance of an aircraft. Using the trigonometric identity

$$\sec^2\phi = 1 + \tan^2\phi$$

and using (3.49) we find

$$\tan \phi = \sqrt{N^2 - 1} \tag{3.59}$$

Then rearranging (3.56) and using (3.59) we find

$$R = \frac{V^2}{g\sqrt{N^2 - 1}} \tag{3.60}$$

This shows that for a given load factor the radius of turn increases with the square of speed and so for supersonic aircraft the radius of turn can become inconveniently large. We note also that the *centripetal* acceleration is $g\sqrt{N^2 - 1}$. From (3.57) we have $L = Nmg$ and introducing the definition of lift coefficient we find

$$V^2 = \frac{Nmg}{\frac{1}{2}\rho SC_L} \tag{3.61}$$

Then on substituting into (3.60)

$$R = \frac{Nm}{\frac{1}{2}\rho SC_L\sqrt{N^2 - 1}} \tag{3.62}$$

which shows that for the minimum radius of turn the aircraft should fly at the incidence for C_{Lmax}. There is no conflict between (3.60) and (3.62), as in (3.60) as the speed increases C_L decreases. Another performance parameter of interest is the rate of turn, $\omega = V/R$. From (3.60) we find

$$\omega = \frac{g}{V}\sqrt{N^2 - 1} \tag{3.63}$$

and on using (3.61)

$$\omega = g\sqrt{\frac{\frac{1}{2}\rho SC_L}{Nmg}}\sqrt{N^2 - 1} = \sqrt{\frac{\frac{1}{2}\rho gSC_L}{m}}\sqrt{\frac{N^2 - 1}{N}} \tag{3.64}$$

showing that the rate of turn increases with C_L.

We now wish to compare conditions for the same aircraft in turning and straight, level flight. For the rest of this section and the next we will denote quantities in the turn with the suffix 't' and those in straight flight will have no suffix. If the aircraft flies at the same lift coefficient in the two cases (i.e. the same incidence) we have from (3.51)

$$C_L = \frac{mg}{\frac{1}{2}\rho V^2 S} = \frac{Nmg}{\frac{1}{2}\rho V_t^2 S}$$

giving

$$V_t = \sqrt{N}.V \tag{3.65}$$

the speed increasing to provide the additional lift required. Since the lift coefficient is unchanged the drag coefficient is unchanged; then using $D \propto V^2$,

$$T_t = ND = NT \text{ and } P_t = N^{3/2}DV = N^{3/2}P \tag{3.66}$$

where T, T_t, P, and P_t are the thrusts and powers required. If the throttle is not opened to provide this increase the aircraft must descend to provide a component of the weight in the forward direction.

If the aircraft flies at the same speed in the two cases we have

$$V^2 = \frac{2mg}{\rho S C_L} = \frac{2Nmg}{\rho S C_{Lt}}$$

or

$$C_{Lt} = NC_L \tag{3.67}$$

This shows that if an aircraft is flying initially in level flight at a speed too close to the stalling speed, an attempt to turn whilst keeping the speed constant could lead to the aircraft stalling. The effect on the thrust and power required depends on the form of the drag polar. In practice for gentle turns, up to a bank angle of about 30°, the effects of turning are small and although pilots tend to maintain the attitude of the aircraft and not open the throttle the rate of loss of height is small. For steep turns the piloting technique is different and opening the throttle is necessary.

Worked example 3.3

An aircraft performs a level turn at a speed of 250 m s^{-1} and a bank angle of 65°. Find the load factor, the radius of turn, the angular rate of turn and the ratio of the power in the turn to that in straight flight.

Solution

Using (3.59) we have $N = \sec\phi = 2.37$. Then from (3.60) $R = V^2 \Big/ \left(g\sqrt{N^2 - 1}\right)$ and substituting gives $R = 250^2 \Big/ \left(9.81\sqrt{2.37^2 - 1}\right) = 2965$ m. The angular rate is $V/R =$ 250/2965 = 0.0843 rad s^{-1} and the power ratio is given by $N^{3/2} = 2.37^{3/2} = 3.65$.

3.4.1 Turns at constant throttle

It is instructive to investigate the rate of loss of height in a turn assuming that the throttle is not opened. For simplicity we assume that the turn takes place at the same lift coefficient as for the initial level flight case and that the engine thrust is independent of speed. The forces acting are illustrated in figure 3.10 and γ is the angle of descent.

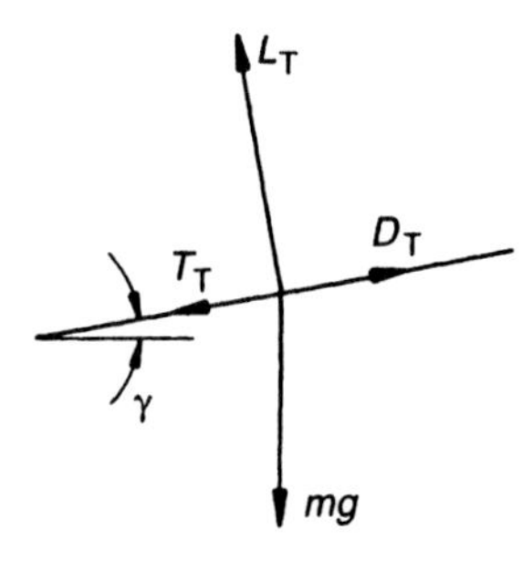

Fig. 3.10 Descent in a turn

In the figure it should be remembered that the lift is also inclined towards the centre of the turn. Resolving normal to the flight path gives

$$L_t \cos\phi - mg\cos\gamma = 0$$

and assuming that γ is a small angle such that $\cos\gamma \simeq 1$ we have

$$L_t \cos\phi = mg \tag{3.68}$$

We can therefore still write $N = \sec\phi$. Resolving along the flight path gives

$$T_t - D_t + mg\sin\gamma = 0 \tag{3.69}$$

Since thrust equals drag in steady level flight we have

$$T = D = T_t$$

Using this and (3.66) in (3.69) gives

$$D - ND + mg\sin\gamma = 0$$

Solving for the descent angle,

$$\sin\gamma = \frac{D(N-1)}{mg} = \frac{N-1}{C_L/C_D} \tag{3.70}$$

showing that the minimum angle occurs at the minimum drag conditions. The rate of losing height is given by $v_s = V_t \sin\gamma$ and on using (3.65) we have

$$v_s = \frac{DV\sqrt{N}(N-1)}{mg} = \frac{V\sqrt{N}(N-1)}{C_L/C_D} \tag{3.71}$$

The first form shows that the minimum rate of loss occurs at the minimum power condition.

3.5 Take-off and landing

The primary objective of this section is to discuss the estimation of the distances required for an aircraft to take off and land. These are not single manoeuvres in the sense that the term has been used up to now but are really a series of manoeuvres and the division into parts can be

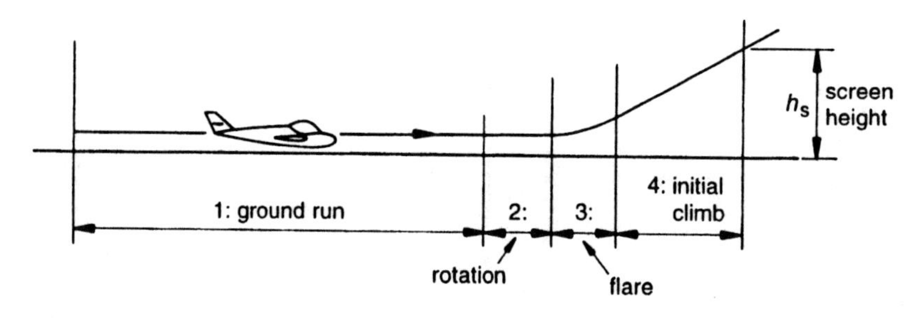

Fig. 3.11 Phases of a take-off

the subject of some debate. Piloting also comes into this discussion as variations in technique have an effect on the distances; the result is that assumptions have to be made which in effect standardize the technique used. We start by discussing take-off; the methods used are easily adapted for the case of landing.

Consider the path of the cg of an aircraft during a take-off as shown in figure 3.11. The take-off is divided into four phases as follows:

1. **Ground run.** During this phase the aircraft accelerates up to the 'rotation speed', V_R, and the attitude of the aircraft is constant so that C_L and C_D are constant.
2. **Rotation.** During this phase the aircraft is rotated (pitched nose-up) by the pilot so that at the end of the phase the lift exceeds the weight and the aircraft lifts off the runway.
3. **Flare.** In this phase the angle of the flight path is changed to match the final climb angle; the cg follows a nearly circular path.
4. **Initial climb.** Here the aircraft climbs at steady angle and speed to the 'screen height' h_s, usually taken to be 15 m. Thrust, C_L and C_D can also be assumed constant.

For a simpler treatment phases 1 and 2 can be combined as can phases 3 and 4; in effect the rotation and flare are assumed to take place instantaneously.

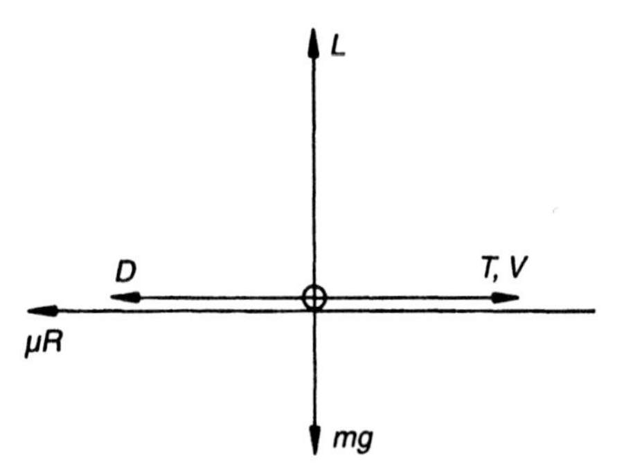

Fig. 3.12 Forces on aircraft in ground run

The forces acting at a point during the ground run are illustrated in figure 3.12, where μR is the rolling friction of the tyres and R is the ground reaction, $(mg - L)$.

The nett forward force then gives the acceleration, a, as

$$m\,a = T - D - \mu R \tag{3.72}$$

This could be integrated once to give the speed and again to find the time taken. However, we require the distance. For this we have

$$a = \frac{dV}{dt} = \frac{dx}{dt}\cdot\frac{dV}{dx} = V\cdot\frac{dV}{dx} = \tfrac{1}{2}\cdot\frac{dV^2}{dx}$$

The acceleration distance is then

$$x_1 = \tfrac{1}{2}\int_0^{V_R}\frac{d(V^2)}{a} \tag{3.73}$$

where from (3.72)

$$a = \frac{T - D - \mu(mg - L)}{m} \tag{3.74}$$

Since the thrust varies with speed in general, the most accurate results are obtained by using numerical integration of (3.73) using equal steps in V^2. The simplest procedure is to calculate the acceleration at say $0.7V_R$ and assume that this is a good approximation to the mean value. Then the acceleration distance is given by

$$x_1 = \frac{m}{\left[T - D - \mu(mg - L)\right]_{\text{mean}}} \cdot \frac{V_R^2}{2} \tag{3.75}$$

To get a feel for the important parameters we introduce some approximations; we can ignore the rolling friction compared to the thrust, and the drag is zero at least initially, so that we approximate the acceleration to T_{mean}/m. We assume that the rotation speed is 15 per cent above the stalling speed so that we have

$$V_R^2 = \frac{1.15^2 mg}{\frac{1}{2}\rho S C_{L\max}}$$

Then

$$x_1 = \frac{1.15^2}{g\rho C_{L\max}} \cdot \left(\frac{mg}{S}\right) \frac{1}{(T_{\text{mean}}/mg)} \tag{3.76}$$

or

$$x_1 \propto \frac{\text{wing loading}}{\text{thrust loading}} \cdot \frac{1}{\sigma C_{L\max}} \tag{3.77}$$

where the thrust loading is defined here as T_{mean}/mg. The ground run is the largest single part of the take-off distance and the rest of it is similarly affected by the above parameters, so that take-off distances correlate well with the right-hand side of (3.77). Weight is a particularly important parameter as increasing weight increases the wing loading and decreases the thrust loading; both of these effects increase the take-off distance. The result is that the ground run is proportional to $(mg)^2$. Evidently increasing altitude increases the take-off distance; there is also a less direct effect of altitude in that the fall-off of air density decreases the thrust. There may also be an effect of airfield temperature, increase of which often decreases engine thrust. The operation of an aircraft from airfields at high altitudes in the tropics usually provides the most testing cases for the design for take-off performance. As a result aircraft manufacturers need to provide curves to operators giving the effects of weight, altitude and temperature on performance, known as 'WAT curves'; in addition the effects of runway slope and wind must be included.

Returning to the subject of estimating the ground run, if we can assume a convenient mathematical form for the effect of speed on the thrust we can then integrate (3.73) analytically to obtain an expression of intermediate accuracy. A suitable form is

$$T = T_0 - kV^2 \tag{3.78}$$

where T_0 is the static thrust and k is a constant. We also write

$$L = \tfrac{1}{2}\rho V^2 S C_{Lg}$$

and

$$D = \tfrac{1}{2}\rho V^2 S C_{Dg}$$

where C_{Lg} and C_{Dg} are the values of C_L and C_D corresponding to the ground attitude of the aircraft with flaps and undercarriage lowered and allowing for the effects of ground proximity. Then substituting into (3.74) and collecting terms,

$$a = \left(\frac{T_0 - \mu mg}{m}\right) - \left(\frac{k + \tfrac{1}{2}\rho S\left(C_{Dg} - \mu C_{Lg}\right)}{m}\right)V^2 \tag{3.79}$$

For simplicity we write this as $a = \mathscr{A} + \mathscr{B}V^2$, where $\mathscr{A}$ and $\mathscr{B}$ are the quantities in the first and second brackets in (3.79) respectively. Then substituting into (3.75) and integrating we find

$$x_1 = \frac{1}{2}\int_0^{V_R} \frac{d(V^2)}{\mathscr{A} + \mathscr{B}V^2} = -\frac{1}{2\mathscr{B}}\left[\ln\left(\mathscr{A} + \mathscr{B}V^2\right)\right]_0^{V_R}$$

or

$$x_1 = \frac{1}{2\mathscr{B}} \ln \frac{\mathscr{A}}{\mathscr{A} + \mathscr{B}V^2} \tag{3.80}$$

Runways are rarely exactly horizontal and the effect of a component of weight along the take-off direction is easily incorporated in the above analysis. Aircraft often take off against a component of the wind; this can be allowed for provided care is taken to distinguish between the aircraft speed relative to the ground, V_g, and that relative to the air, V_a say. The former is related to the distance along the ground and the latter to the aerodynamic forces; the two are related by $V_a = V_g + w$, where w is the headwind component.

The distance travelled during the rotation phase is estimated from the time taken for rotation. For transport aircraft a rotation rate of 3° per second is assumed; typically the aircraft will need to rotate through 12°, giving a time of four seconds. The distance can then be estimated by numerical integration as for the ground run or by assuming a mean speed and multiplying by the time. The aircraft will lift off when the lift exceeds the weight but rotation may continue as during the flare lift must exceed the weight to provide a centripetal acceleration.

The last two phases, the flare and the initial climb, are dealt with together and are shown in more detail in figure 3.13.

The flare ends when the flight path reaches the climb angle given by (3.26) or (3.28), repeated here:

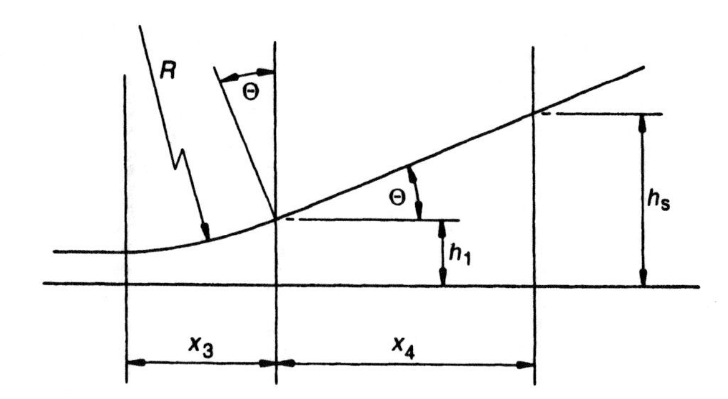

Fig. 3.13 Detail of flare and initial climb

$$\sin \Theta = \frac{T - D}{mg} \tag{3.81}$$

where the thrust and drag are calculated using values appropriate to the aircraft speed and condition. This angle is needed later to determine x_3. Then assuming that h_1 is the height at the end of the flare we have

$$\tan \Theta = \frac{h_s - h_1}{x_4}$$

or

$$x_4 = \frac{h_s - h_1}{\tan \Theta} \tag{3.82}$$

where h_1 has yet to be determined. The distance taken in the flare is fixed effectively by the centripetal acceleration available at the end of rotation, $\Delta n.g$, then

$$\Delta n.g = \frac{V^2}{R}$$

or

$$R = \frac{V^2}{\Delta n.g} \tag{3.83}$$

and from figure 3.11 we have, if the climb angle is not too large,

$$x_3 = R\Theta \tag{3.84}$$

We still need to find h_1, which from figure 3.13 is

$$h_1 = R(1 - \cos \Theta)$$

Then using the series for cos Θ we have

$$h_1 = R\left(1 - \left\{1 - \Theta^2/2! + \Theta^4/4! - \ldots\right\}\right) \approx \frac{R\Theta^2}{2} \tag{3.85}$$

to second order in Θ and enabling us to find x_4 using R from (3.83). In some cases the height gained in the flare found in this way is greater than the screen height. As the take-off is taken to end at the screen height there is no climb-away phase in this case and the angle Θ is found

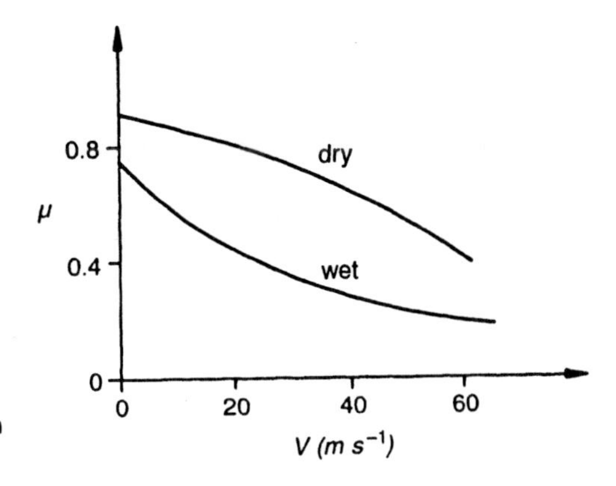

Fig. 3.14 Variation of maximum braking friction with speed

from (3.85) putting $h_1 = h_s$; x_3 is found from (3.84).

3.5.1 Landing

Landing can be divided into four phases which correspond to those used to analyze the take-off, as shown in figure 3.14, or into two phases.

The distance taken in this case is much more affected by piloting technique. Student pilots tend to end the flare with the aircraft above the runway and the aircraft 'floats' along in the ground effect, decelerating until it finally stalls onto the runway. Experienced pilots generally touch down at the end of the flare at a speed higher than the stalling speed and it is this distance that is required. 'Derotation' follows very shortly after touchdown.

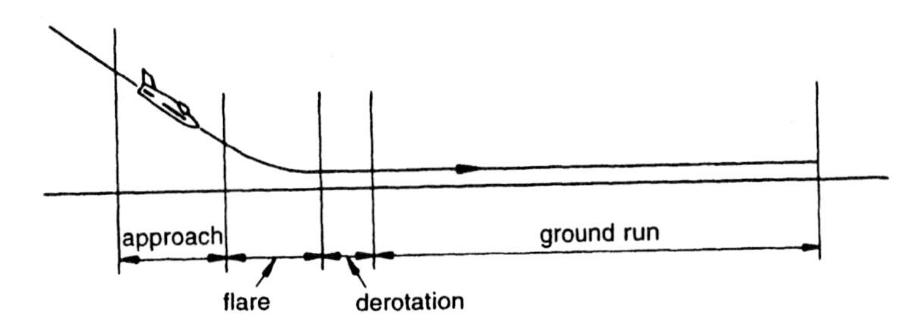

Fig. 3.15 Phases during landing

There are a number of factors which can make the estimation of the ground run distance more complicated than that of take-off. Energy is dissipated in flexing the tyres and in the brakes, so that the coefficient of friction is now the sum of the rolling and braking coefficients and is strongly affected by the speed and whether the runway is dry or wet, as sketched in figure 3.15 (71026).

These values could only be achieved if all the wheels have brakes and with perfect antiskid systems. The maximum deceleration may be limited by passenger comfort or other reasons and there may be a limit to the energy which can be absorbed by a braking system; possibly

80 per cent of these values is a fair assumption. The brakes can only be applied after touch-down and so there is a delay of about two seconds before they become effective. The same applies to the use of reverse thrust or 'lift dumpers' (a form of spoiler) which are found on sophisticated aircraft. The braking friction is a much more important factor than the rolling friction is in take-off distance estimation and it is more necessary to use numerical integration.

3.5.2 Balanced field length

Safety is a prime consideration in all aircraft design and in this context it is necessary to consider the effect of engine failure during take-off. In the case of a single-engined aircraft, failure just after take-off is mainly a matter of looking for a suitable field to land in and landing safely. Failure on the ground simply means aborting the take-off. In the case of multi-engined aircraft, if failure occurs during the take-off the pilot has the options of either continuing or

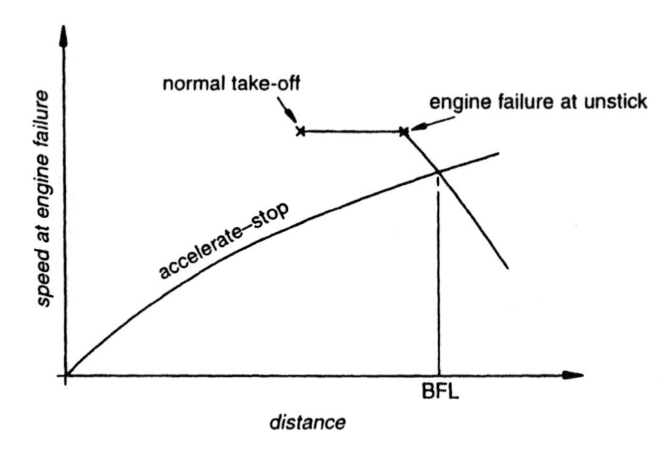

Fig. 3.16 Determination of balanced field length

throttling back the remaining engines and applying the brakes. The latter is known as an 'accelerate–stop' manoeuvre. At any given speed one or other of these options will require the smaller amount of runway; at low speeds the aircraft should stop whilst at speeds near to that for lift-off the better option is to continue. The pilot needs a simple criterion on which to judge the correct option and this is provided by the speed at which the engine fails. Calculations are therefore made for both distances and plotted against speed of engine failure as shown in figure 3.16.

In the calculation of accelerate–stop distance allowances must be made for the time taken for the pilot to make the decision to stop, the time taken for the remaining engines to run down and for brakes and lift-dumpers to become effective. Where the curves intersect defines the decision speed, V_1, and the corresponding distance is known as the 'balanced field length' (BFL). The minimum field length for the safe operation of an aircraft for a specified take-off weight is the larger of the BFL and the landing distance at the maximum permitted landing weight. Since the airworthiness regulations require that multi-engined aircraft are capable of climbing with one engine failed, engine failure during the climb-away will result in climb at a reduced angle. The pilot is likely in that case to abort the mission, jettisoning fuel to reduce the weight to the maximum for landing, and land.

Rate of climb with one engine failed is a safety issue at any altitude and the airworthiness regulations specify minimum rates of climb which depend on the number of engines.

3.5.3 Reference speeds during take-off

There are a number of speeds attained by an aircraft during take-off between which the airworthiness regulations require certain relations to be satisfied. The basic speeds are the stalling speed V_{s1}, the minimum control speeds in the air V_{MCa} and on the ground V_{MCg}, the decision speed V_1 and the minimum airspeed at which the aircraft can take off, V_{MU}. The latter may be related to the maximum angle the aircraft can reach without contacting the runway. The minimum control speeds are discussed in Section 6.3.3.

The principal requirements are that:

1. The speed at which rotation is initiated, V_R, shall be at least 5 per cent greater than V_{MCg} and that rotation at V_R followed by the maximum rate of rotation shall not result in a lift-off speed less than $1.05V_{MU}$ with one engine failed or $1.1V_{MU}$ with all engines operative.
2. V_1 shall be equal to or greater than $1.05V_{MCa}$.
3. The target speed at the end of take-off, V_2, which is chosen by the designer, must provide a specified rate of climb and be equal to or greater than $1.2V_{s1}$ for two- or three-engined aircraft and $1.15V_{S1}$ for four-engined aircraft. Also V_2 must be equal to or greater than $1.1V_{MCa}$.

For a complete statement of the requirements for large civil aircraft the European Joint Airworthiness Requirements (JAR)–25 or the equivalent US Federal Aviation Regulations (FAR) Part 25 should be consulted.

Student problems

3.1 An aircraft having a mass of 36 290 kg and a wing area of 93 m^2 has jet engines having a thrust of 57.8 kN, assumed not to vary with speed. The drag coefficient is given by $C_D = 0.014 + 0.05C_L^2$. Find, at sea-level:
(a) the angle and steady rate of climb at a speed of 92 m s^{-1};
(b) the maximum steady angle of climb, the corresponding speed and rate of climb;
(c) the speed for maximum steady rate of climb, the corresponding angle and rate of climb. (A)

3.2 The aircraft of the previous question is fitted instead with turboprop engines of total power 10 000 kW having propellers of 87 per cent efficiency. Repeat parts (a) and (c) of the question. (A)

3.3 At sea-level a jet aircraft has its maximum steady rate of climb at a speed of 152 m s^{-1}, which occurs at a shallow angle. The aircraft has a mass of 22 200 kg, a wing loading of 2.4 kN m^{-2} and a drag coefficient given by $C_D = 0.018 + 0.065C_L^2$. Estimate the angle and steady rate of climb when climbing at the same incidence but using rocket giving 65 kN of additional thrust. (A)

3.4 An aircraft has the following lift/drag characteristics:

C_L	0.2	0.3	0.4	0.6
C_D	0.018	0.0195	0.024	0.036

The engine has a thrust of 26 kN at an altitude where the relative density is 0.45 and the speed of sound is 310 m s^{-1}. If the aircraft mass is 18 000 kg and the wing area is 50 m^2, find the speed for the maximum rate of climb at this altitude if the aircraft is constrained to fly at constant Mach number, allowing for the effects of acceleration. Also find the rate of climb. The temperature lapse rate is 6.5 K km^{-1}. (Hint: in this case $C_D \neq a + bC_L^2$.) (A)

3.5 A sailplane with a wing loading of 250 N m^{-2} has a best gliding angle of 1.5° at 25 m s^{-1}.

Determine the coefficients a and b in the equation for the drag coefficient $C_D = a + bC_L^2$. Also determine the minimum sinking speed and the glide angle at which it is achieved assuming standard sea-level conditions. (A)

3.6 An aircraft has a wing area of 260 m^2, a mass of 44 000 kg and an engine power given by $5000\sigma^{0.65}$ kW. Find the absolute ceiling if the drag coefficient is $C_D = 0.016 + 0.052C_L^2$ and the propeller efficiency is 88 per cent. Assume that the relative density is given by

$$\sigma = \frac{20 - H}{20 + H}$$

where H is the height in km. (A)

3.7 A twin engine turboprop aircraft having engines which give a maximum power of 5000 kW has been modified to be used as a 'flying test-bed' for a turbojet engine giving a maximum thrust of 30 kN, both figures at sea-level and assumed to be independent of speed. Find the maximum steady rate of climb at sea-level with all three engines working and the speed at which it occurs. The drag coefficient is given by $C_D = 0.016 + 0.055C_L^2$, the aircraft mass is 48 000 kg, the wing area is 260 m^2 and the propeller efficiency is 87 per cent. (A)

3.8 Show that the steady rate of climb of a jet engine aircraft is given by

$$v_{co} = \frac{2}{3}\sqrt{\frac{mg}{3a\rho S}}\left(\frac{\tau^2 + \tau\tau^1 - 12ab}{\sqrt{\tau + \tau^1}}\right)$$

where $\tau = T/mg$, $\tau^1 = \sqrt{\tau^2 + 12ab}$, $C_D = a + bC_L^2$ and the thrust has been assumed constant with speed. Try to find two other equivalent expressions.

3.9 Calculate the radii of turns of load factor 2.6 for aircraft at Mach numbers $\mathbf{M} = 0.85$, 1.5 and 2.5 at a height where the speed of sound is 300 m s^{-1}, find the bank angle and comment. (A)

3.10 A propeller-driven aircraft of mass 7500 kg is flying straight and level at 110 m s^{-1} at an altitude where $\sqrt{\sigma} = 0.776$. It then executes a correctly banked turn at a rate of 8° s^{-1} whilst flying at the same incidence. If the wing loading is 1.8 kN m^{-2} find the radius of turn and the load factor. (A)

3.11 An aircraft of mass 10 000 kg has jet engines giving a thrust of 40 kN. The wing area is 50 m^2, the wing lift curve slope is 4.5 and the no-lift incidence is $-2.5°$. The wing chord line is parallel to the engine thrust line. Find the radius of a correctly banked turn in which the total load factor is 4 at an altitude where the relative density is 0.74, the wing incidence then being 8°. (A)

3.12 A jet aircraft is flying straight and level at 100 m s^{-1} true airspeed; it then executes a turn at the same incidence and throttle setting. It is observed to lose 200 m in altitude in 15 s. If the lift/drag ratio at this incidence is 14 find the angle of bank and the radius of turn, neglecting any changes of thrust or air density. (A)

3.13 A jet aircraft is flying straight and level at a speed 20 per cent greater than the minimum drag speed when it executes a correctly banked turn at the same incidence and throttle setting. The bank angle is 50° and it is observed to lose 100 m in altitude in 14 s. It is known that the maximum speed at that altitude is 240 m s^{-1}, the air density is 0.8 kg m^{-3} and the maximum thrust is 25 kN. The wing area is 50 m^2 and the mass is 15 000 kg. Find the constants a and b in the parabolic drag law $C_D = a + bC_L^2$, neglecting any changes of air density or thrust at constant throttle.

3.14 As part of a two-stage take-off calculation find the ground run distance for an aircraft with the following characteristics: mass = 21 000 kg, wing loading = 1.9 kN m^{-2}, C_{Lg} = 0.4, unstick C_L = 1.8, drag coefficient in the air and on the ground $C_D = 0.058 + 0.05C_L^2$, total engine power = 6000 kW, propeller efficiency at take-off speed = 58 per cent, static thrust = 92 kN, $\mu = 0.035$. Assume a parabolic law for the thrust variation between zero speed and unstick. (A)

3.15 Repeat the last problem including a headwind of 10 m s^{-1}.

3.16 As part of a two-stage calculation find the ground run distance for an aircraft with the following characteristics landing on a runway with a 1° downward slope: mass = 20 000 kg, wing area = 100 m^2, reverse thrust = 30 kN (assumed constant with speed), $C_{Lmax} = 2.8$, $C_{Lg} = 0.2$, $C_{Dg} = 0.17 + 0.06C_{Lg}^2$, coefficient of braking friction = 0.2. Assume that the touch-down speed is 30 per cent above the stalling speed. (A)

3.17 An aircraft has the following lift/drag characteristics in the ground run configuration:

C_{Lg}	0.1	0.2	0.3	0.4	0.5
C_{Dg}	0.0586	0.0598	0.0625	0.0676	0.0741

Find the lift coefficient which gives the shortest ground run in taking off from a level runway, and find the distance (two-stage calculation). Other characteristics of the aircraft are: mass = 30 000 kg, thrust = 90 kN, C_L at unstick = 1.7, coefficient of rolling friction = 0.04. (A)

3.18 Using a two-stage calculation find the total landing distance from 15 m for an aircraft with the following characteristics: mass = 28 000 kg, wing loading = 1.8 kN m^{-2}, C_{Lg} = 0.3, C_L on approach and touchdown = 1.7, $C_{Dg} = 0.06 + 0.62C_{Lg}^2$. The coefficient of braking friction is given by

$$\mu = \frac{2650}{2915 + V^2}$$

where V is in m s^{-1}. Integrate analytically or use Simpson's rule using equal intervals in V^2. (A)

3.19 Find the distance required in the last two phases of a four-stage take-off calculation for an aircraft with the following characteristics: mass = 77 000 kg, constant thrust = 110 kN, wing area = 160 m^2, $C_{Lmax} = 1.5$, $C_D = 0.056 + 0.04C_L^2$. Assume that the unstick speed is 20 per cent above the stalling speed and that the increment in normal acceleration in the circular arc is 0.1g and the screen height is 15 m. (A)

3.20 Show that the increment in ground run distance during take-off caused by using a runway of adverse slope β is approximately

$$\Delta x = \frac{\beta V_{us}^{\ 2}}{2g\tau\{\tau - (\mu + \beta)\}}$$

where $\tau = T_0/mg$. Hint: use the series ln $(1 + x) \simeq x + \ldots$.

3.21 Show that the ground run distance given by (3.81) can be written approximately as

$$x = \frac{w}{\rho g C_{Lus}(\tau - \mu')}$$

where $w = mg/S$, $\tau = T/mg$, C_{Lus} is the unstick C_{L} and

$$\mu' = \mu\left(1 - C_{\mathrm{Lg}}/2C_{\mathrm{Lus}}\right) + C_{\mathrm{Dg}}/2C_{\mathrm{Lus}}$$

Hint: an expression of the form $(1 + x)$ can be written in the form $\left(\dfrac{1+u}{1-u}\right)$ and $\ln\left(\dfrac{1+u}{1-u}\right) \simeq 2\{u + \text{terms of third and higher order}\}$.

Notes

1. It is important for the student beginning this subject not to confuse the climb and incidence angles.
2. Theoretically there could be more than one tangent.

4
Introduction to stability and control

4.1 Aims of study

Suppose we have an aircraft in some state of steady flight. If it is disturbed, by a gust say, or by the pilot, it is regarded as stable if it returns to a sensibly steady state within a finite time. The final state, however, does not have to be identical to the initial state, although it often will be. Depending on circumstances we may be able to tolerate a small degree of instability or even deliberately design an aircraft to be quite unstable; in the latter case, however, a reliable automatic stabilization system will be required. We normally require more than mere stability; the response to gusts must not make the pilot's task difficult, produce an uncomfortable ride for passengers, impose excessive loads on the aircraft, or make the aircraft unsuitable as an aiming platform. The pilot must be able to control the aircraft accurately without having to perform excessive feats of skill or strength.

Our first aim then is to study the dynamics of the aircraft and its interaction with the aerodynamics in order to be able to assess and possibly improve the dynamic characteristics. A further aim is to understand the physics of the processes involved. If necessary we make approximations as, while better numerical results can generally be found using a computer, little real understanding follows its use alone. With a good understanding of the physics involved, solutions to design problems can be put forward.

4.2 First thoughts on stability

There are a few matters to be discussed as a preliminary to the remaining chapters, consisting of some notation and a little basic theory mainly to obtain some insight.

4.2.1 Choice of axes

In solving a problem involving the dynamics of a body it is usually necessary to set up axes of reference and define variables of motion; however, aircraft present some problems in this respect. We have to choose between (i) axes fixed relative to the ground, (ii) axes which translate with the aircraft but do not rotate and (iii) axes fixed relative to the aircraft so that they both rotate and translate with it. A little thought shows that in the first case an aircraft flying in a straight line has moments of (constant) forces and moments of inertia about the axes which will vary with time. In the second case simple rotations of the aircraft have the same effect. We are then forced to adopt axes which remain fixed relative to the aircraft so that they move with it and recognize that our equations of motion will need to be modified accordingly. Such axes are known as 'body axes'. The origin is best placed at the cg as then there are no moments of the weight to allow for. Our axes are then chosen as follows. The Ox axis points in the direction of flight, Oy points at right angles to the plane of symmetry and towards the starboard wing tip (the right-hand one looking forwards), and Oz points downwards and completes a right-handed set. Figure 4.1 shows the axes.

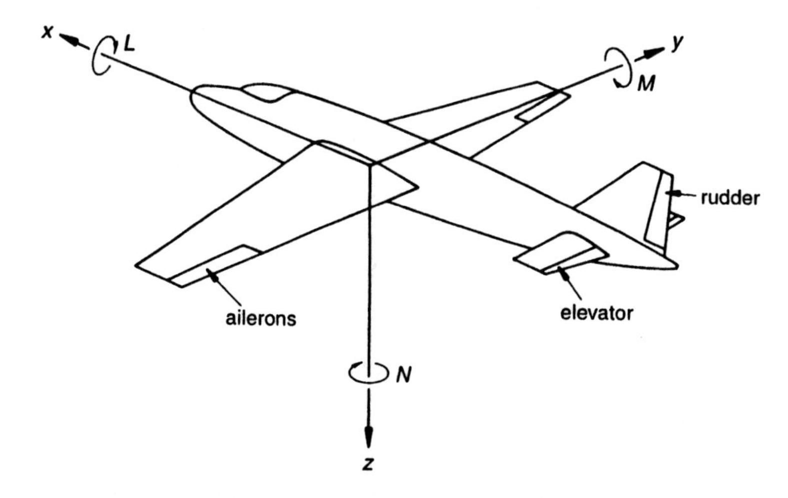

Fig. 4.1 Stability axes and control surfaces of an aircraft

The dynamics of the aircraft are then dealt with using these axes; for other purposes such as finding the change of height in a disturbance we will have to introduce further sets of axes.

In the next two chapters we shall be mostly concerned with the moments about the axes just defined; these are L about the Ox axis, M about the Oy axis, and N about the Oz axis, as shown in figure 4.1. They are known as the rolling, pitching and yawing moments respectively. Like much of the terminology these have been borrowed from nautical parlance. It is slightly unfortunate that L is used for rolling moment as it is also used for lift. The practice is, however well established, has merit and causes no difficulty. This notation will be elaborated upon in Chapter 8 to deal with the requirements of subsequent chapters.

The velocity components, forces and moments and other similar quantities will change when the aircraft is disturbed but we need to be able to define nondimensional quantities. We therefore define a datum state and use the corresponding values of speed, density and so on for this purpose. The datum state will be the initial flight condition if it was a steady state and therefore in equilibrium. The subscript 'e' is used to denote a datum quantity; in particular the airspeed is written V_e, replacing the V for true airspeed used in earlier chapters.

4.2.2 Static and dynamic stability

An aircraft in flight is completely unconstrained and so has six degrees of freedom, three in translation and three rotational ones. This means that there are many ways for instability to appear and due to the high speeds serious consequences may ensue. To assist in dealing with these problems we first discuss a weaker form of stability, known as 'static stability'. A body is said to be statically stable if a disturbance generates a force or moment, as appropriate, which tends to restore the body to the initial state.

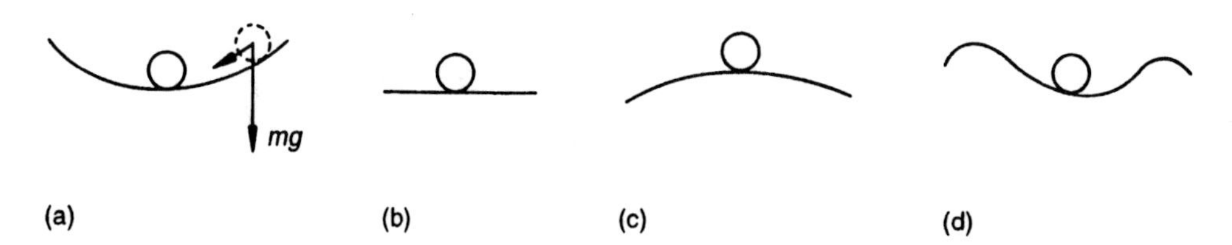

Fig. 4.2 Illustration of various terms in static stability: (a) stable, (b) neutral stable, (c) unstable, (d) conditional stability

Consider the ball in a groove as shown in figure 4.2(a). If the ball is displaced to the right as shown then a component of the weight appears towards the initial position; initially therefore the ball was statically stable. By contrast the ball in figure 4.2(c), when displaced, has a force on it tending to increase the displacement and is therefore statically unstable. The ball in figure 4.2(b) has no force on it produced by a displacement and so is in a state of neutral static stability. When an investigation using the full dynamic equations finds that an aircraft is in a stable state it is said to be 'dynamically stable'. It can be shown that positive static stability is a necessary but not sufficient condition for dynamic stability. A further possibility is shown in figure 4.2(d) where the ball is stable to small displacements but unstable if displaced too far; this can be described as a case of 'conditional stability'. An aircraft can be statically stable to small changes in incidence but unstable if the change of incidence results in the wing stalling.

In the case of aircraft it is useful to quantify the static stability; the restoring factor is usually a moment so the quantity is the rate of change of a moment with displacement, and so has the form of a stiffness. We also need to change the same displacements using the controls; the latter apply moments which are resisted by the same restoring moments. The consequence is that an aircraft with a high degree of static stability also requires large control movements and vice versa. We often find this sort of relationship between stability and control in mechanical systems.

4.2.3 Approximate treatment of response to gusts

A very important feature of the dynamics of aircraft is the response to gusts and in particular to vertical gusts; an approximate treatment of the latter case yields useful information. Suppose that an aircraft, flying horizontally at steady speed V_e, flies into a region of the atmosphere where the air has a constant upward velocity w_g. We will assume that the aircraft does not change its forward speed or its attitude in space and that the lift changes instantly to correspond to the incidence change. The latter implies, for instance, that we can ignore the time taken for the aircraft to enter fully into the region of rising air. Initially the vertical velocity of the aircraft is zero; let the downward velocity (i.e. along Oz) of the aircraft after a time t from entering the gust be w. Both w and w_g are taken to be small compared with the forward velocity. The increment of incidence on the wing at time t is then

$$\Delta\alpha = \frac{w + w_g}{V_e} \tag{4.1}$$

and hence the upward force on the aircraft is

$$\Delta L = \tfrac{1}{2}\rho V_e^2 Sa\Delta\alpha$$

where a is the lift curve slope. Then from (4.1)

$$\Delta L = \tfrac{1}{2}\rho V_e Sa(w + w_g) \tag{4.2}$$

The downward acceleration is given by

$$m\frac{dw}{dt} = -\Delta L$$

and on substituting from (4.2) and rearranging we obtain

$$\frac{\tau}{a}\frac{dw}{dt} + w = -w_g \tag{4.3}$$

where

$$\tau = \frac{m}{\frac{1}{2}\rho V_e S} \tag{4.4}$$

The quantity τ is a constant having the units of time and will be discussed further in the next section. The solution to (4.3) is

$$w = -w_g[1 - \exp(-at/\tau)] \tag{4.5}$$

The *upward* velocity of the aircraft ($-w$) therefore tends to the gust velocity. On differentiating (4.5) we find the acceleration is

$$\frac{dw}{dt} = -\frac{w_g a}{\tau}\exp(-at/\tau) \tag{4.6}$$

The acceleration is a maximum at zero time and its maximum *upward* value as a fraction of g is

$$n_{\max} = -\frac{1}{g}\frac{dw}{dt} = \frac{\rho V_e S w_g a}{2mg} \tag{4.7}$$

This result shows that the maximum normal acceleration is proportional to the air density, the airspeed and the lift curve slope; it is also inversely proportional to the wing loading.

Suppose that we choose to use τ as a unit of time, so that we write $t = \tau\hat{t}$ and we measure the velocities in terms of V_e so we write $w = V_e\hat{w}$ and $w_g = V_e\hat{w}_g$. Then on substituting, the vertical velocity becomes

$$\hat{w} = -\hat{w}_g[1 - \exp(-a\hat{t})] \tag{4.8}$$

and introducing the lift coefficient in the undisturbed state, C_{Le}, the maximum normal acceleration, in 'g' units, from (4.7) is

$$n_{\max} = \frac{a}{C_{Le}}\hat{w}_g \tag{4.9}$$

So that to the degree of accuracy of this theory the motion of all aircraft with the same value of lift curve slope is the same in these variables. If also the initial lift coefficient is the same, then so is the maximum normal acceleration. We will return to the problem of the response of aircraft to gusts in Chapter 10.

4.2.4 The natural time scale

We turn now to determining what we can learn from the use of dimensional analysis. Suppose we have two geometrically similar aircraft performing similar manoeuvres and wish to find how the times for the manoeuvres depend on the various relevant parameters. Specifically we define similar manoeuvres such that the changes of the same velocity component are the same as a percentage of the undisturbed velocity in the two cases. The relevant parameters are the aircraft mass m, the density of the air ρ, the aircraft initial velocity V_e and a characteristic length l. The introduction of the viscosity of the air or the speed of sound would simply result in the appearance of the Reynolds number and the Mach number so we need not include these. We then assume that the time is given by a relation of the form

$$t = \text{function}\,(l, V_e, \rho, m) \tag{4.10}$$

The dimensions of these quantities are as follows: t: T; l: L; V_e: LT^{-1}; ρ: ML^{-3}; m: M. Using Buckingham's π-theorem we have five quantities with three fundamental dimensions; the result we require must then be one involving two nondimensional groups. If we keep all parameters constant except the mass, then since the acceleration is inversely proportional to mass, as in (4.7), the time taken for a given change of a velocity component must be proportional to the mass. Consequently we choose the group which contains t to include the ratio t/m; the form chosen is then

$$\rho V_e l^2 . \frac{t}{m} = f\left(m / \rho l^3\right)$$

or

$$t = \frac{m}{\rho V_e l^2} f\left(m / \rho l^3\right) \tag{4.11}$$

If we replace l^2 by the wing area as usual, the time taken is proportional to the quantity τ found in the previous section regardless of the manoeuvre. This is the magnitude of the time unit in the nondimensionalized system to be introduced in Chapter 8.

The quantity in the brackets in (4.11) is also usually written in terms of wing area and is regarded as the ratio of a fictitious aircraft density m/Sl to the density of air. The quantity is known as the 'aircraft relative density parameter' and written as

$$\mu = \frac{m}{\rho S l} \tag{4.12}$$

It is also sometimes known as 'normalized mass'. It appears naturally as the result of analysis of the motion of aircraft in later chapters where the characteristic length, l, is chosen to suit the particular case.

4.2.5 Simple speed stability

Suppose that an aircraft is flying steadily at a speed above that for minimum drag, so that the thrust and drag are in equilibrium. If it receives a disturbance which increases the forward speed the thrust will in general increase slower than the drag and there will be a nett rearward force on the aircraft: see, for instance, figure 3.5 or 3.6. This force will decrease the speed and the aircraft is stable. Conversely if the initial speed is below that for minimum drag the nett force is forwards and the aircraft is unstable. In fact an aircraft does not normally behave like this as the speed change also generates pitching moments which alter the incidence and other forces appear on the aircraft. However if the aircraft is constrained to maintain its attitude, as for instance if it is forced to follow a glide slope as a preliminary to landing, then the problem may appear.

4.3 Controls

To control the aircraft we give the pilot the means to produce forces along the axes or moments about them, although normally not all possible forces and moments are provided. A short list of the methods used includes the engine throttle(s), the ailerons, elevator and rudder as shown in figure 4.1, the all-moving tailplane or foreplane and spoilers. Ailerons, elevators and rudders are known as flap type controls; they differ from the flaps discussed in Chapter 1 in that they are required to increase or decrease the lift and so are designed to move in both directions from the neutral position. The principle of operation is, however, the same.

Moving the pilot's stick forwards produces a downward movement of the elevator, increasing the lift on the tailplane and producing a nose-down pitching moment. Movement of the stick to the right, or a clockwise movement of a control-wheel, raises the starboard aileron and lowers the port one, producing a clockwise rolling moment about a forward axis. Forward movement of the pilot's right foot on the rudder bar moves the rudder trailing edge to starboard giving a clockwise yawing moment about a downward axis. Essentially these controls produce a moment about an axis and hence an angular acceleration, but what we usually require are displacements which only appear later as a result of the response of the aircraft.

4.3.1 Flap type controls

A typical flap type control is shown in figure 4.3(a) and a typical pressure distribution along the chord with and without deflection of the flap in figure 4.3(b).

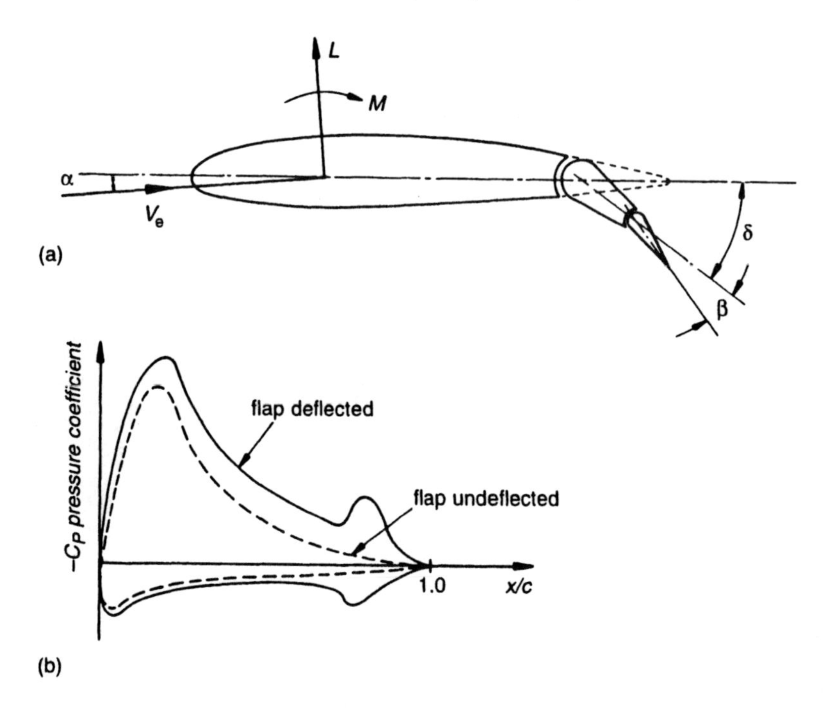

Fig. 4.3 (a) Definition of angles for a control surface. (b) Pressure distribution on a wing with deflected flap

Apart from changing the lift from the main surface, a pitching moment M is produced and a moment H about the hinge. These are taken as positive in the nose-up or trailing edge down sense as shown. Control surfaces are usually placed as far from the cg as possible and so the pitching moment produced is usually much smaller than the moment of the lift about the cg. Except on some aircraft which have powered controls, control surfaces are usually provided with a 'trim tab' which the pilot can adjust independently of the main control to vary the hinge moment. The trim tab is a small flap mounted on the main flap and is used to reduce the pilot's stick force to zero in steady flight.

Theory and experiment for incidence and flap angles less than about 15° agree on a linear variation of lift with these angles so that we can write the lift in the form

$$\frac{L}{\frac{1}{2}\rho V_e^2 S} = C_L = a_1(\alpha - \alpha_0) + a_2\delta + a_3\beta \tag{4.13}$$

where α_0 is the no-lift angle, and δ and β are the flap and tab angles. The constants a_1, a_2 and a_3 are functions of the main surface geometry, the ratio (flap chord)/(wing chord), and Reynolds and Mach numbers (Aero C.01.01.03 and .04, 74011, 74012). The hinge moment is similarly expressed as

$$\frac{H}{\frac{1}{2}\rho V_e^2 S_\delta c_\delta} = C_H = b_0 + b_1(\alpha - \alpha_0) + b_2\delta + b_3\beta \tag{4.14}$$

where S_δ and c_δ are the plan area of the flap aft of the hinge line and the mean chord of that area. The constants b_0, b_1, b_2 and b_3 are functions of the same parameters as for lift (Aero C.04.01.01, .02 and .06). The constant b_0 appears because on a cambered aerofoil section even at zero mean lift there are pressure differences between the upper and lower surfaces. The sections of tailplanes and fins are often symmetrical and then b_0 is zero. Unless special measures are taken both b_1 and b_2 are negative, as might be expected from the pressure distribution, figure 4.3(b).

The symbol δ is only used for the general case; for the specific cases of the aileron, elevator and rudder we use the symbols ξ, η and ζ respectively; the subscripts on the flap area and chord are similarly changed.

On a few aircraft the control surfaces are expected to provide control about more than one axis. Examples of this are the elevons on tail-less aircraft which deflect in opposite directions to act as ailerons and in the same sense to perform the function of elevators. A further example is that of the vee-tail in which antisymmetric deflections of the elevators give control in yaw and symmetric ones control in pitch. Another possible complication on some aircraft is that the ailerons are split with the inboard section only being used at high speed.

It is worth noting that the American notation differs in some respects from that just described; much more use is made of angles as subscripts. Thus the wing or aircraft-less-tail lift slope is written as $C_{L\alpha}$. The aileron, elevator and rudder angles are denoted by δ_a, δ_e and δ_r. The slopes a_1, a_2 and a_3 are denoted by using the control angles as a subscript to C_L, thus the equivalents for the tailplane and elevator are $C_{L\alpha_t}$, $C_{L\delta_e}$ and $C_{L\beta}$. A similar notation is used for the hinge moments.

4.3.2 Balancing of flap type controls

In smaller aircraft the control surfaces are connected directly to the pilot's controls and to move the control surfaces he or she has to apply forces to overcome the hinge moments. In larger aircraft powered controls are used, but in either case it makes sense to reduce the hinge moments by aerodynamic means. There is, however, a limit to the degree of balance to be aimed for; in no circumstances can b_2 be allowed to become positive in an aircraft that is ever likely to be controlled manually. In such a case the control is unstable and would require constant attention from the pilot. There are a number of methods of balancing controls, some of which are now obsolescent. We will briefly describe four methods of which the first is probably the most important: these are (a) the sealed nose balance, (b) the horn balance, (c) the geared tab and (d) the set back hinge. We are not discussing here the mass balancing of control surfaces which is necessary for the avoidance of some flutter problems.

The sealed nose balance is shown in figure 4.4. In this form of balance a 'beak' extends forward from the control surface and is joined to the structure inside the wing by a flexible but impervious sheet of material. Suppose the control surface is deflected downwards, then the pressures on the wing surface increase on the lower surface and decrease on the upper. These pressures are vented through to the surfaces of the beak and act to reduce the hinge moment; a similar action takes place for an increase of incidence. This method of balance therefore reduces the magnitude of both b_1 and b_2 (Aero C.04.01.00 and .04). Any leakage of air from the lower surface to the upper reduces the effectiveness of a control surface (Aero C.01.01.04) and in this respect this method is clearly superior; another advantage is that the mechanism is contained within the aerofoil section.

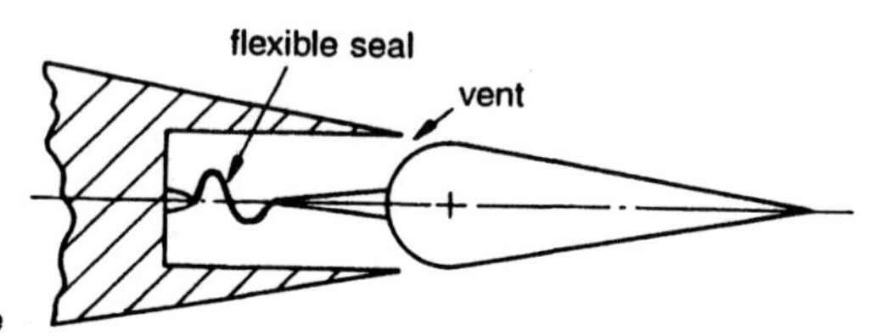

Fig. 4.4 Sealed nose balance

The horn balance is shown in figure 4.5 and consists of continuing the flap surface forward over a small part of its span to include the whole wing tip. This is an effective method of balancing which reduces the magnitude of both b_1 and b_2 (88003) but is not very suitable for high speed aircraft.

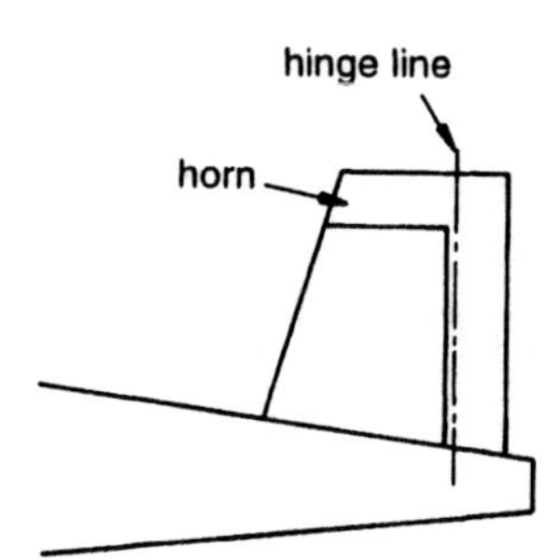

Fig. 4.5 The horn balance

The principle of the geared tab is shown in figure 4.6. In this form of balance a lever attached to the tab is connected via a rigid link *AB* to a bracket attached to the control surface. Suppose that the control surface is deflected downwards, then the link will push the tab upwards which then produces a lift on the control having a moment assisting the main surface deflection. Only the magnitude of b_2 is reduced.

The principle of the set back hinge is perhaps the simplest of all the methods. The standard position of the hinge is taken to be at the centre of the circle forming the nose of the control as shown in figure 4.7(a). A set back hinge position is shown in figure 4.7(b); there also have to be modifications to the rear of the main part of the wing to provide clearance. This method has a

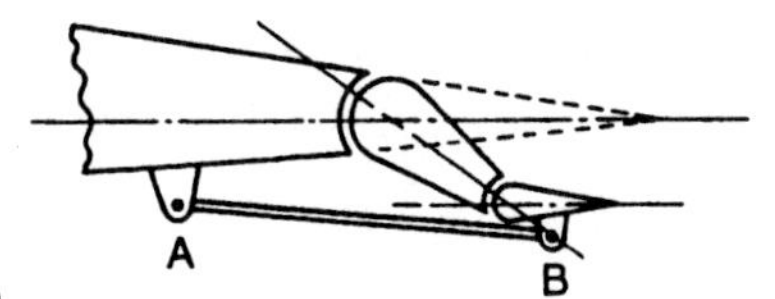

Fig. 4.6 The geared tab

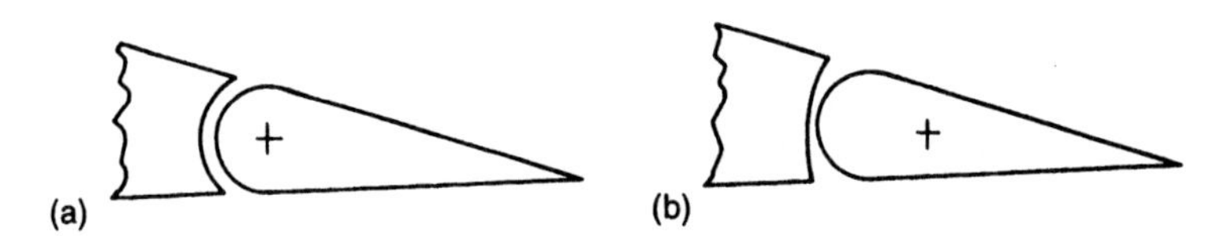

Fig. 4.7 (a) Standard position of hinge. (b) The set back hinge

fairly strong effect on the balance as the load on the control surface is concentrated towards the nose as shown in figure 4.3. The magnitudes of both b_1 and b_2 are reduced (Aero C.04.01.03).

4.3.3 Spoilers

Although spoilers have not been used very extensively, they have been used successfully for roll control on large airliners at low speed in conjunction with the ailerons. For the purposes of roll control the spoiler on one wing is raised, thus reducing the lift on that wing and producing a rolling moment. Figure 4.8(b) shows a sketch of the lift reduction due to a spoiler as a function of the distance d which it projects from the aerofoil section. Their disadvantages are that they cause a loss in the total lift and that their characteristics are non-linear; this latter

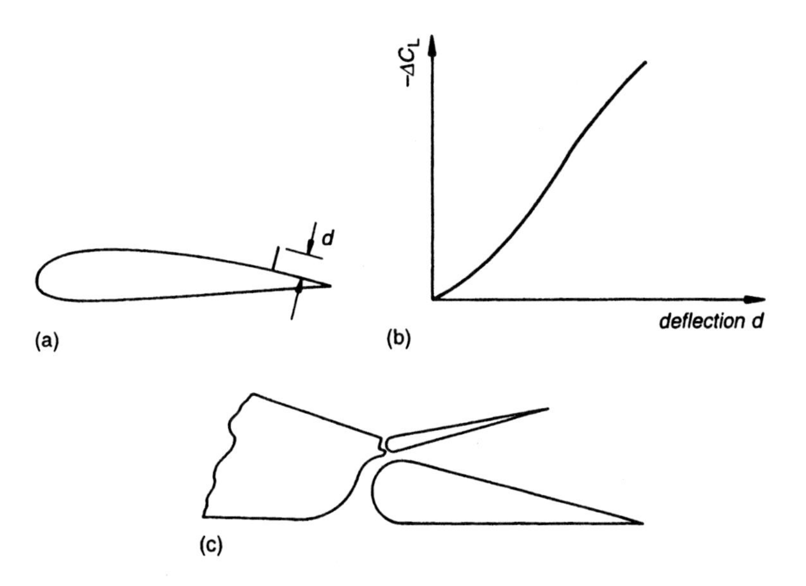

Fig. 4.8 The spoiler: (a) definition of deflection, (b) effect on lift coefficient, (c) the lift dumper

problem has prevented their common use until the coming of modern control systems. Other characteristics can be found from ESDU Data Items 92002 and 92030.

Another version of the spoiler is the 'lift dumper' shown in figure 4.8(c). It is only operated on the ground during landing and is used to reduce the lift and so increase the load on the wheels to improve the effectiveness of the brakes.

Student problem

4.1 Think of three other methods of control which have been or could be used on aircraft and reasons why they are not used or not often used.

5
Elementary treatment of pitching motion

5.1 Introduction

In this chapter we discuss the behaviour of an aircraft in pitch. Although we shall think of the aircraft moving in pitch we shall neglect the effects of inertia and unsteady aerodynamics, a procedure often described as 'quasi-static', since it is correct only in the limit of very slow motion. Our purpose is to determine such things as the elevator angles and stick forces required for steady flight as a function of the properties of the aircraft and to consider the static stability of the aircraft. To do this we require an equation relating the pitching moment to such quantities as speed, elevator angle and cg position. In some sense we model the aircraft behaviour in pitch and the resulting equation is known as the 'pitching moment equation' of the aircraft. The reader is reminded of the general assumptions given in Section 1.2 to which is added that of symmetric flight, i.e. there is no sideslip.

5.2 Modelling an aircraft in slow pitching motion

There are various ways in which the pitching moment equation of an aircraft might be found, such as windtunnel testing or computer modelling of the flow about the aircraft. These methods are not available until the design of an aircraft is fairly well fixed; prior to this the pitching moment equation has to be found from data correlated from theory, windtunnel test results and earlier designs. The process is one of synthesis with the pitching moment equation being built up by considering the various components of the aircraft in turn. We need first to examine two concepts related to the pitching moment characteristics of a wing.

5.2.1 Centre of pressure and aerodynamic centre

For simplicity we first consider a rectangular wing of moderate to high aspect ratio. The various theories for wings generally find that the plot of aerodynamic pitching moment against lift is a linear one and windtunnel experiments at reasonably high Reynolds numbers mostly agree except at the stall. In nondimensional terms the line has a slope b and intersects the C_L-axis at C_{M_0} as shown in figure 5.1. Usually C_{M_0} and b are negative as shown. The axis about which pitching moments have been measured is at a distance $x_1.c$ from the leading edge.

The pitching moment equation for the rectangular wing can then be expressed as

$$\frac{M_1}{\frac{1}{2}\rho V_e^2 Sc} = C_{M_1} = C_{M_0} + bC_L \tag{5.1}$$

Now taking moments we find the pitching moment about a general point at a distance $x.c$ from the leading edge as

$$M = M_1 - (x_1 - x)\,Lc\cos\alpha - (x_1 - x)Dc\sin\alpha$$

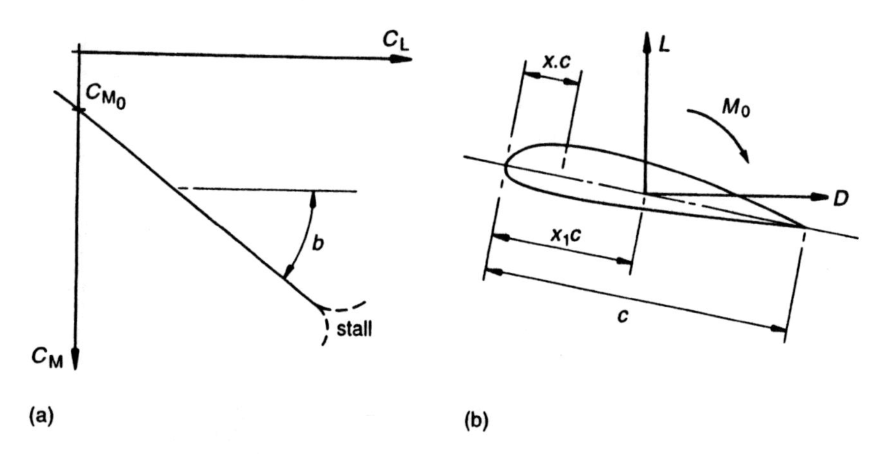

Fig. 5.1 Terms defining (a) the pitching moment curve, (b) positions along the chord

or in nondimensional terms

$$C_M = C_{M_1} - (x_1 - x)C_L \cos\alpha - (x_1 - x)C_D \sin\alpha$$

This expression is rather inconvenient but can be simplified if we remember that for practical wings $C_L \gg C_D$ and we restrict the incidence to values below the stall, i.e. $\alpha < 15°$ say, so that $\cos\alpha \simeq 1$ and $\sin\alpha \simeq \alpha$. We can then neglect the last term as the product of two small quantities and so on substituting for C_{M_1} from (5.1) and putting $\cos\alpha = 1$ we find

$$C_M = C_{M_0} + [b - (x_1 - x)]C_L \tag{5.2}$$

At this stage we can define two points along the wing chord.

- The 'centre of pressure', x_{cp}, is the point for which $C_M = 0$, i.e.

$$x_{cp} = -\frac{C_{M_0}}{C_L} + (b - x_1) \tag{5.3}$$

  In fact this point is of limited practical use, as when C_L passes through zero, x_{cp} goes through infinity (unless $C_{M_0} = 0$ as it does for a symmetrical section).
- The 'aerodynamic centre', x_0, is the point for which $dC_M/dC_L = 0$. On differentiating (5.2) with respect to C_L we have

$$\frac{dC_M}{dC_L} = b - (x_1 - x)$$

  Equating to zero and solving for $x = x_0$, say, then we find

$$x_0 = x_1 - b \tag{5.4}$$

Interpreted, this means that the aerodynamic centre (in wing chords) is at a distance equal to the slope of the C_M against C_L curve behind the point about which pitching moments have been measured. The pitching moment about the aerodynamic centre is constant at C_{M_0} and the pitching moment equation of the wing in terms of x_0 is, on substituting for b from (5.4) into (5.2),

$$C_M = C_{M_0} + (x - x_0)C_L \tag{5.5}$$

The value of x_0 is approximately 0.25 for a rectangular wing in low speed flow and 0.5 in supersonic flow.

5.2.2 The reference chord

In the last section pitching moment coefficients were defined using the wing chord; however, for most planforms the choice of reference chord is not so obvious. We are free to choose any dimension we please and so attempt to choose one in a logical and useful way; we also choose a longitudinal position in the aircraft for it. Consider first the calculation of the overall no-lift pitching moment for a wing given values for each streamwise section c_{M0}, defined as

$$c_{M0} = c_{M0}(y) = \frac{1}{\frac{1}{2}\rho V_e^2 c^2} \frac{\partial M_0(y)}{\partial y} \tag{5.6}$$

where y is the spanwise distance of the section from the aircraft centreline. Bearing in mind that no-lift pitching moment is a pure torque and assuming that the wing is untwisted, the no-lift pitching moment for the whole wing is

$$\tfrac{1}{2}\rho V_e^2 \int_{-s}^{s} c_{M0} c^2 \mathrm{d}y = \tfrac{1}{2}\rho V_e^2 \bar{\bar{c}} C_{M0} \tag{5.7}$$

say, where s is the semispan and C_{M0} is the overall no-lift pitching moment. In this equation $\bar{\bar{c}}$ is the reference chord which has still to be defined; the magnitude of C_{M0} will be fixed by that choice. Now consider the case of an untwisted wing of constant section; often a fairly accurate estimate of the no-lift pitching moment coefficient will be obtained by assuming c_{M0} to be constant at its two-dimensional value. From (5.7) this estimate is given as

$$S\bar{\bar{c}} C_{M0} = 2c_{M0} \int_0^s c^2 \mathrm{d}y$$

We now choose to define $\bar{\bar{c}}$ such that in this case $C_{M0} = c_{M0}$, so that

$$\bar{\bar{c}} = \frac{2}{S}\int_0^s c^2 \mathrm{d}y = \int_0^s c^2 \mathrm{d}y \Bigg/ \int_0^s c \mathrm{d}y \tag{5.8}$$

This length is known as the 'aerodynamic mean chord' (amc) and it is often convenient, but not essential, to use it as the reference chord throughout stability work.

A wing can be effectively twisted in two ways: it can be of constant wing section and the chord lines be rotated with spanwise position; or the camber of the wing sections can change, changing the local no-lift angles; conceivably both methods could be used. The total effect is known as 'aerodynamic twist' and produces a further contribution to the no-lift pitching moment (87001).

We turn now to the question of locating the aerodynamic mean chord longitudinally and define a distance $\bar{\xi}_0$ for any planform by

$$\bar{\xi}_0 = \frac{2}{S}\int_0^s \xi_0 c dy \tag{5.9}$$

where ξ_0 is the distance from the leading edge of the centreline chord (the 'apex') to the leading edge of the local chord as shown in figure 5.2.

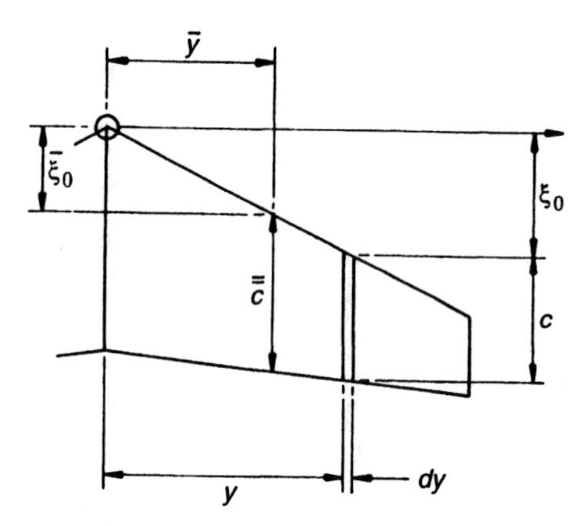

Fig. 5.2 Determination of reference chord

For a wing with straight leading and trailing edges as shown it is mathematically simple, if a little tedious, to show that the wing chord at a distance $\bar{\xi}_0$ from the apex has a length of $\bar{\bar{c}}$. Another way of stating this is to say that the aerodynamic mean chord 'fits' on the wing at this distance back and at the spanwise position $\bar{y}$ shown. If we consider a line at a constant fraction k of the chord then for any wing

$$\bar{\xi}_k = \frac{2}{S}\int_0^s \xi_k c dy = \frac{2}{S}\int_0^s (\xi_0 + kc)c dy = \frac{2}{S}\left(\int_0^s \xi_0 c dy + k\int_0^s c^2 dy\right) = \bar{\xi}_0 + k\bar{\bar{c}} \tag{5.10}$$

showing that $\bar{\xi}_1 - \bar{\xi}_0 = \bar{\bar{c}}$, but we cannot in general expect the chord to fit the wing.

When discussing the pitching moment characteristics of a wing it is necessary to have a moment reference point. For low speeds, for which the aerodynamic centre of a thin wing section in two-dimensional flow is theoretically a quarter of the chord behind the leading edge, the obvious point to take is the quarter-chord point of the aerodynamic mean chord. If it could be assumed that the lift on a plane wing were proportional to the local chord and located at the quarter-chord point then the aerodynamic centre would be $\bar{x}_{0.25}$. For supersonic flow the equivalent point is at half-chord. The use of this choice of reference points removes the gross effects of wing geometry on overall aerodynamic centre of a wing (70011, 70012). The variation of aerodynamic centre of a typical wing with Mach number is sketched in figure 5.3. We note that there is a rapid rearward movement of aerodynamic centre at transonic Mach numbers.

5.2.3 The aircraft-less-tailplane

We proceed with our programme to build up the pitching moment equation of the aircraft by adding various components, but will leave the tailplane until Section 5.2.5 because of its importance and the influence on it from the wing.

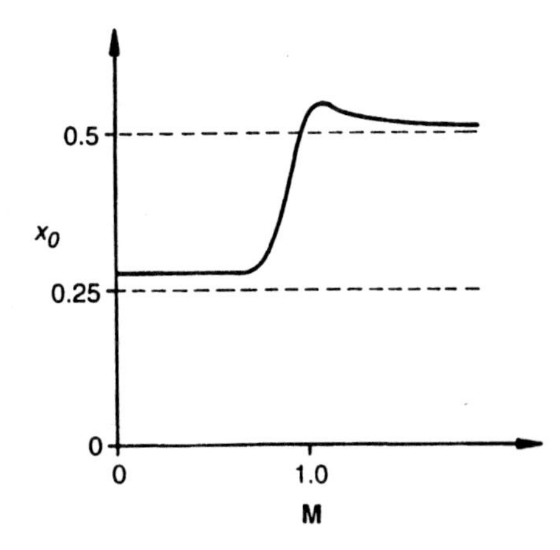

Fig. 5.3 Typical variation of aerodynamic centre with Mach number

Adding a fuselage to a wing reduces the lift over the part of the gross wing included in the fuselage as it is a less efficient lift-producer. The lift produced by the fuselage is mostly concentrated near the nose, so the nett effect is a very small change in the total lift and a shift forward of the aerodynamic centre of the combination compared with wing alone (76015). Also because the two components have different no-lift angles there is an effect on the no-lift pitching moment (89042). Engine nacelles have a similar effect, whether located on the wing (76015) or on pylons (77012, 78013). If a propeller is placed in a flow at a small angle of incidence to its axis of rotation it produces a lift force normal to the axis and a pitching moment (89047); both will contribute to a shift in the aerodynamic centre of the combination.

We can write the results of this process in the following form:

$$C_{M0}^{wb} = C_{M0}^{wing} + \Delta C_{M0}^{fuse} + \Delta C_{M0}^{nac} + \ldots \tag{5.11}$$

and

$$h_0 = x_0^{wing} + \Delta x_0^{fuse} + \Delta x_0^{nac} + \ldots \tag{5.12}$$

where h_0 is the aircraft-less-tail aerodynamic centre position. The reference quantities used throughout are gross wing area and a wing chord, e.g. the amc, and the distances are measured from a point such as the leading edge of the amc as defined in the previous section.

5.2.4 The pitching moment equation of the complete aircraft

We are now in a position to consider the complete aircraft but still leave detailed consideration of the tailplane for later. Figure 5.4 shows the major forces and moments acting on the aircraft. It will be noted that the thrust and drag vectors, T and D, have been drawn as co-linear; this will not usually be the case but the error is small and of little consequence to our present purpose. Level or nearly level flight has also been assumed.

We now need to take moments about a suitable axis. Normally in the dynamics of rigid bodies the centre of gravity (cg) is the obvious choice; however, for an aircraft this is a point which might vary rather randomly by a significant amount even over a short time due to the consumption of fuel, the jettisoning of items or for other reasons. A better choice is the aerodynamic centre which is a fixed point at speeds for which compressibility is unimportant and above these speeds is a determinable quantity. Consider the aircraft shown in figure 5.4.

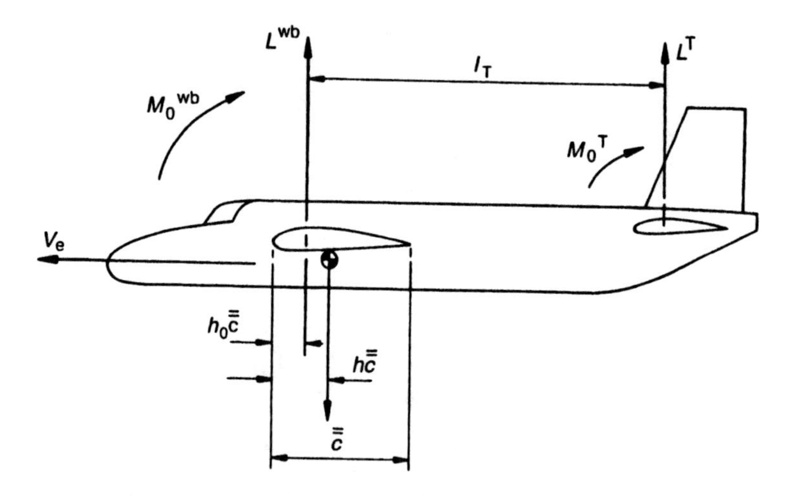

Fig. 5.4 Definition of terms for the pitching moment equation

Taking moments about the aerodynamic centre we find

$$M = M_0^{\mathrm{wb}} + M_0^{\mathrm{T}} + (h - h_0)\bar{\bar{c}}mg - L^{\mathrm{T}}l_{\mathrm{T}} \tag{5.13}$$

Also we have

$$mg = L^{\mathrm{wb}} + L^{\mathrm{T}} \tag{5.14}$$

Then dividing through by $\frac{1}{2}\rho V_e^2 S\bar{\bar{c}}$ and defining

$$C_{\mathrm{L}} = \frac{mg}{\frac{1}{2}\rho V_e^2 S},\ C_{\mathrm{L}}^{\mathrm{T}} = \frac{L^{\mathrm{T}}}{\frac{1}{2}\rho V_e^2 S_{\mathrm{T}}},\ C_{\mathrm{M}_0} = \frac{M_0^{\mathrm{wb}} + M_0^{\mathrm{T}}}{\frac{1}{2}\rho V_e^2 S\bar{\bar{c}}},\ \bar{V}_{\mathrm{T}} = \frac{S_{\mathrm{T}}l_{\mathrm{T}}}{S\bar{\bar{c}}}$$

we find

$$C_{\mathrm{M}} = C_{\mathrm{M}_0} + (h - h_0)\bar{\bar{c}}C_{\mathrm{L}} - \bar{V}_{\mathrm{T}}C_{\mathrm{L}}^{\mathrm{T}} \tag{5.15}$$

and

$$C_{\mathrm{L}} = C_{\mathrm{L}}^{\mathrm{wb}} + \frac{S_{\mathrm{T}}}{S}\cdot C_{\mathrm{L}}^{\mathrm{T}} \tag{5.16}$$

The quantity $\bar{V}_{\mathrm{T}}$ is known as the 'tailplane volume coefficient'. Note that in defining $C_{\mathrm{L}}^{\mathrm{T}}$ we have divided the tailplane lift by the same dynamic pressure as for the wing. The wing lift is produced by deflecting the oncoming airflow downwards and, provided that there are at most only small regions of separated flow, there is no reduction in the speed of the flow except in a narrow wake region. The result is that the tailplane operates generally in an airflow having the same dynamic pressure as the wing. There are, however, reasons why the tailplane may not be as effective as we expect; they will be discussed in Section 5.8.1.

5.2.5 Tailplane contribution to the pitching moment equation

The main purpose of this section is to find an expression for the tailplane lift allowing for the interference from the wing; a secondary purpose is to discuss the hinge moments on the elevator. Using the general expression (4.13) and writing the elevator angle η in place of the general flap angle δ, we can write the tailplane lift coefficient as

$$C_L^T = a_1\alpha^T + a_2\eta + a_3\beta \tag{5.17}$$

Here α^T and β are the tailplane incidence and tab angle, a_1 is the tailplane lift curve slope, a_2 is the elevator lift curve slope and a_3 is the tab lift curve slope. It is convenient in this work to measure incidence, not relative to the chord line as is normal, but to a line parallel to the flow direction which results in zero lift. This line is known as the 'no-lift line', and eliminates the no-lift angle from our equations even for cambered aerofoil sections. The result is that we write $C_L = a\alpha$ to find the lift from the incidence. Figure 5.5 shows the angles of the flow relative to the wing and tailplane, where 'nll' stands for no-lift line.

Here ε is the 'downwash angle' i.e. the mean angle at the tailplane that the airflow is deflected through by the wing, and η_T is the 'tailplane setting angle' defined as the angle between the wing and tailplane no-lift lines. In figure 5.5 the magnitude of the angles has been exaggerated for clarity.

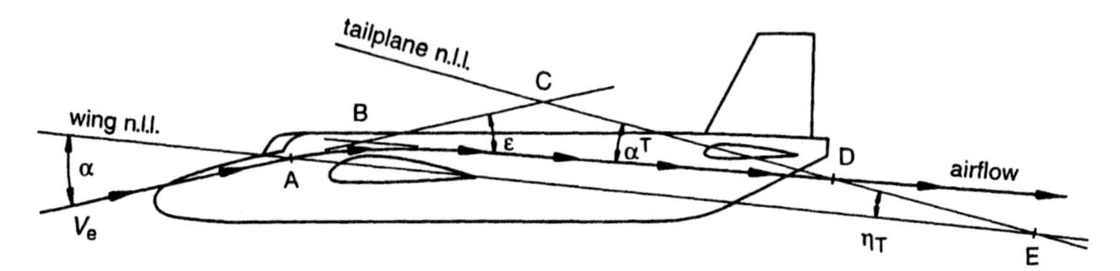

Fig. 5.5 Determination of tailplane incidence (nll = no-lift line)

To find the tailplane incidence α^T we consider the triangles ACE and BCD. The angle at C can be expressed as $\pi - \alpha - \eta_T$ or as $\pi - \varepsilon - \alpha^T$. Then solving gives

$$\alpha^T = \alpha + \eta_T - \varepsilon \tag{5.18}$$

Now the downwash angle is proportional to the wing lift produced by incidence so we can write

$$\varepsilon = C_L^{wb}\frac{d\varepsilon}{dC_L^{wb}} \tag{5.19}$$

which is a form which also makes some allowance for lift produced by lowering the flaps, for instance. It is usually more convenient to work in terms of the aircraft-less-tail incidence α and write

$$\varepsilon = \alpha . \frac{d\varepsilon}{d\alpha} \tag{5.20}$$

which assumes that the lift results entirely from the incidence. The downwash derivative $d\varepsilon/d\alpha$ is a function of the wing geometry and position of the tailplane relative to the wing (80020). We can now substitute (5.20) into (5.18) to obtain

$$\alpha^T = \alpha\left(1 - \frac{d\varepsilon}{d\alpha}\right) + \eta_T \tag{5.21}$$

To find the incidence α we use the aircraft-less-tail lift curve slope a. Neglecting the small tailplane lift so that $C_L^{wb} \simeq C_L$, see (5.16), we express the incidence as C_L/a. We then substitute into (5.21) to find

$$\alpha^T = \frac{C_L}{a}\left(1 - \frac{d\varepsilon}{d\alpha}\right) + \eta_T \tag{5.22}$$

Finally we can substitute this result into (5.17) to obtain the tailplane lift as

$$C_L^T = \frac{a_1}{a}\left(1 - \frac{d\varepsilon}{d\alpha}\right)C_L + a_1\eta_T + a_2\eta + a_3\beta \tag{5.23}$$

We now turn our attention to the hinge moment produced by the elevator. From (4.14) and using the results above we find

$$C_H = \frac{b_1}{a}\left(1 - \frac{d\varepsilon}{d\alpha}\right)C_L + b_1\eta_T + b_2\eta + b_3\beta \tag{5.24}$$

where we have assumed that the tailplane has a symmetrical aerofoil section so that $b_0 = 0$. If the aircraft has fully powered controls information on the hinge moments will assist the aircraft designer in specifying the actuators required. With manual controls the control circuit will transmit a force back to the pilot at the stick.

5.2.6 The pitching moment equation, 'stick fixed'

The final task remaining to us in the development of the pitching moment equation is to substitute for C_L^T in (5.19) using (5.23), giving

$$C_M = C_{M_{fix}} = C_{M_0} + (h - h_0)C_L - \bar{V}_T\left[\frac{a_1}{a}\left(1 - \frac{d\varepsilon}{d\alpha}\right)C_L + a_1\eta_T + a_2\eta + a_3\beta\right] \tag{5.25}$$

This equation is known as the 'pitching moment equation, stick fixed', and is the basis of most of the work of the rest of this chapter. The label 'stick fixed' relates to the fact that the elevator angle η is controlled by the pilot through the control circuit. We later meet the situation where it is determined by the pilot in a less direct manner. It can be helpful to remember that the first two terms on the right-hand side are generated by the aircraft-less-tail and the part in square brackets by the tailplane.

5.3 Trim

The term 'trim' was originally a nautical one. If a ship does not float at the required attitude, then the cargo or ballast must be moved to bring the cg and centre of buoyancy into the same vertical line. If an out-of-trim ship is held at the desired attitude and then released, it will move to an attitude which satisfies that condition. In our case we use the controls to trim. In any steady flight condition the moments about any axis must total to zero. For the specific case of pitching we have

$$\boxed{M = 0 = C_M} \tag{5.26}$$

When this condition is satisfied the aircraft is said to be trimmed. This condition has at least three uses for us.

- Used in conjunction with (5.13) we can solve for the tailplane lift for use in stressing the tailplane and rear fuselage; in this case it may be necessary to deal correctly with the moments due to the thrust and drag.
- Used with (5.15) we can find the tailplane lift coefficient and check that the stalling value is never exceeded.
- We can find the elevator angle to trim, which is the subject of the next section.

5.3.1 Trim, 'stick fixed'

Proceeding immediately to use (5.26) with (5.25) and solving for the elevator angle gives

$$\eta_{\mathrm{trim}} = \frac{C_{M_0}}{\overline{V}_T a_2} - \frac{a_1 \eta_T}{a_2} - \frac{a_3 \beta}{a_2} + \frac{C_L}{\overline{V}_T a_2}\left[(h - h_0) - \overline{V}_T \frac{a_1}{a}\left(1 - \frac{d\varepsilon}{d\alpha}\right)\right] \tag{5.27}$$

For an aircraft for which the effects of compressibility are negligible many of the terms are constant and this is better thought of in the form

$$\eta_{\mathrm{trim}} = \eta_0 + C_L.\mathrm{function}(h) \tag{5.28}$$

where η_0 is a constant and $h\bar{\bar{c}}$ is the cg position. As so often happens we can think of C_L as a stand-in for speed, so that this equation gives the variation of elevator angle to trim with speed. A plot of η_{trim} against C_L is known as an 'elevator trim curve'. A set of trim curves consists of straight lines of slope depending linearly on cg position and all passing through the point $\eta = \eta_0$ at $C_L = 0$. Figure 5.6 shows the form of typical trim curves.

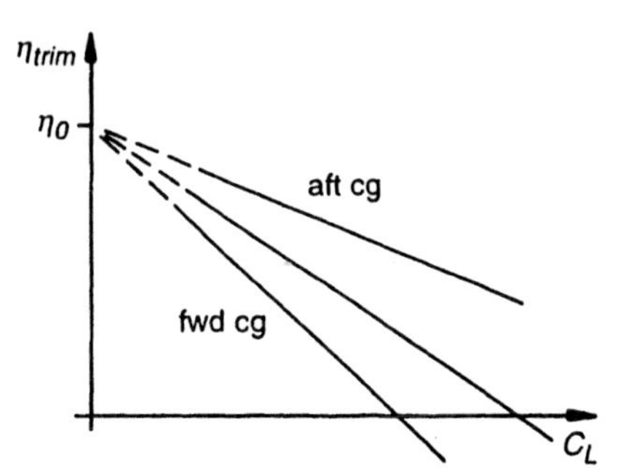

Fig. 5.6 Typical elevator trim curves

The corresponding hinge moment coefficient and hence stick force can be found by substituting $\eta = \eta_{\mathrm{trim}}$ into (5.24). Writing for brevity

$$\bar{e} = (1 - d\varepsilon / d\alpha)$$

we find

$$C_{H_{\mathrm{trim}}} = \frac{b_1}{a} C_L \bar{e} + b_1 \eta_T + b_2\left(\frac{C_{M_0}}{\overline{V}_T a_2} - \frac{a_2}{a_1}\eta_T - \frac{a_3}{a_2}\tau\right) + \frac{b_2 C_L}{\overline{V}_T a_2}\left((h - h_0) - \overline{V}_T \frac{a_1}{a}\bar{e}\right) + b_3 \beta$$

Rearranging and writing for brevity

$$\bar{b}_1 = b_1 - \frac{a_1}{a_2} b_2 \text{ and } \bar{b}_3 = b_3 - \frac{a_3}{a_2} b_2 \tag{5.29}$$

we find

$$C_{H_{trim}} = \frac{b_2 C_{M_0}}{\bar{V}_T a_2} + \bar{b}_1 \eta_T + \bar{b}_3 \tau + \frac{b_2 C_L}{\bar{V}_T a_2} \left[(h - h_0) - \bar{V}_T \frac{a_1}{a} \left(1 - \frac{d\varepsilon}{d\alpha} \right) \left(1 - \frac{a_2 b_1}{a_1 b_2} \right) \right] \tag{5.30}$$

Again this can be written in the form

$$C_{H_{trim}} = C_{H_0} - \text{function}(h).C_L \tag{5.31}$$

where C_{H_0} is a constant. This also leads to trim curves, which in this case are known as 'hinge moment trim curves' and are of the form shown in figure 5.7.

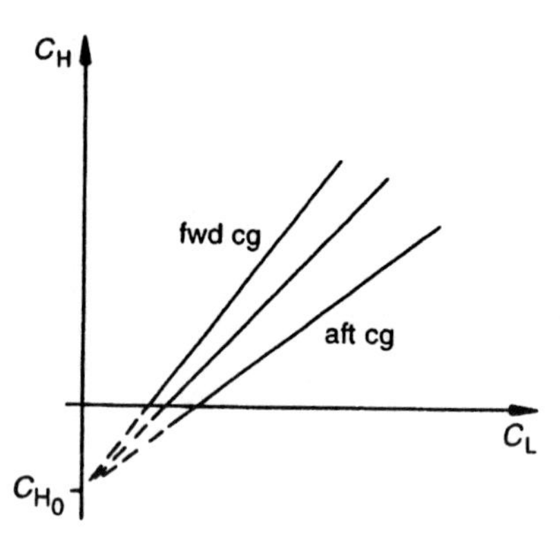

Fig. 5.7 Typical hinge moment trim curves

5.3.2 Trim, 'stick free'

Continuously reacting against the stick force represented by the hinge moment coefficient derived above would be very tiring for a pilot of an aircraft equipped with manual controls. For this purpose trim tabs are provided so that the pilot can reduce the stick force to zero, and in principle the pilot can then take his hands off the stick. The aircraft is then said to be trimmed 'stick free' or 'hands off', the conditions for which are therefore

$$\boxed{C_M = 0 \text{ and } C_H = 0} \tag{5.32}$$

This is what a pilot would regard as trimming an aircraft, rather than the simpler condition (5.26) used earlier, because as we shall see later the first condition is satisfied by a stable aircraft responding automatically to the out-of-balance moment. Using this second condition with (5.24) and solving for the elevator angle η, and writing $\eta = \eta'$, gives

$$\eta' = -\left[\frac{b_1 C_L}{b_2 a} \left(1 - \frac{d\varepsilon}{d\alpha} \right) + \frac{b_1}{b_2} \eta_T + \frac{b_3}{b_2} \tau \right] \tag{5.33}$$

Providing the pilot applies no restraining force to the stick, the elevator is now effectively geared to the tab, with gear ratio $-b_3/b_2$ and with an offset depending on speed, via C_L. The behaviour of the aircraft as a whole is affected, and we have a new pitching moment equation, obtained by substituting for η' from (5.33) above into the pitching moment equation stick fixed (5.25). Writing $C_M = C_{M\text{free}}$, the result is

$$C_{M\text{free}} = C_{M_0} + (h - h_0)C_L - \bar{V}_T\left[\frac{\bar{a}_1}{a}\left(1 - \frac{d\varepsilon}{d\alpha}\right)C_L + \bar{a}_1\eta_T + \bar{a}_3\tau\right] \tag{5.34}$$

where we have written

$$\bar{a}_1 = a_1 - \frac{a_2 b_1}{a_2} \text{ and } \bar{a}_3 = a_3 - \frac{a_2 b_3}{b_2} \tag{5.35}$$

This equation is known as the 'pitching moment equation stick free'; the similarity to the pitching moment equation stick fixed (5.25) should be noted.

Applying the first of the conditions (5.32) we could solve (5.34) for the tab angle to trim stick free; this leads to a set of tab angle to trim curves. These resemble the elevator angle to trim curves of figure 5.6.

The quantities $\bar{a}_1$ and $\bar{a}_3$ can be shown to be the tailplane lift curve slopes with the elevator left free to float. Consider an isolated tailplane; we can write

$$C_L^T = a_1\alpha^T + a_2\eta + a_3\beta$$

and

$$C_H = b_1\alpha^T + b_2\eta + b_3\beta$$

Putting $C_H = 0$ and solving for η gives

$$\eta = -\frac{b_1}{b_2}\alpha^T - \frac{b_3}{b_2}\beta$$

then substituting this into the equation for C_L gives

$$C_L^T = \left(a_1 - \frac{a_2 b_1}{b_2}\right)\alpha^T - \left(a_3 - \frac{a_2 b_3}{a_2}\right)\beta$$

Then differentiating with respect to C_L^T gives $\dfrac{\partial C_L}{\partial \alpha^T} = \bar{a}_1$ and $\dfrac{\partial C_L}{\partial \beta} = \bar{a}_3$.

5.3.3 Trim near the ground

When an aircraft flies close to the ground a number of changes occur to its aerodynamic characteristics, some directly due to ground proximity and others due to changes in its configuration such as lowering flaps. The lift and downwash are changed due to the ground effect (72023) and lowering the flaps gives a change in the no-lift pitching moment (Aero F.08.01.01 and .02). Some changes can be accommodated in the previous work without difficulty, others cannot. Whilst we could derive an expression for, say, the elevator angle to trim it would be even more complicated than the expression we already have; we choose therefore to illustrate calculating η_{trim} by an example.

Worked example 5.1

An aircraft is flying close to the ground at a speed of 50 m s^{-1}. Determine the elevator angle to trim the aircraft, with zero tab angle, if the aerodynamic characteristics in this condition are as follows: Wing loading = 360 kg m^{-2}, $a = 4.7$, $a_1 = 3.4$, $a_2 = 2.0$, $\overline{V}_T = 0.48$, ΔC_L due to flap = 0.9, C_{M_0} flaps down = -0.162, $d\varepsilon/dC_L = 0.11$. The cg is $0.03\,\bar{\bar{c}}$ aft of the aircraft-less-tail aerodynamic centre. The reduction in downwash due to ground effect is 1.6° and the tailplane setting angle is $-3°$. All slopes are expressed per radian.

Solution

We first find the lift coefficient and then use the pitching moment equation in the form (5.15) to find the tailplane lift coefficient, thus

$$C_L = \frac{mg}{\frac{1}{2}\rho V^2 S} = \frac{360 \times 9.81}{0.5 \times 1.225 \times 50^2} = 2.306$$

Then substituting into (5.15), i.e.

$$C_M = C_{M_0} + (h - h_0)C_L - \overline{V}_T C_L^T$$

with $C_M = 0$, for trim, gives

$$0 = -\ 0.162 + 0.03 \times 2.306 - 0.48 \times C_L^T$$

and solving for C_L^T gives the value as -0.1934. We now use (5.17) to find the elevator angle but we must first calculate the tailplane incidence from (5.18). This requires the wing incidence and the downwash. The part of the wing lift coefficient due to incidence is 2.306 – 0.9 = 1.406, then the wing incidence from no lift is

$$\frac{1.406}{4.7} = 0.2991 \text{ rad} = 17.14°$$

The downwash angle in the absence of ground effect would be

$$\varepsilon = \frac{d\varepsilon}{dC_L}.C_L = 0.11 \times 2.306 = 0.2537 \text{ rad} = 14.53°$$

and with ground effect this becomes 14.53 – 1.6 = 12.93°. Then using (5.18)

$$\alpha^T = \alpha - \varepsilon - \eta_T = 17.14 - 12.93 + (-3) = 1.21°$$

Finally using (5.17), i.e.

$$C_L^T = a_1\alpha^T + a_2\eta + a_3\beta$$

or

$$-0.1934 = 3.4 \times \frac{1.21}{57.3} + 2.0 \times \eta$$

and solving for η gives $\eta_{trim} = -0.1326$ rad $= -7.6°$.

5.4 Static stability

We turn now to our second main purpose in this chapter which is to discuss static stability. We shall do this using an intuitive argument, but the results will be confirmed in Section 9.3.5. In Section 4.2.2 we stated the static stability was related to measuring the tendency for an aircraft to return to its initial flight condition after a disturbance. Consider the aircraft shown in figure 5.8, which has been disturbed from the trimmed state by a small angle $\delta\alpha$.

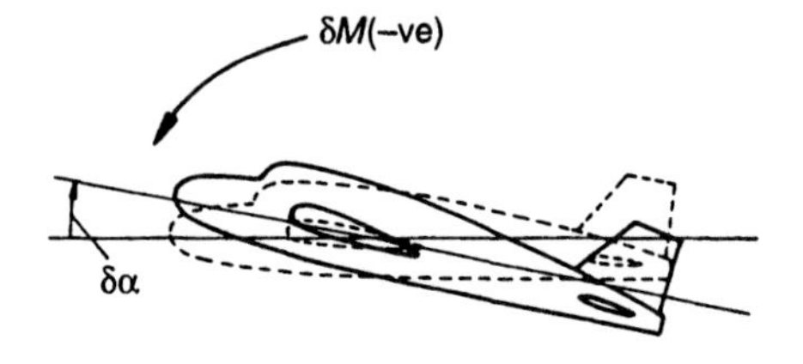

Fig. 5.8 Illustration of sense of required pitching moment required in a pitch disturbance

A pitching moment δM will be generated. This is required to be in the opposite sense to $\delta\alpha$, as shown, if the aircraft is to have a tendency to return to its initial state. As we measure both α and M positive in the nose-up sense we can write the condition for static stability as $\delta M/\delta\alpha < 0$ or in the limit of small quantities $\mathrm{d}M/\mathrm{d}\alpha < 0$. For later application we will find it more convenient to work in terms of C_L and C_M; providing that C_L is an increasing function of incidence the condition for static stability can be expressed as

$$\boxed{\frac{\mathrm{d}C_M}{\mathrm{d}C_L} < 0} \tag{5.36}$$

If we assume that the motion of disturbance has been a slow one as described in Section 5.2, we can use this condition with the pitching moment equation that we have derived and use the results to discuss aircraft stability.

5.4.1 Static stability, 'stick fixed'

We proceed immediately to use condition (5.36) with the pitching moment equation, stick fixed (5.25), assuming that h_0, a, a_1 and $\mathrm{d}\varepsilon/\mathrm{d}\alpha$ are independent of C_L. The result (which we now write as a partial derivative as so many parameters have been kept constant) is

$$\frac{\partial C_{M_{fix}}}{\partial C_L} = (h - h_0) - \bar{V}_T \frac{a_1}{a}\left(1 - \frac{\mathrm{d}\varepsilon}{\mathrm{d}\alpha}\right) \tag{5.37}$$

Except for the cg position, h, all the quantities on the right-hand side of the above equation are constant provided that compressibility effects are negligible. As cg position is varied a position will be reached such that the right-hand side is zero. At this position no restoring moment will be generated by a disturbance and the aircraft will remain in the new attitude; it is then in a state of neutral stability. Putting the left-hand side to zero and solving for $h = h_n$, say, gives

$$h_n = h_0 + \bar{V}_T \frac{a_1}{a}\left(1 - \frac{d\varepsilon}{d\alpha}\right) \tag{5.38}$$

This possible cg position is known as the 'neutral point stick fixed'. The process used to obtain it is the same as that used in Section 5.3.1 to obtain the aerodynamic centre, so that the difference between these quantities is simply a matter of terminology. The first term on the right-hand side is the aircraft-less-tail aerodynamic centre and the second can be regarded as the shift of aerodynamic centre due to the tailplane.

As mentioned in Section 4.2.2 it is convenient to have a measure of the degree of stability. The quantity $-dM/d\alpha$ is in effect a torsion spring rate and we take a nondimensional version as our measure and write $H_n = -\partial C_M/\partial C_L$, where H_n is known as the 'cg margin stick fixed'. Then from (5.37) we find

$$H_n = \frac{\partial C_M}{\partial C_L} = (h_0 - h) + \bar{V}_T \frac{a_1}{a}\left(1 - \frac{d\varepsilon}{d\alpha}\right) \tag{5.39}$$

and using (5.38) gives

$$H_n = h_n - h \tag{5.40}$$

In other words the cg margin is the distance between the neutral point and the actual cg position, expressed in terms of the reference chord. It is positive when the cg lies ahead of the neutral point and the aircraft is then stable. The condition (5.36) for stability in this case then becomes

$$H_n > 0 \tag{5.41}$$

The quantity on the right-hand side of (5.39) occurred previously in (5.27) in square brackets; reference to Section 5.3.2 shows that (5.28) can then be written as

$$\eta_{trim} = \eta_0 - \frac{C_L H_n}{\bar{V}_T a_2} \tag{5.42}$$

or rearranging

$$H_n = -\bar{V}_T a_2 \frac{d\eta_{trim}}{dC_L} \tag{5.43}$$

This states that the cg margin is proportional to the slope of the elevator trim curve. As noted in Section 4.4.2 this is entirely reasonable as the stiffer the aircraft in pitch as measured by the cg margin the more elevator angle will be required to change the pitch attitude, as measured by the lift coefficient. Note also that the sense of the stick movement will change with the sign of the cg margin.

Equation (5.43) leads to a method of finding the neutral point of an aircraft experimentally. Given an aircraft equipped with means to measure equivalent airspeed and elevator angle we trim the aircraft in level flight at a series of speeds. From the speed and known mass we can find the lift coefficient and plot elevator angle against it, i.e. we find the elevator trim curve. We then find the slope and repeat the experiment with various cg positions (which have to be carefully controlled) and plot the slopes against cg position. A graph like figure 5.9 results. Extrapolating back the curve to the cg axis we find a cg position for which $\mathrm{d}\eta_{\mathrm{trim}}/\mathrm{d}C_{\mathrm{L}} = 0$, i.e. $H_{\mathrm{n}} = 0$ or $h = h_{\mathrm{n}}$.

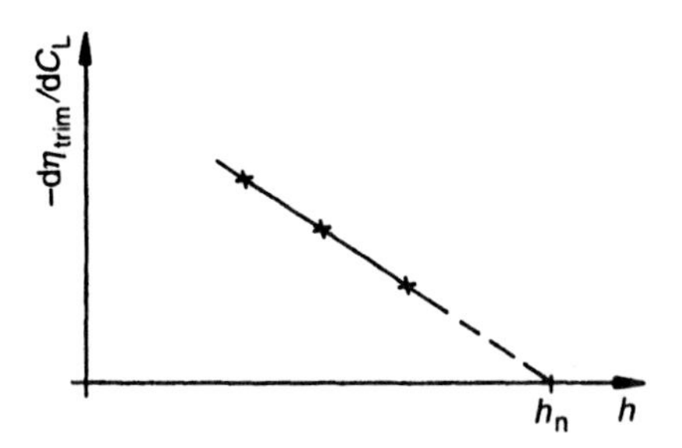

Fig. 5.9 Determination of neutral point fixed from flight test results

5.4.2 Static stability, 'stick free'

We now apply the condition (5.36) for static stability to the pitching moment equation 'stick free' (5.34). In fact the similarity between this equation and that for stick fixed conditions (5.25) enables us to define the 'neutral point stick free' and its position h'_{n} similarly and find that

$$h'_{\mathrm{n}} = h_0 + \bar{V}_{\mathrm{T}} \frac{\bar{a}_1}{a}\left(1 - \frac{\mathrm{d}\varepsilon}{\mathrm{d}\alpha}\right) \tag{5.44}$$

Using the same arguments as before we define the 'cg margin stick free' as

$$H'_{\mathrm{n}} = -\frac{\partial C_{\mathrm{M\,free}}}{\partial C_{\mathrm{L}}} \tag{5.45}$$

and find that

$$H'_{\mathrm{n}} = h_0 + \bar{V}_{\mathrm{T}} \frac{\bar{a}_1}{a}\left(1 - \frac{\mathrm{d}\varepsilon}{\mathrm{d}\alpha}\right) \tag{5.46}$$

and also

$$H'_{\mathrm{n}} = h'_{\mathrm{n}} - h \tag{5.47}$$

As before the cg margin, stick free, measures the corresponding degree of stability and is equal to the distance the cg is ahead of the neutral point, stick free, measured in reference chords.

We can put (5.44) in a slightly different form by writing $\bar{a}_1$ as

$$\bar{a}_1 = a_1 - \frac{a_2 b_1}{b_2} = a_1\left(1 - \frac{a_2 b_1}{a_1 b_2}\right) \tag{5.48}$$

and substituting to give

$$h_n' = h_0 + \bar{V}_T \frac{a_1}{a}\left(1 - \frac{a_2 b_1}{a_1 b_2}\right)\left(1 - \frac{d\varepsilon}{d\alpha}\right)$$

Subtracting the neutral point stick fixed distance from (5.38) we find the distance between the two as

$$h_n' - h_n = \bar{V}_T \frac{a_2}{a(-b_2)}\left(1 - \frac{d\varepsilon}{d\alpha}\right)b_1 \tag{5.49}$$

Now $d\varepsilon/d\alpha$ always lies between zero and one, also a, a_2 and $(-b_2)$ are always positive. Then if b_1 is negative, as it often is, the neutral point stick fixed lies behind the cg margin, stick free. We have seen in Chapter 4 that b_1 and b_2 can be adjusted by the designer, and it can often be advantageous to make $h_n' > h_n$ by making b_1 positive.

Again we find a connection between the stability and control. If we substitute for $\bar{a}_2$ from (5.48) into (5.46) and compare with (5.31) we find

$$C_{H_{trim}} = C_{H_0} - \frac{b_2 C_L}{\bar{V}_T a_2} H_n' \tag{5.50}$$

or

$$H_n' = -\frac{\bar{V}_T a_2}{b_2} \cdot \frac{dC_{H_{trim}}}{dC_L} \tag{5.51}$$

The remarks at the end of the last section can be echoed here. These equations state that the slope of the hinge moment trim curve is proportional to the cg margin, stick free, and in particular the sense of the stick force to change speed will change with the sign of H_n'. Also if we measure the hinge moment at a series of speeds, perhaps by measuring stick force, we can plot C_H against C_L, i.e. finding the trim curve, and determine its slope. Then repeating at a series of cg positions and plotting the slope against cg position will determine the cg position for which the slope vanishes. This point will be the neutral point, stick free.

Worked example 5.2

An aircraft with wings of rectangular planform and the characteristics given below is in steady level flight at a lift coefficient of 0.3. Find the elevator angle to trim with zero tab angle and the cg margin stick fixed. Wing area = 25 m^2, aspect ratio = 6, cg at 0.6 m aft of the leading edge, tail arm = 6 m, tailplane setting angle = $-1°$, $d\varepsilon/d\alpha = 0.46$, $a_1 = 4.6$, $a_2 = 3.1$, $a_3 = 1.6$, $C_{M_0} = -0.036$, tailplane area = 3.7 m^2, aerodynamic centre at $0.25c$.

Solution

We first calculate the wing chord from (1.2) and for a rectangular wing $c = \bar{\bar{c}} = \sqrt{S/A} = \sqrt{25/6} = 2.041$ m Then the fractional cg position is 0.6/2.041 = 0.294 and the tail volume coefficient is

$$\bar{V}_T = \frac{S_T l_T}{S\bar{\bar{c}}} = \frac{3.7 \times 6}{25 \times 2.041} = 0.4351$$

The cg margin stick fixed is then, from (5.39),

$$H_n = (h_0 - h) + \bar{V}_T \frac{a_1}{a}\left(1 - \frac{d\varepsilon}{d\alpha}\right)$$
$$= (0.25 - 0.294) + 0.4351 \times \frac{3.1}{4.6}(1 - 0.46) = 0.1143 \text{ (Ans)}$$

Using (5.27) and (5.42) we have

$$\eta_{trim} = \frac{C_{M0}}{\bar{V}_T a_2} - \frac{a_1 \eta_T}{a_2} - \frac{a_3 \beta}{a_2} - \frac{C_L H_n}{\bar{V}_T a_2}$$
$$= \frac{-0.036}{0.4351 \times 1.6} - \frac{3.1 \times (-1)}{57.3 \times 1.6} - 0 - \frac{0.3 \times 0.1143}{0.4351 \times 1.6}$$
$$= -0.067\,15 \text{ rad} = -3.85^\circ \text{ (Ans)}$$

Worked example 5.3

Find the maximum lift coefficient which can be trimmed by the aircraft with the following characteristics, when the cg is at its forward limit of 0.13 of the amc, assuming the tab angle to be zero: $\bar{V}_T = 0.48$, $a = 4.5$, $a_1 = 2.8$, $a_2 = 1.2$, $d\varepsilon/d\alpha = 0.4$, aerodynamic centre position $h_0 = 0.18$, tailplane setting angle $\eta_T = -1.8^\circ$, $C_{M0} = -0.018$. The elevator angle travel limits are ±30°.

Solution

We first find the cg margin as in the above example:

$$H_n = (0.18 - 0.13) + \frac{0.48 \times 2.8}{4.5}(1 - 0.4) = 0.2292$$

Then on solving the trim equation for the lift coefficient

$$C_L = \frac{1}{H_n}(C_{M0} - \bar{V}_T a_1 \eta_T - \bar{V}_T a_2 \eta_{trim})$$

and on substituting numbers with $\eta = -30^\circ$, since with the cg ahead of the aerodynamic centre a download will be required on the tail,

$$C_L = \frac{1}{0.2292}\left(-0.018 + \frac{0.48 \times 2.8 \times 1.8}{57.3} + \frac{0.48 \times 1.2 \times 30}{57.3}\right) = 1.42 \text{ (Ans)}$$

5.5 Actions required to change speed

We have rather left the pilot out of our considerations up to now. His or her appreciation of the pitching stability of an aircraft is related to the 'actions' required of him at the stick and he knows little of neutral points as such. He expects to move the stick forward to increase the speed and to have to push to achieve this movement. The more the stick has to be moved or pushed the greater his assessment of the stability. He will, of course also have to move the throttle lever if the flight path is to remain at the same angle to the horizontal and we are assuming that the aircraft does not have powered controls.

To calculate the stick movements and forces we define a 'stick gearing', G_η, as

$$G_\eta = \frac{\text{elevator deflection}}{\text{stick movement}} \tag{5.52}$$

where the units are metres and radians, or in symbols

$$\eta = G_\eta s \tag{5.53}$$

Here the stick movement s measured at the pilot's handgrip is taken as positive in the forward direction, corresponding to a positive elevator deflection. To relate the pilot's stick force to the hinge moment we note that the work done at the elevator in a small displacement $\delta\eta$, corresponding to a stick displacement δs, is $H\delta\eta$. This is equal to the work done by the pilot on the stick, neglecting friction. If P_η is the backward force exerted by the pilot, then

$$H\delta\eta = HG_\eta\,\delta s = P_\eta\,\delta s$$

or

$$P_\eta = HG_\eta \tag{5.54}$$

Stick position and stick force can now be found using (5.26) and (5.30) for any combination of speed, height, mass and so on. However, we are more interested in the slopes of the stick movement and force with speed curves. We have assumed that the pilot's stick is connected to the elevator; however, with the more sophisticated aircraft it may well be connected to an 'all-moving tailplane', though the alterations to the theory given here are minor and fairly obvious. In these cases the automatic control system is often connected to the elevator.

5.5.1 Stick movement and force to change speed

Now the wing loading and height are constant so that $C_L V^2$ = constant, then taking logarithms and differentiating we find

$$\frac{\mathrm{d}C_L}{\mathrm{d}V} = -\frac{2C_L}{V}$$

Then using (5.43) and (5.53) and putting $V = V_e$ and C_L and C_{Le}

$$\frac{\mathrm{d}s}{\mathrm{d}V} = \frac{1}{G_\eta}\cdot\frac{\mathrm{d}\eta}{\mathrm{d}C_L}\cdot\frac{\mathrm{d}C_L}{\mathrm{d}V} = \frac{2C_{Le}}{G_\eta V_e \bar{V}_T a_2}\cdot H_n \tag{5.55}$$

Hence the slope of the curve is positive if the cg margin is positive, i.e. the stick moves forward with speed if the aircraft is stable, stick fixed.

In a similar manner using (5.50) we find

$$\frac{\mathrm{d}C_{\mathrm{H}}}{\mathrm{d}V} = \frac{2C_{\mathrm{L}}b_2}{V\overline{V}_{\mathrm{T}}a_2} \cdot H'_{\mathrm{n}}$$

and the stick force slope, using (5.54), is

$$\frac{\mathrm{d}P_\eta}{\mathrm{d}V} = G_\eta \cdot \tfrac{1}{2}\rho V_{\mathrm{e}}^2 S_\eta c_\eta \cdot \frac{2C_{\mathrm{Le}}b_2}{V_{\mathrm{e}}\overline{V}_{\mathrm{T}}a_2} \cdot H'_{\mathrm{n}} = \frac{mg}{S} G_\eta S_\eta c_\eta \cdot \frac{2b_2}{V_{\mathrm{e}}\overline{V}_{\mathrm{T}}a_2} \cdot H'_{\mathrm{n}} \qquad (5.56)$$

It can be seen that the slope is inversely proportional to the speed. Because the control derivative b_2 is always negative, the slope of the curve is negative if the cg margin is positive, i.e. the stick has to be pushed forward with speed increase if the aircraft is stable, stick free. Hence in both cases the pilot's expectations of a stable aircraft correspond with our criteria. It should also be noted that as C_{L} is proportional to wing loading, both slopes are proportional to the wing loading, that is as the aircraft loading is increased the aircraft gets stiffer to control.

5.6 Manoeuvre stability

The other main use the pilot makes of his control over the elevator angle is to apply normal acceleration to the aircraft, as in a pullout or turn. He also uses the actions required to assess the stability of the aircraft, and so we choose to study the stick action required to 'pull' normal acceleration. We are free to choose the particular manoeuvre studied so that the theory is arbitrary to some degree. Both the pullout and the steady turn have advantages and disadvantages from our point of view. In the turn the stick action is nonlinear as a function of the normal acceleration, and the power plant may introduce gyroscopic terms. However, the latter can be corrected for by, for instance, performing turns in opposite directions and averaging. The pullout is linear in normal acceleration, but is a more difficult manoeuvre for a pilot to perform accurately. The pullout theory also omits unsteady aerodynamic effects which can be significant. However, with these reservations in mind, we proceed to discuss the pullout manoeuvre.

5.6.1 The pullout manoeuvre

We assume that the flight path is an arc of a circle and compare an aircraft performing a steady pullout at the horizontal position with the same aircraft flying level at the same speed, as shown in figure 5.10.

We will study the stick actions required to produce a normal acceleration on the aircraft of ng, where g is the acceleration due to gravity. This will require us to produce pitching moment equations for the pullout case, which we derive from those for the level flight case by considering the differences between them. Let suffix 1 denote conditions in level flight, then equating nett vertical force to mass (m) times the acceleration

$$L - mg = mng$$

or

$$L = mg(1 + n)$$

we also have $L_1 = mg$ and so define $C_{\mathrm{L}_1} = mg/\tfrac{1}{2}\rho V_{\mathrm{e}}^2 S$

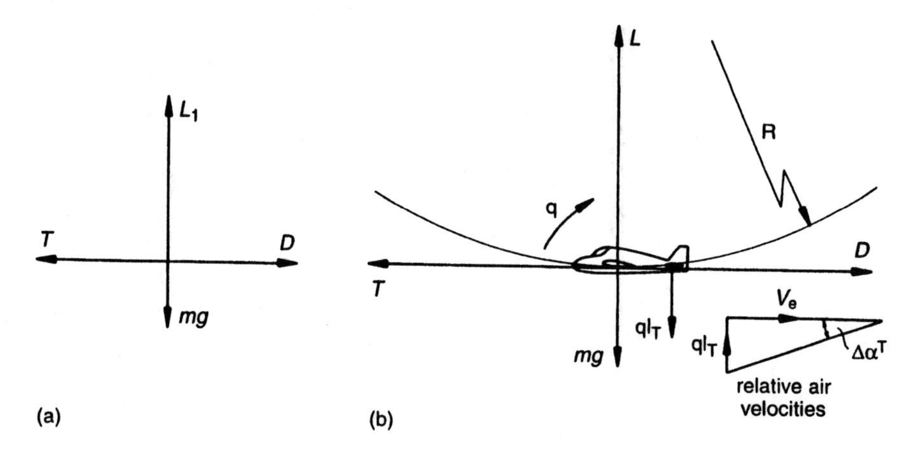

Fig. 5.10 The pullout manoeuvre: (a) forces in level flight at the same speed, (b) forces in pullout

Then

$$C_L = \frac{L}{\frac{1}{2}\rho V_e^2 S} = (1 + n)C_{L_1} \tag{5.57}$$

As the aircraft follows the circular flight path the aircraft must pitch. Let the rate be q, and so the tailplane has a downward velocity ql_T as shown in figure 5.10. Then considering the triangle of relative air velocities shown, there is an increment in tailplane incidence $\Delta\alpha_T$, assumed small, given by

$$\tan \Delta\alpha_T = \frac{ql_T}{V_e} \simeq \Delta\alpha_T$$

To express this in terms of the normal acceleration, we have

$$ng = \frac{V_e^2}{R} \text{ and } q = \frac{V_e}{R}$$

where R is the radius of the flight path. So $q = ng/V_e$, giving

$$\Delta\alpha_T = \frac{ngl_T}{V_e^2} \tag{5.58}$$

This must be added to the tailplane incidence as found in (5.22) and clearly the tailplane lift is increased, giving an extra, nose-down pitching moment. The wing also gives a pitching moment proportional to the rate of pitch, but provided that the wing is not of low aspect ratio or highly swept the effect can be neglected. This property of the wing is discussed further in Section 9.2.3.

5.6.2 Manoeuvre stability, 'stick fixed'

We repeat here our pitching moment equation, stick fixed in the level flight case (5.25), adapted for the slight change of notation

$$C_{M_{\text{fix},1}} = C_{M_0} + (h - h_0)C_{L_1} - \bar{V}_T\left[\frac{a_1}{a}\left\{1 - \frac{d\varepsilon}{d\alpha}\right\}C_{L_1} + a_1\eta_T + a_2\eta + a_3\beta\right] \quad (5.59)$$

We now modify this for the case of the pullout. Using (5.57) and (5.58) and writing the elevator angle as $\eta = \eta_1 + \Delta\eta$, then

$$C_{M_{\text{fix},n}} = C_{M_0} + (h - h_0)(1+n)C_{L_1} - \bar{V}_T\left[\frac{a_1}{a}\left\{1 - \frac{d\varepsilon}{d\alpha}\right\}(1+n)C_{L_1} + a_1\left\{\eta_T + \frac{ngl_T}{V_e^2}\right\} + a_2(\eta_1 + \Delta\eta) + a_3\beta\right] \quad (5.60)$$

Subtracting (5.59) from (5.60), solving for $\Delta\eta$ and dividing by n, we find

$$\frac{\Delta\eta}{n}\cdot\frac{\bar{V}_T a_2}{C_{L_1}} = (h - h_0) - \bar{V}_T\left[\frac{a_1}{a}\left\{1 - \frac{d\varepsilon}{d\alpha}\right\} + \frac{a_1 g l_T}{V_e^2 C_{L_1}}\right] \quad (5.61)$$

We simplify the last term in the square brackets as follows:

$$\frac{V_e^2 C_{L_1}}{gl_T} = \frac{V_e^2}{gl_T}\cdot\frac{mg}{\frac{1}{2}\rho V_e^2 S} = \frac{2m}{\rho S l_T} = 2\mu \quad (5.62)$$

say, where μ is a form of the aircraft relative density parameter, referred to in Section 4.2.4. The elevator angle per unit increment of normal acceleration, $\Delta\eta/n$, is also known briefly, if a little confusingly, as the 'elevator angle per g'. It will vanish when the cg position is h_m where

$$h_m = h_0 + \bar{V}_T\left[\frac{a_1}{a}\left\{1 - \frac{d\varepsilon}{d\alpha}\right\} + \frac{a_1}{2\mu}\right] \quad (5.63)$$

h_m is known as the 'manoeuvre point, stick fixed'. Comparing with the expression for the neutral point, stick fixed (5.38), we see that

$$h_m = h_n + \frac{\bar{V}_T a_1}{2\mu} \quad (5.64)$$

so that the manoeuvre point, stick fixed always lies a small distance behind the neutral point, stick fixed. As altitude increases, μ also increases and $h_m \rightarrow h_n$. By analogy with the cg margin we define the manoeuvre margin, stick fixed as

$$H_{\mathrm{m}} = h_{\mathrm{m}} - h = (h_0 - h) + \bar{V}_{\mathrm{T}}\left[\frac{a_1}{a}\left\{1 - \frac{\mathrm{d}\varepsilon}{\mathrm{d}\alpha}\right\} + \frac{a_1}{2\mu}\right] \tag{5.65}$$

Then stick movement per g using (5.60) and (5.53) is

$$\frac{\Delta s}{n} = -\frac{C_{\mathrm{L}_1}}{\bar{V}_{\mathrm{T}} a_2 G_\eta} \cdot H_{\mathrm{m}} \tag{5.66}$$

i.e. the stick is moved back to apply positive normal acceleration, as expected. This expression should be compared with that for stick movement to change speed (5.55). By differentiating (5.60) with respect to n, keeping elevator and tab angles fixed and comparing with (5.65), we see that

$$H_{\mathrm{m}} = -\frac{1}{C_{\mathrm{L}_1}} \cdot \frac{\partial C_{\mathrm{M_{fix,n}}}}{\partial n} \tag{5.67}$$

and this expression should be compared with the definition of cg margin, stick fixed (5.41). Similarly the condition for positive manoeuvre stability is

$$H_{\mathrm{m}} > 0 \tag{5.68}$$

5.6.3 Manoeuvre stability, 'stick free'

We could proceed to discuss the stick force required to pull normal acceleration in a similar manner to the previous section but clearly the equations involved would be still longer. Little enlightenment would be achieved; fortunately, however, we can make use of the similarities between static and manoeuvre stability that we have noted. The fundamental reason that they arise is that both change of speed and increase of normal acceleration involve changing the lift coefficient. We require the increase in hinge moment coefficient and so consider (5.51). The term $\bar{V}_{\mathrm{T}} a_2$, known as the 'elevator power', is $\mathrm{d}C_{\mathrm{m}}/\mathrm{d}\eta$ and H_{n}' is defined as $-\mathrm{d}C_{\mathrm{m}}/\mathrm{d}C_{\mathrm{L}}$ under stick free conditions. We can then rewrite (5.51) in the form

$$\left(\frac{\mathrm{d}C_{\mathrm{m}}}{\mathrm{d}C_{\mathrm{L}}}\right) \Big/ \left(\frac{-\mathrm{d}C_{\mathrm{m}}}{\mathrm{d}\eta}\right) \cdot \left(\frac{\partial C_{\mathrm{H}}}{\partial \eta}\right) = -\frac{\mathrm{d}C_{\mathrm{H}}}{\mathrm{d}C_{\mathrm{L}}}$$

which gives (5.51) the appearance of being self evident. We anticipate finding a manoeuvre margin, stick free, H_{m}' which is related to the rate of change of hinge moment with C_{L} in a similar manner, that is

$$H_{\mathrm{m}}' = -\frac{\bar{V}_{\mathrm{T}} a_2}{b_2} \cdot \frac{\mathrm{d}C_{\mathrm{H}}}{\mathrm{d}C_{\mathrm{L}}}$$

where in this case the change in C_{L} is caused by increase in normal acceleration and so $-\mathrm{d}C_{\mathrm{m}}/\mathrm{d}C_{\mathrm{L}}$ is interpreted as H_{m}'. Since this is a linear relation we can replace the differentials by differences, and then replacing ΔC_{L} by nC_{L_1} we find

$$\frac{\Delta C_H}{n} \cdot \frac{C_{L_1} b_2}{\bar{V}_T a_2} = -H'_m \tag{5.69}$$

Bearing in mind that we have shown in Section 5.3.2 that the effective lift curve slope of a tailplane with free elevator is $\bar{a}_1$ we take the manoeuvre margin stick free as

$$H'_m = (h_0 - h) + \bar{V}_T \left[\frac{\bar{a}_1}{a} \left\{ 1 - \frac{d\varepsilon}{d\alpha} \right\} + \frac{\bar{a}_1}{2\mu} \right] \tag{5.70}$$

where we have been guided by the form of (5.65). We also define the manoeuvre point, stick free, H'_m such that

$$H'_m = h'_m - h \tag{5.71}$$

Using (5.54) a change in stick force is related to the change in hinge moment coefficient by

$$\Delta P_\eta = G_\eta \tfrac{1}{2} \rho V_e^2 S_\eta c_\eta \Delta C_H$$

Then the stick force to pull g is

$$\frac{\Delta P_\eta}{n} = G_\eta \tfrac{1}{2} \rho V_e^2 S_\eta c_\eta \cdot \frac{C_{L_1} b_2}{\bar{V}_T a_2} \cdot (-H'_m)$$

and on substituting for C_{L_1} we obtain

$$\frac{\Delta P_\eta}{n} = G_\eta S_\eta c_\eta \frac{mg}{S} \cdot \frac{(-b_2)}{\bar{V}_T a_2}) \cdot H'_m \tag{5.72}$$

The quantity on the right-hand side is positive as b_2 is always negative, indicating that a pull force is required for positive normal acceleration provided that $H'_m > 0$. The relations (5.69)–(5.72) can also be derived in a completely formal manner.

These manoeuvre points can be found experimentally using a technique similar to that for the neutral points. A series of pullouts are performed with different values of normal acceleration and readings of elevator angle or hinge moment are recorded. The slopes against normal acceleration are then obtained. This is repeated using different cg positions and the slopes plotted against cg position. Extending the plot to the cg axis enables the cg positions for the vanishing of H_m or H'_m to be found. These are the respective manoeuvre points.

5.7 The centre of gravity range and airworthiness considerations

We have seen that the cg position plays a significant role in the aircraft pitching characteristics. If the cg is well forward the margins are large as are the forces and moments required. The result on a manually controlled aircraft may be excessive stick movements or forces that are larger than we can expect a pilot to exert for any length of time. If the cg is aft of a neutral point the aircraft will be unstable and stick movements or forces to change speed will be in the 'wrong' sense. We will see in Chapter 9 that if the cg is not too far back the aircraft will only diverge slowly and may well be controllable by a good pilot. If the cg falls behind a manoeuvre point the aircraft will diverge rapidly and be uncontrollable. These facts are summarized in figure 5.11.

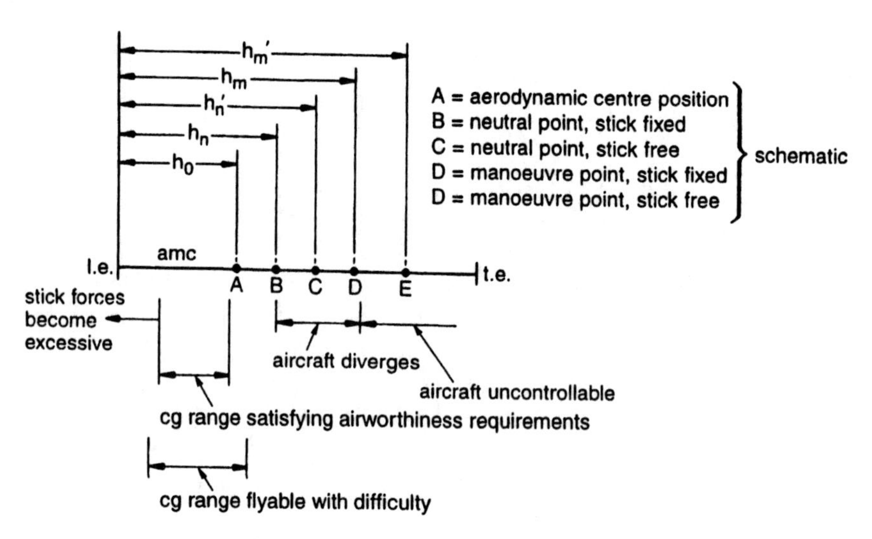

Fig. 5.11 Schematic of various points of importance in static stability

These matters have to be attended to by the designer during the design process by specifying the permissible limits of travel for the cg. On a manually controlled aircraft the forward limit will probably be fixed by the stick movements or forces, the aft limit by the cg margin, stick free. An aircraft with powered controls (without provision for reversion to manual control) can avoid the limit on stick force. An aircraft in which the pilot's control operates on force rather than displacement (possibly a 'sidestick') can avoid the restriction on stick movement. Aircraft equipped with auto-stabilization can cope with a degree of inherent instability; the aft cg limit will then be fixed by the minimum manoeuvre margin that the system can deal with. Other considerations may also operate, for instance the aft limit may be fixed by the minimum load on the nose-wheel on the ground to provide adequate steering, or the forward limit by stressing of the nose undercarriage.

There are a number of airworthiness requirements relating to static longitudinal stability, the most important of which will be outlined. The slope of the stick force against airspeed curve must be stable and have a gradient not less than 1 lb per 6 knots (0.741 N per knot or 1.44 N per m s^{-1}) over a range of speeds within 15 per cent of the trimmed speed. With regard to stalling it must be demonstrated that as speed is reduced warning of the approach of the stall begins at not less than the greater of $1.05V_{s_1}$ and V_{s_1} + 5 knots. This may be provided by natural aerodynamic effects or artificially. At the stall no abnormal nose-up pitching may occur and the stick force must be positive up to and throughout the stall; it must be possible to prevent the stall promptly and to recover by normal means. The action of the lateral controls must remain normal up to stall. If the stall is unsatisfactory the designer may take precautions to prevent stalling such as installing a 'stick-pusher' or make provisions within the Automatic Flight Control System.

5.8 Some further matters[1]

There are various useful extensions and improvements to this theory which can be made. We shall consider three.

5.8.1 More accurate expression for the cg margin, 'stick fixed'

The theory for the cg margin as laid out in Section 5.4.1 frequently does not quite agree with the results of experiments, and a 'tailplane efficiency' is often introduced to explain the difference. Typically this has a value of about 0.9 and is used to factor the tailplane volume coefficient. Apart from errors in estimating the terms in the expression (5.39) for cg margin there are several possible sources of discrepancy, such as bending of the fuselage or twisting of the tailplane and variations of downwash over the tailplane. However one certain error exists which is the result of an approximation made in Section 5.2.5. We neglected the tailplane lift in finding the aircraft-less-tail incidence which was used to obtain (5.22), the expression for the tailplane incidence. In short, we should have used C_L^{wb} instead of C_L and written

$$\alpha^T = \frac{C_L^{wb}}{a}\left(1 - \frac{d\varepsilon}{d\alpha}\right) + \eta_T$$

and on substituting for C_L^{wb} from (5.16) this becomes

$$\alpha^T = \frac{1}{a}\left(C_L - \frac{S_T}{S}C_L^T\right)\left(1 - \frac{d\varepsilon}{d\alpha}\right) + \eta_T$$

Following the previous analysis, we now substitute into (5.17) to obtain

$$C_L^T = \frac{a_1}{a}\left[\left\{C_L - \frac{S_T}{S}C_L^T\right\}\left\{1 - \frac{d\varepsilon}{d\alpha}\right\}\right] + a_1\eta_T + a_2\eta + a_3\beta$$

and solving for C_L^T we get

$$C_L^T = \frac{1}{1+F}\left[\frac{a_1}{a}\left\{1 - \frac{d\varepsilon}{d\alpha}\right\}C_L + a_1\eta_T + a_2\eta + a_3\beta\right] \tag{5.73}$$

where

$$F = \frac{a_1 S_T}{aS}\left(1 - \frac{d\varepsilon}{d\alpha}\right) \tag{5.74}$$

We now use (5.76) in place of (5.23) in (5.15) to obtain an improved pitching moment equation, stick fixed:

$$C_{M_{fix}} = C_{M_0} + (h - h_0)C_L - \frac{\bar{V}_T}{1+F}\left[\frac{a_1}{a}\left\{1 - \frac{d\varepsilon}{d\alpha}\right\}C_L + a_1\eta_T + a_2\eta + a_3\beta\right] \tag{5.75}$$

It is evident that the divisor $(1 + F)$ will appear beneath the tailplane volume coefficient throughout all subsequent analysis based on (5.75), and in particular in the expression for cg margin, stick fixed. If we substitute the typical approximate values $a_1/a = 0.7$, $S_T/S = 0.25$ and $d\varepsilon/d\alpha = 0.4$ into (5.77) we obtain the result $1/(1 + F) \simeq 0.9$, in agreement with the value of tailplane efficiency quoted above. This correction is significant because the tailplane term in the cg margin, stick fixed (5.39) is always positive whilst the term $(h_0 - h)$ will be negative near the aft cg limit. In the case of the derivation for the pitching moment equation, stick free, we will need to modify (5.24) in a similar manner, with the result that $\bar{a}_1$ will appear instead of a_1 thereafter in the expression for F in this case. It is shown in Section 9.2.2 that $(1 + F)a$ is the lift curve slope of the whole aircraft, stick fixed.

5.8.2 Canard aircraft

For various good reasons a few aircraft having a canard layout have appeared in recent years and it is worth seeing briefly how the ideas in this chapter can be applied. Figure 5.12 shows the forces and moments involved and some of the special notation needed in this section.

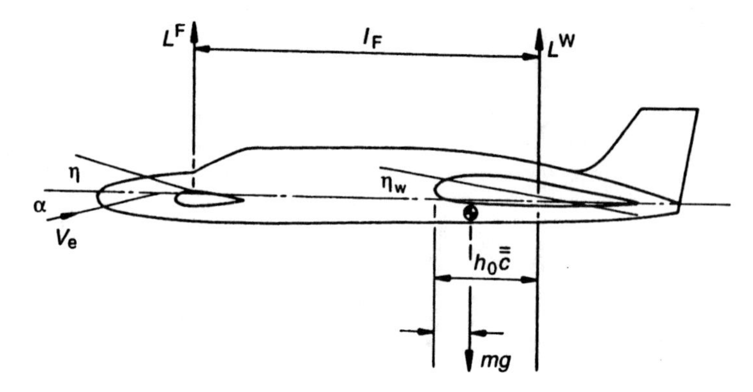

Fig. 5.12 Definition of terms used in determination of pitching moment equation for canard aircraft

One of the advantages of this layout can be seen from this figure: suppose the pilot wishes to initiate a pullout and pulls the stick back to increase the foreplane[2] incidence. The lift on the foreplane appears almost immediately and starts the pullout; then as the aircraft responds in pitch the wing incidence increases and with it the wing lift and the pullout develops fully. With a conventional layout, the tailplane has to have a downward lift increment to pitch the aircraft nose-up; this lift increment therefore opposes the effect of the wing. The result is that canard aircraft are likely to have better manoeuvre characteristics.

Now equating vertical forces we have

$$mg = L^W + L^F$$

and taking moments about the aerodynamic centre

$$M = M_0 - (h_0 - h)mg\bar{\bar{c}} + l_F C_L^F$$

Then rendering these equations nondimensional we have

$$C_L = \frac{S_F}{S} C_L^F + C_L^W \tag{5.76}$$

and

$$C_M = C_{M_0} + (h - h_0)C_L + \bar{V}_F C_L^F \tag{5.77}$$

where S_F is the area of the foreplane. We now need to express the foreplane lift in terms of the incidence of the fuselage and the foreplane angle to the fuselage. Unlike tailplanes, foreplanes are usually placed on the fuselage at a point where the fuselage width is a significant fraction of the span; the result is that the foreplane lift curve slope with fuselage incidence and the slope with angle relative to the fuselage are rather different and we must distinguish between them. At this stage we divide canard aircraft into two types:

- 'close coupled', where the upwash in front of the wing appreciably affects the foreplane average incidence due to its proximity;
- 'loose coupled' where we can ignore the upwash effect.

In both cases we must allow for the downwash effect on the wing. We assume here that we are dealing with a loose coupled layout. Foreplanes are commonly all-moving surfaces so accordingly we write

$$C_L^F = a_1\alpha + a_2\eta \tag{5.78}$$

where η is the angle of the foreplane relative to the fuselage datum. To proceed we require the fuselage incidence α. We will obtain this from (5.76), but as foreplanes normally carry a relatively larger lift force than tailplanes it is most unlikely that we can ignore the foreplane lift term in this equation. The wing incidence, allowing for the downwash, is

$$\alpha^W = \alpha + \eta_W - \varepsilon$$

where ε is now the average downwash angle over the wing. Writing

$$\varepsilon = \alpha\frac{d\varepsilon}{d\alpha} + \eta\frac{d\varepsilon}{d\eta}$$

the wing incidence becomes

$$\alpha^W = \alpha\bar{e} - \eta\varepsilon_\eta + \eta_W \tag{5.79}$$

where

$$\bar{e} = 1 - \frac{d\varepsilon}{d\alpha} \text{ and } \varepsilon_\eta = \frac{d\varepsilon}{d\eta} \tag{5.80}$$

Then as $C_L^W = a\alpha^w$ we find, on writing $s_F = S_F/S$ and substituting in (5.76),

$$C_L = s_F(a_1\alpha + a_2\eta) + a(\alpha\bar{e} - \eta\varepsilon_\eta + \eta_W) \tag{5.81}$$

Then solving for α we have

$$\alpha = \frac{C_L}{a_1 s_F + a\bar{e}} + \text{terms in } \eta \text{ and } \eta_W \tag{5.82}$$

Finally on substituting for α into (5.78) and then substituting for C_L^F into (5.77) we find

$$C_M = C_{M_0} + (h - h_0)C_L + \frac{a_1 \bar{V}_F}{a_1 s_F + a\bar{e}} C_L + \text{terms in } \eta \text{ and } \eta_W \tag{5.83}$$

Then following the arguments of Section 5.4.1 we find the cg margin, stick fixed, to be

$$H_n = (h_0 - h) - \frac{a_1 \bar{V}_F}{a_1 s_F + a\bar{e}} \tag{5.84}$$

As might have been expected the foreplane has a destabilizing effect, with the result that foreplanes normally are rather smaller in area than tailplanes and so are usually more heavily loaded.

5.8.3 Effects of springs or weights in the control circuit

In Chapter 4 we discussed how the aircraft designer can modify the hinge moment with incidence slope, b_1, and we have seen in Section 5.4.2 how this influences the cg margin, stick free. An alternative method of adjusting this margin is to place a weight or spring in the control circuit, which has the advantage of leaving the aerodynamics of the control unaltered. The elevator must still be mass balanced, for flutter avoidance reasons, so that the weight is placed in the control circuit. It is assumed that we are dealing with an aircraft with a manual control system.

To see how the stability is affected, suppose we place a weight in the control circuit which gives a trailing edge down (positive) moment to the elevator. If the aircraft slows down, the dynamic pressure decreases allowing the elevator to increase its deflection. This increases the tailplane lift giving a nose-down pitching moment resulting in a nose-down attitude. The forward component of the weight then accelerates the aircraft and hence the effect is a stabilizing one. A similar effect will be produced by a spring. In both cases only the stick free stability can be affected.

Let us now proceed to investigate this effect analytically. Suppose that a weight is placed in the control circuit giving a moment K about the elevator hinge. To the right-hand side of (5.24) we must add a term

$$\frac{K}{\frac{1}{2}\rho V_e^2 S_\eta c_\eta} = \frac{KS}{S_\eta c_\eta mg} \cdot C_L = \Lambda \cdot C_L, \text{ say}$$

Then when solving (5.24) to find the elevator angle in the stick free condition, as we did in Section 5.3.2, an extra term $-\Lambda C_L/b_2$ appears on the right-hand side of (5.33). This leads to an extra term on the right-hand side of (5.34), the pitching moment equation, stick free, of $\Lambda \bar{V}_T a_2 C_L/b_2$. The final result, after following the same arguments as in Section 5.4.2, is that the neutral point stick free is further aft by a distance, measured in reference chords, of $\Lambda \bar{V}_T a_2 C_L/(-b_2)$, increasing the stability. The same result applies to the manoeuvre point stick free, the moment due to the weight being nK in this case.

Student problems

5.1 A straight tapered wing has a gross area S, an aspect ratio A and taper ratio λ. Show that the centreline chord is given by

$$c_0 = \frac{2}{1+\lambda}\sqrt{S/A}$$

If the leading edge sweep is Λ_0 show that the sweep on the kth chord line is given by

$$\tan \Lambda_n = \tan \Lambda_k - \frac{4n(1-\lambda)}{A(1+\lambda)}$$

Show that the aerodynamic chord (amc) is given by

$$\bar{\bar{c}} = \frac{2c_0(1+\lambda+\lambda^2)}{3(1+\lambda)}$$

and if the wing is part of an aircraft with fuselage of diameter d, show that the chord at the wing/fuselage junction is

$$c_b = c_0\{1 - d(1-\lambda)/b\}$$

Hint: see figure 1.2, produce leading and trailing edges to meet and use similar triangles.

Find the amc, mean chord, fuselage-side chord and quarter-chord sweep for a wing of area 175 m^2, aspect ratio 7.5, taper ratio 0.4, body diameter 4 m and leading edge sweep 45°. (A)

5.2 A model wing in a windtunnel has the following characteristics: area = 0.25 m^2, flap area = 0.06 m^2, flap mean chord = 40 mm, $a_1 = 3.5$, $a_2 = 1.7$, $a_3 = 0.3$, $b_1 = 0.09$, $b_2 = -0.014$, $b_3 = -0.03$. Estimate the lift and flap hinge moment at a wind speed of 60 m s^{-1} if the incidence, flap and tab angles are 3°, −4° and 5° respectively. What angle would the flap take up if it were perfectly free to move (the other angles remaining the same) and what would be the lift in this case? (A)

5.3 Find the cg margin, stick free, for an aircraft with the following characteristics: wing area = 95 m^2, tailplane area = 15 m^2, tail arm = 10 m, amc = 3.2 m, $a = 4.5$, $a_1 = 3.7$, $a_2 = 2.0$, $b_1 = 0.01$, $b_2 = -0.06$, $d\varepsilon/d\alpha = 0.47$. The cg lies a distance of 1.1 m, and the aircraft-less-tail aerodynamic centre 0.57 m aft of the leading edge of the amc. (A)

5.4 An aircraft, with the characteristics given below, is to be designed such that it is in trim at a C_L of 0.3 with elevators in the neutral position (i.e. $\eta = 0$). Find the tailplane setting to achieve this, assuming zero tab angle.

$\bar{V}_T = 0.48$, $a = 4.5$, $a_1 = 2.8$, $a_2 = 1.2$, $d\varepsilon/d\alpha = 0.4$, aerodynamic centre position $h_0 = 0.18$, cg position $h = 0.27$, $C_{M0} = -0.016$. Also find the most forward position for the cg for which the trimmed lift coefficient is 1.4. The elevator has travel limits of ±30°. (A)

5.5 A glider has the following characteristics: tail arm = 8 m, wing area = 42 m^2, amc = 2 m, tailplane area = 5 m^2, $a = 4.6$, $a_2 = 1.19$, $b_1 = -0.008$, $b_2 = -0.013$, $d\varepsilon/d\alpha = 0.39$. Find the movement of neutral point on freezing the stick. (A)

5.6 An aircraft is flying close to the ground with full flaps deployed giving a lift coefficient of 2.0 which includes an increment of 0.8 due to the flaps. Find the elevator angle to trim and the pilot's stick force if the characteristics are as follows: wing loading = 3.925 kN m^{-2}, $\bar{V}_T = 0.5$, $a = 4.4$, $a_1 = 3.1$, $a_2 = 2.5$, $b_1 = -0.045$, $b_2 = -0.075$, C_{M0} in take-off configuration = −0.18, elevator area = 6.7 m^2, elevator chord = 0.7 m, stick gearing = 1.45 rad m^{-1}, $d\varepsilon/dC_L = 0.11$, tailplane setting angle = −3°. The cg is 0.05$\bar{\bar{c}}$

aft of the aerodynamic centre and the reduction of downwash due to the ground is 1.8°. Assume zero tab angle. (A)

5.7 An aircraft has the following characteristics: aerodynamic centre position $h_0 = 0.22$, $a = 3.6$, $a_1 = 3.0$, $a_2 = 1.5$, $d\varepsilon/d\alpha = 0.4$. It is to be designed to satisfy the following conditions:

(a) at the forward cg limit the change of elevator angle for a C_L change of 1.0 is 10°;
(b) at the aft cg limit the cg margin stick fixed is to be 0.05;
(c) the cg range is $0.05\bar{\bar{c}}$.

Find the forward cg position and the tail volume coefficient. (A)

5.8 Determine the tailplane area required to given an aircraft a cg margin, stick free, of 0.12. The other characteristics are: aircraft-less-tail aerodynamic centre at $0.18\bar{\bar{c}}$, cg at $0.23\bar{\bar{c}}$, $\bar{\bar{c}} = 1.5$ m, $a = 0.48$, $a_1 = 3.0$, $a_2 = 1.8$, $b_1 = 0.01$, $b_2 = -0.06$, wing area = 10 m^2, $d\varepsilon/d\alpha = 0.47$.

5.9 Show that the tab angle to trim an aircraft, stick free, is given by

$$\beta_{\text{trim}} = \left(\frac{C_{M0}}{\bar{V}_T\bar{a}_3} - \frac{\eta_T\bar{a}_1}{\bar{a}_3}\right) - \frac{C_L H_n'}{\bar{V}_T\bar{a}_3}$$

The cg margin, stick free, of an aircraft is 0.06. At an altitude where the relative density is 0.61, the tab angle to trim, stick free, is zero at a speed of 180 m s^{-1}. Find the tab angle to trim, stick free, at an altitude where the relative density is 0.74 when the speed is 85 m s^{-1}, given that the wing loading is 3.0 kN m^{-2} and $\bar{V}_T = 0.55$, $a_2 = 2.3$, $a_3 = 0.5$, $b_2 = -0.15$, $b_3 = -0.003$. (A)

5.10 Show that the change in tab angle to trim, stick free, due to a small change in speed ΔV is

$$\Delta\beta_{\text{trim}} = \frac{2C_{Le}}{\bar{V}_T\bar{a}_3} H_n' \frac{\Delta V}{V_e}$$

5.11 An aircraft with the following characteristics performs a pullout manoeuvre at 150 m s^{-1} at a low level, pulling an excess normal acceleration of $2.5g$: mass = 65 000 kg, wing area = 190 m^2, $\bar{V}_T = 0.49$, tail arm = 14 m, $d\varepsilon/d\alpha = 0.49$, $a = 4.5$, $a_1 = 3.7$, $a_2 = 2.0$, $h_0 = 0.16$, $h = 0.25$. The stick gearing is 1.2 rad m^{-1}. Find the elevator angle change from level flight. (A)

5.12 Estimate the stick force required to pull an excess normal acceleration of $3.5g$ at the altitude where $\sigma = 0.74$ at a speed of 150 m s^{-1} for an aircraft with the following characteristics: stick fixed cg margin = 0.09, $a = 5.0$, $a_1 = 3.7$, $a_2 = 1.6$, $b_1 = -0.06$, $b_2 = -0.13$, wing loading = 1.96 kg m^{-2}, elevator area = 1.8 m^2, elevator chord = 0.6 m, $\bar{V}_T = 0.55$, tail arm = 9.1 m, $d\varepsilon/d\alpha = 0.4$, stick gearing = 2 rad m^{-1}.

5.13 Show that in a steady level correctly banked turn at a speed V, the rate of pitch, q, of an aircraft is given by

$$q = \frac{g}{V}(N - 1/N)$$

where $N = L/mg$, and that the change of incidence of the tailplane is

$$\Delta\alpha_T = \frac{C_{L1}}{2\mu}(N - 1/N)$$

where C_{L_1} is the lift coefficient at the same speed in level flight and $\mu = m/\rho S l_T$.

An aircraft is in level flight at a speed of 120 m s^{-1} at sea-level; it is then put into a level turn at the same speed with a bank angle of 70°. Find the change in elevator angle required to trim given the following particulars: wing loading = 2.5 kN m^{-2}, tail arm = 6 m, $\bar{V}_T = 0.55$, $d\varepsilon/d\alpha = 0.45$, $h - h_0 = 0.05$, $a = 4.5$, $a_1 = 3.4$, $a_2 = 1.8$. (A)

Notes

1. This section can be omitted when first studying this chapter.
2. The author prefers the term 'foreplane' for the forward lifting surface by analogy with 'tailplane'; 'canard' is used here only as a description of the layout.

6
Lateral static stability and control

6.1 Introduction

One of the reasons that the Wright brothers were successful in designing and constructing the first man-carrying aircraft was their realization that it was necessary to provide control about all three axes. It is all too evident from cinefilm of many of the early attempts to fly that control in roll was desperately needed, not least to react the propeller torque. The Wrights used wing warping and, for good measure, used coupled contrarotating propellers. Shortly after their first flight, ailerons were invented and are almost universally used today.

In this chapter we consider control and stability about the roll and yaw axes. We also introduce a notation which will be made much use of in later chapters.

6.2 Simple lateral aerodynamics

We first consider some of the simple background aerodynamics, assuming that the aircraft has conventional flap type controls and a conventional layout.

6.2.1 Aileron and rudder controls

To start our discussion we look at the relation between the rudder angle and the resulting yawing moment. Consider the aircraft shown in figure 6.1, where ζ is the rudder deflection angle from the neutral position, positive as shown.

The sideways lift on the fin will be $Y_{fin} = \frac{1}{2}\rho V_E^2 S_F a_2^F \zeta$ where S_F is the fin area and a_2^F is the rudder lift curve slope. Assuming that the centre of pressure of the lift on the fin due to rudder deflection is a distance l_R aft of the cg, the yawing moment produced will be negative (see figure 4.1) and can be written

$$N_R = \overset{\circ}{N}_\zeta \cdot \zeta$$

where

$$\overset{\circ}{N}_\zeta = \frac{\partial N_R}{\partial \zeta} = -\tfrac{1}{2}\rho V_e^2 S_F l_R a_2^F \tag{6.1}$$

In this expression we have used ζ as a suffix to indicate differentiation with respect to rudder angle, a practice we often use. The superscript '°' is to emphasize that the quantity is dimensional; it can be pronounced as 'ord'. The side force will also produce a rolling moment as shown in figure 6.1 and we introduce a derivative $\overset{\circ}{L}_\zeta$ to represent this effect.

The ailerons are designed to produce a rolling moment which we write in a similar manner as $L_A = \overset{\circ}{L}_\xi . \xi$. Here $\overset{\circ}{L}_\xi$ also will be a negative quantity as positive aileron angle is defined as

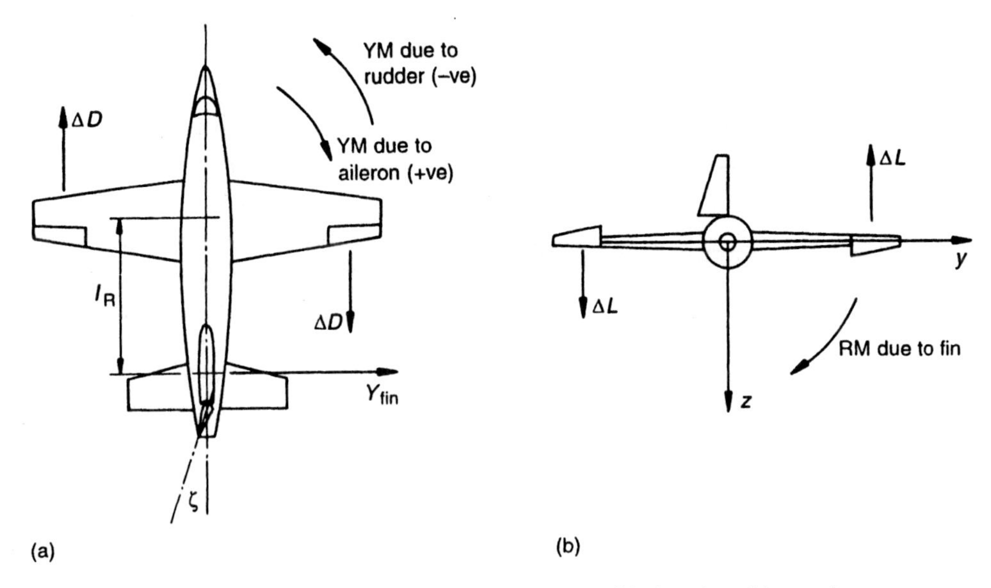

Fig. 6.1 Effects of aileron and rudder deflection: (a) plan view, (b) rear view

starboard aileron down, giving a positive lift on that wing and hence a negative rolling moment. The increase of lift on the starboard wing increases the trailing vortex drag, giving a yawing moment in the positive sense. Similarly the drag on the port wing is decreased adding to the positive yawing moment. From figure 6.1 it can be seen that the yawing moment due to aileron opposes the turn. This effect is known as the 'adverse yawing moment due to aileron', and particular measures may be taken to reduce it. The rolling moment from the rudder is adverse at low incidence, but helpful at high incidence, depending on whether the centre of pressure is above or below the x-axis. However, the moment arm of the fin side force is generally small and this is therefore usually a much less serious effect.

It should be noted that both the aileron and the rudder produce rolling and yawing moments, in different amounts, and so usually have to be used in a coordinated manner.

6.2.2 Sideslip

A pilot, by suitable adjustment of the controls, can fly an aircraft steadily in a straight line but with its longitudinal axis at an angle to the direction of flight. This manoeuvre is a 'straight sideslip', the drag is increased and so it is occasionally used to lose height. Normally the aircraft will be at a small roll angle so that there is a component of the weight to balance aerodynamic sideforces. Consider the aircraft shown in figure 6.2(a) which is in a steady sideslip, but with the wings level. The aircraft has velocity components V_e along its X-axis and v along its Y-axis, resulting in a sideslip angle, β, between its longitudinal axis and the direction of flight.

If we neglect any interference effects from the wing or fuselage, the fin incidence angle will be β, as shown, given by

$$\tan\beta = \frac{v}{V_e} \simeq \beta \tag{6.2}$$

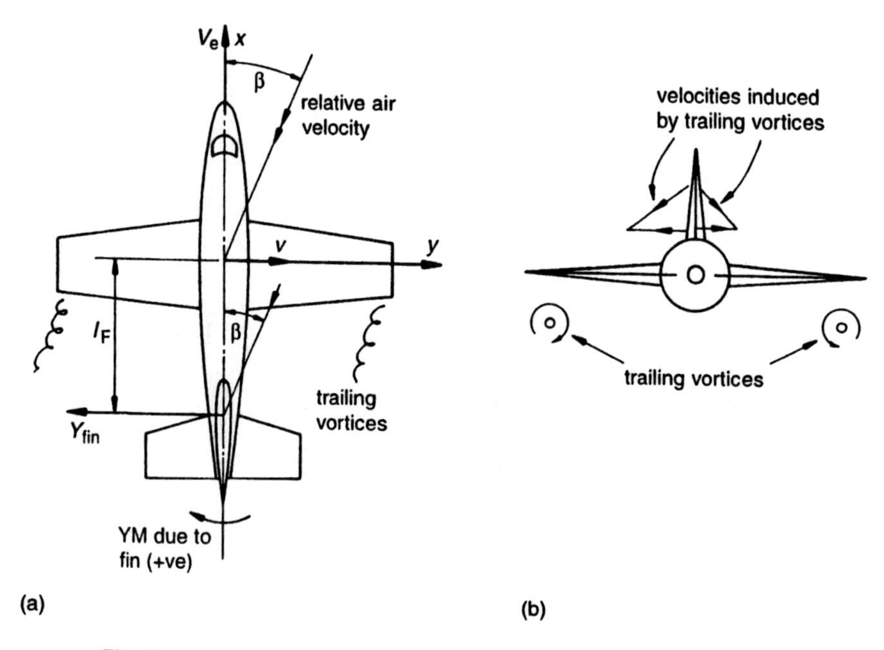

Fig. 6.2 Velocities and angles in a sideslip: (a) plan view, (b) rear view

assuming that β is a small angle. This will give a sideways lift on the fin of

$$Y_{\text{fin}} = \tfrac{1}{2}\rho V_e^2 S_F a_1^F \cdot \beta = \tfrac{1}{2}\rho V_e S_F a_1^F \cdot v \tag{6.3}$$

where a_1^F is the fin lift curve slope. Assuming that the fin lift acts at a distance l_F aft of the cg, there is a positive yawing moment which we write

$$N_{\text{fin}} = \mathring{N}_{v,\text{fin}} \cdot v$$

where

$$\mathring{N}_{v,\text{fin}} = \tfrac{1}{2}\rho V_e S_F l_F a_1^F \tag{6.4}$$

Other components of the aircraft produce yawing moments; in particular the fuselage produces a moment of the opposite sign. The latter is the result of a sideways lift on it which has a centre of pressure near the nose, similar to that produced in the pitching case. The result is that the yawing moment due to sideslip derivative for the aircraft as a whole is rather less than the fin contribution.

We now consider some further effects of sideslip, namely those due to dihedral and sweep-back on the wings. Figure 6.3(a) shows the rear view of an aircraft having dihedral in a sideslip.

Resolving the relative air velocity into the normal to the starboard wing mean plane gives an upward velocity component of $v\Gamma$, where Γ is the dihedral angle, positive as shown and assumed small. The result is an increase in incidence of amount $v\Gamma/V_e$, which increases the lift on the wing and gives a negative rolling moment. The opposite effect occurs on the other wing, which again produces a negative rolling moment.

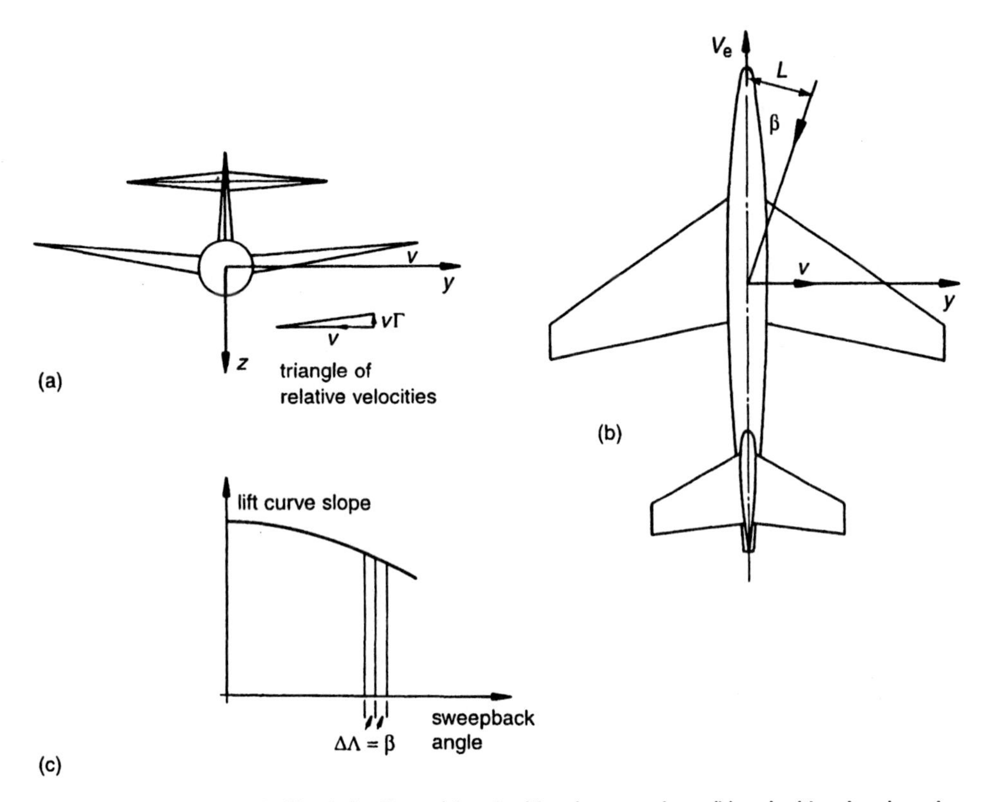

Fig. 6.3 Determination of dihedral effect: (a) velocities in rear view, (b) velocities in plan view, (c) variation of lift curve slope with sweepback angle

Figure 6.3(b) shows an aircraft with sweptback wings in a sideslip. We see that the sweepback angle is decreased by β on the starboard wing and increased by β on the port wing. Now the lift curve slope of wings decreases as the sweep is increased. Assuming that this wing is at some incidence, the lift on the starboard side therefore increases with sideslip giving a negative rolling moment. The opposite occurs on the port side, adding to the magnitude of the rolling moment.

There are still more mechanisms for producing rolling moment on an aircraft in a sideslip; we will discuss only one more, that due to wing position on the fuselage. In a sideslip the airflow past the fuselage can be thought of as composed of two flows, the flow in unsideslipped flight and a flow from one side of the fuselage to the other, the 'crossflow'. Figure 6.4(a) shows the ideal flow past a circular cylinder, which resembles the crossflow expected past a circular fuselage. Figure 6.4(b) shows the cross-section of a high wing aircraft in the region of the wing. Near the point A the air is deflected upwards relative to the wing, increasing the incidence locally; similarly the incidence is decreased near B. The resulting local changes to the wing lift gives a negative rolling moment from both wings.

We shall see later that the rolling moment due to sideslip effect is a stabilizing one provided that it is not too large. The effects on aircraft layout can be seen in actual designs. With unswept wings, low wing layouts usually have noticeable dihedral whilst high wing ones little or none. Highly swept, low wing aircraft have little dihedral whilst high wing ones often have negative dihedral angle, known as 'anhedral'.

The three effects discussed above all depend on changes of lift distribution over the span of the wing, and so are accompanied by spanwise changes in the trailing vortex drag. The result

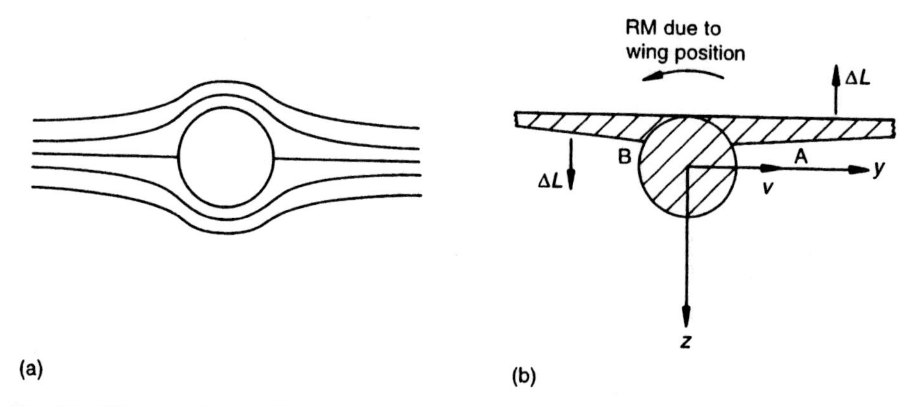

Fig. 6.4 Effect of wing position on L_v: (a) ideal flow past a circular cylinder, (b) forces produced by crossflow

in cases of dihedral and sweep is a contribution to yawing moment due to sideslip in the same sense as the fin contribution. In the case of dihedral the effect is usually small, but for sweep the effect is proportional to C_L^2, and so can become important at high incidence.

We write the sideforce, rolling and yawing moments due to sideslip for the whole aircraft using derivatives thus:

$$\left.\begin{aligned} Y &= \mathring{Y}_v \cdot v \\ L &= \mathring{L}_v \cdot v \\ N &= \mathring{N}_v \cdot v \end{aligned}\right\} \qquad (6.5)$$

6.2.3 Effect of rate of yaw

In any turn, in order for the aircraft axis to be continually tangential to the path of the cg, the aircraft must have an angular velocity in yaw, and there are other cases in which the aircraft is yawing. Let us consider the effect of a small rate of yaw on an aircraft flying in a straight line and wings level, i.e. without the extra complications of a turn, as shown in figure 6.5.

As a consequence of the yaw rate r, the fin has a sideways velocity of rl_F. Then considering the triangle of relative air velocities, the fin has an incidence of $\alpha_F = rl_F/V_e$, which gives rise to a sideforce of

$$Y_{\text{fin}} = \tfrac{1}{2}\rho V_e^2 S_F a_1^F \cdot \alpha_F = \tfrac{1}{2}\rho V_e S_F l_F a_1^F \cdot r$$

This in turn gives a yawing moment which opposes the yawing motion and which we write in the form

$$N_{\text{fin}} = \mathring{N}_{r,\text{fin}} \cdot r$$

where

$$\mathring{N}_{r,\text{fin}} = -\tfrac{1}{2}\rho V_e S_F l_F^2 a_1^F \qquad (6.6)$$

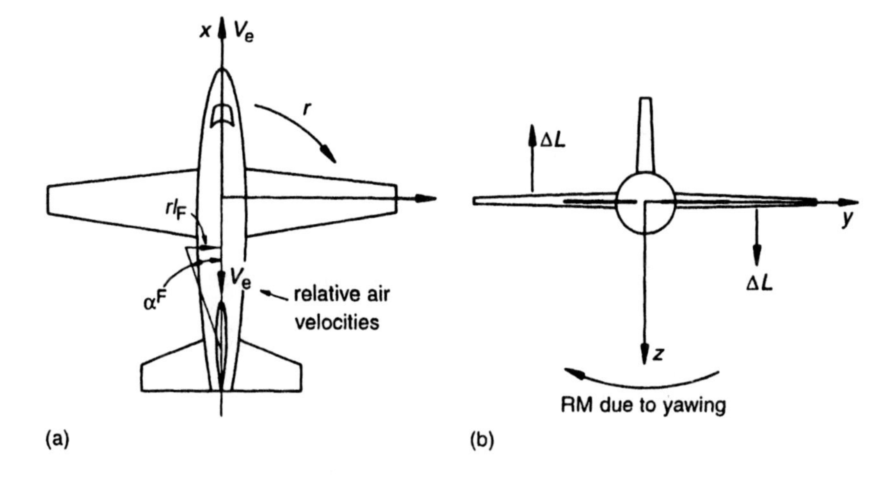

Fig. 6.5 Effect of rate of yaw: (a) velocities in plan view, (b) lift forces

Other parts of the aircraft also give rise to yawing moments as a response to rate of yaw.

Rate of yaw has another effect, which is the production of a rolling moment; to see how this may happen consider again the aircraft of figure 6.5. The starboard wing tip is moving backwards relative to the cg and so its nett velocity is reduced, and hence also is the lift. This gives rise to a positive rolling moment; the reverse effect appears on the port wing and so this contributes to the rolling moment in the same sense. Other parts of the aircraft also contribute to this effect.

We write the sideforce, rolling and yawing moments due to rate of yaw using derivatives thus:

$$\left.\begin{aligned} Y &= \mathring{Y}_r \cdot r \\ L &= \mathring{L}_r \cdot r \\ N &= \mathring{N}_r \cdot r \end{aligned}\right\} \tag{6.7}$$

In fact the sideforce due to yaw rate is very small and we frequently neglect it.

6.3 Trimmed lateral manoeuvres

In this section we consider the aileron and rudder angles required to perform two simple manoeuvres, the correctly banked turn and the straight sideslip. Strictly speaking the angles found are the changes from straight and level flight, since even in that condition an aircraft may need aileron or rudder angles to counter power effects or other asymmetries. It is cumbersome and unnecessary to use the superscript ord in this section since it is evident that all the terms in the equations are dimensional.

6.3.1 The correctly banked turn

Consider again the aircraft shown in figure 3.9, which is performing a turn at bank angle ϕ to the vertical. We repeat the relation (3.55) here:

$$L \sin \phi = m \frac{V_e^2}{R}$$

and rewrite it in terms of the rate of turn ω as

$$L \sin \phi = m V_e \omega \tag{6.8}$$

where ω is the angular velocity of the aircraft about the vertical; we have again neglected the side force generated by rate of yaw. We need to resolve this velocity along the z-axis to find the rate of yaw and hence express the aerodynamic effects of yaw rate. Considering the triangle of angular velocities shown in figure 3.9 we find

$$r = \omega \cos \phi \tag{6.9}$$

We also find a rate of pitch, $q = r \sin \phi$. This means that there will be a change in the trim in pitch: see Section 5.5. From (3.46) we have $L \cos \phi = mg$, then using (6.9) to eliminate $\cos \phi$ in this we have

$$Lr/\omega = \mathrm{mg}$$

and using this to eliminate ω from (6.8) gives

$$r = (g \sin \phi)/V_e \tag{6.10}$$

Since the flight condition is a steady one the total moments about the roll and yaw axes must be zero (see for comparison Section 5.2). Then adding the moments due to yaw rate and control action we have

$$\left.\begin{aligned} L_r r + L_\xi \xi + L_\zeta \zeta &= 0 \\ N_r r + N_\xi \xi + N_\zeta \zeta &= 0 \end{aligned}\right\} \tag{6.11}$$

Solving these simultaneous equations for ξ and ζ in terms of r and then substituting r from (6.10) gives

$$\xi = \frac{r}{\Delta}\left(L_r N_\zeta - L_\zeta N_r\right) = \frac{g \sin \phi}{V_e \Delta}\left(L_r N_\zeta - L_\zeta N_r\right) \tag{6.12}$$

$$\zeta = \frac{r}{\Delta}\left(L_\xi N_r - L_r N_\xi\right) = \frac{g \sin \phi}{V_e \Delta}\left(L_\xi N_r - L_r N_\xi\right) \tag{6.13}$$

where

$$\Delta = L_\zeta N_\xi - L_\xi N_\zeta \tag{6.14}$$

These expressions are dominated by the direct effects of the controls and the rate of yaw derivatives and can be approximated as

$$\xi \simeq -\frac{L_r \cdot r}{L_\xi} \text{ and } \zeta \simeq -\frac{N_r r}{N_\zeta} \tag{6.15}$$

This shows that the aileron is primarily used to balance the rolling moment due to yaw rate and the rudder to balance the yawing moment, as would be expected. Once the aircraft has achieved a steady turn the aileron and rudder angles required are usually quite small. Having found these angles it is possible to find the stick forces from equations of the form (4.17) and (5.54).

6.3.2 Steady straight sideslip

Consider the aircraft shown in figure 6.6, which is performing a steady straight sideslip, with sideslip velocity v.

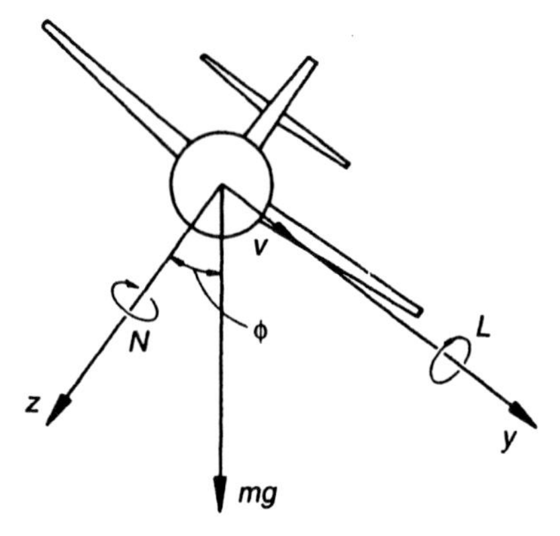

Fig. 6.6 Forces on aircraft in steady sideslip

Resolving forces along the Y direction and neglecting any sideforces generated by the control surfaces we have

$$Y_v . v + mg \sin \phi = 0 \tag{6.16}$$

The rolling and yawing moments due to sideslip and from the controls must be in balance giving

$$\left.\begin{aligned} L_v \cdot v + L_\xi \cdot \xi + L_\zeta \cdot \zeta = 0 \\ N_v \cdot v + N_\xi \cdot \xi + N_\zeta \cdot \zeta = 0 \end{aligned}\right\} \tag{6.17}$$

Solving these simultaneously for ξ and ζ in terms of v gives

$$\xi = \frac{v}{\Delta}\left(L_v N_\zeta - L_\zeta N_v\right) \tag{6.18}$$

$$\zeta = \frac{v}{\Delta}\left(L_\xi N_v - L_v N_\xi\right) \tag{6.19}$$

where Δ is given by (6.14) and the roll angle can be found from (6.16). Again these expressions are dominated by the direct effects of the control angles and the effects of sideslip and can be approximated as

$$\xi \simeq -\frac{L_v \cdot v}{L_\xi} \text{ and } \zeta \simeq -\frac{N_v \cdot v}{N_\zeta} \tag{6.20}$$

This shows that in this case the aileron is primarily used to balance the rolling moment due to sideslip and the rudder to balance the yawing moment, as would be expected. The aileron and rudder angles increase rapidly with sideslip angle, whilst the roll angle increases less rapidly. Figure 6.7 shows a sketch of typical roll, aileron and rudder angles as a function of sideslip angle.

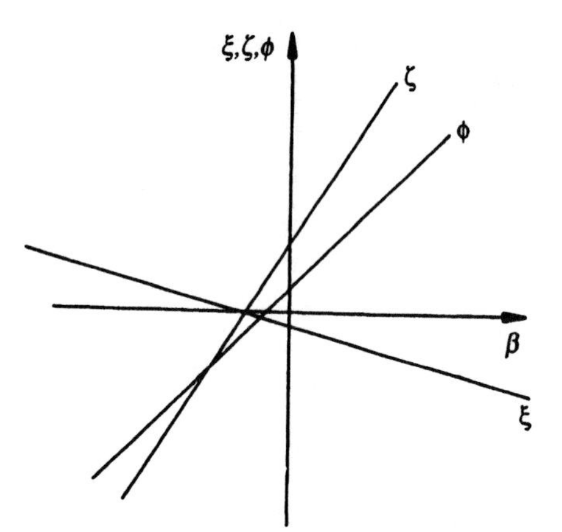

Fig. 6.7 Variation of roll angle, ϕ, aileron angle, ξ, and rudder angle, ζ, with sideslip, β, in a steady sideslip

6.3.3 Minimum control speeds

If an engine fails on a multi-engined aircraft then the pilot must be able to manoeuvre the aircraft. At low speeds the dynamic pressure may not be sufficient to produce sufficient yawing moment from the rudder. There is therefore a minimum speed at which the pilot can maintain straight flight. The minimum control speed in the air, V_{MCa}, is defined in the airworthiness requirements roughly as the airspeed at which, when the critical engine is made inoperative, it is possible to maintain control of the aircraft in straight flight with a bank angle of not more than 5°. This has to be demonstrated by flight test when the aircraft is built. The rudder pedal force may not exceed 150 lb (= 667 N) and the change in heading may not be more than 20°. Also V_{MCa} may not exceed $1.2V_{\mathrm{S_1}}$. In designing an aircraft the designer will choose a value for V_{MCa} with reference to the desired take-off performance and use it as one factor in the design of the fin and rudder. At an assumed value for V_{MCa} in the calculations, the yawing moment from the remaining engines will balance that from the fully deflected rudder. The minimum control speed on the ground, V_{MCg}, is similarly defined; no use of a steerable nose-wheel may be assumed.

6.4 Static stability

In this section we consider what preliminary insights we may obtain from applying the ideas of static stability. Consider an aircraft disturbed in sideslip as in figure 6.2. The yawing moment produced by the fin tends to turn the aircraft into the direction of the resultant velocity, i.e. it tends to reduce the sideslip. Therefore this is a stable response and is known as 'directional' or 'weathercock' stability. The condition for this is then

$$\boxed{N_v > 0} \tag{6.21}$$

The fin provides positive directional stability, whilst the fuselage and any engine nacelles or propellers ahead of the cg are destabilizing. We note that a rolling moment will also be generated through the derivative L_v so that the aircraft will be given both roll and yaw accelerations.

Now let us consider an aircraft which has been given a small angle of rotation in roll around its velocity vector. No restoring moment in roll will appear because no surface of the aircraft has changed its incidence to the flow. However, there is a component of the weight along the Y-axis which will produce a sideslip; this in turn will produce a rolling moment through the derivative L_v. Positive roll angle, as shown in figure 3.9, will produce a positive sideslip velocity, and so a negative rolling moment is required for static stability. This effect is known as 'static lateral stability' and the condition for it to be positive is therefore

$$\boxed{L_v < 0} \tag{6.22}$$

This sideslip will also produce a yawing moment bringing the mechanism of the directional stability into play. We see therefore that these static stability conditions are coupled and, although necessary, are not sufficient and only analysis of the dynamic stability gives the complete picture.

As in the case of static stability in pitch the measurement of the trim curves indicates the static stability. The procedure is to fly the aircraft in a series of straight sideslips, and measure the sideslip, roll, aileron and rudder angles. Then, assuming that the direct effects of the aileron and rudder are in the usual senses, (6.20) shows that for positive static lateral stability aileron angle decreases with sideslip. For positive directional stability rudder angle increases with sideslip. Figure 6.7 shows the aileron, rudder and roll angles in a sideslip for an aircraft with positive directional and lateral static stability. If stick and rudder forces are also measured then these can be used to infer the static stabilities, stick free. These are positive when the forces are in the direction to produce these stick and rudder movements. The airworthiness requirements ask that an aircraft is laterally stable as discussed above and require the trim curves and the rudder pedal force curve not to have any reversal of slope up to the maximum angles available or up to a pedal force of 180 lb (= 800 N).

Student problem

6.1 A twin-engine aircraft is flying at low altitude when one engine fails. Find the slowest speed at which it can fly steadily without sideslip. Assume that the thrust of the remaining engine just balances the drag and that engine thrust is independent of speed. The maximum rudder angle is 30°, the rudder lift curve slope is 0.55 and the rudder moment arm about the cg is 16 m. Other information is as follows: mass = 200 000 kg, wing area = 300 m^2, fin area = 42 m^2, the engines are 5 m from the centreline, $C_D = 0.036 + 0.062C_L^2$. (A)

7
Revision and extension of dynamics

7.1 Introduction

In this chapter we first discuss some simple aircraft motions which can be treated approximately using the familiar first- and second-order ordinary differential equations. We continue by deriving equations for dealing with the dynamics of a body using axes fixed in the body itself as discussed in Section 4.2.1, and use them to discuss the motion of a bifilar pendulum. Finally a generalized method of writing systems of linear differential equations is introduced.

7.2 Some simple aircraft motions

In this section we consider three simple aircraft motions, arranged in order of increasing complexity. The second and third involve second-order ordinary differential equations with oscillatory solutions.

7.2.1 Pure rolling

We saw in the last chapter that when the ailerons on an aircraft are deflected a rolling moment is produced; the aircraft will then accelerate in roll and quite quickly reach a steady rate of roll. There will also be a yawing moment from the ailerons and when a roll angle has built up a sideways component of the weight will appear; these and other effects will lead to motion in these freedoms. However, it is evident from flying displays that a pilot can fly an aircraft such that the motion is almost pure rolling and we idealize the situation and just consider moments about the roll axis.

Rolling an aircraft will produce changes in the local angles of incidence on the wing, as shown in figure 7.1.

A section of the wing a distance y from the roll axis will have a downward velocity py, where p is the angular velocity in roll, and considering the triangle of relative air velocities an approximate increase in incidence of py/V_c is produced. This, however, is altered by induced effects and no simple expression can be found for the change in the local lift. The moment of the lift about the roll axis is also proportional to y and opposes the roll. The final result is that the overall rolling moment is proportional to p and to (span)^2, provided stalling of the wing is avoided. This moment is in the opposite sense to the rolling velocity, i.e. it is a damping moment. Although other parts of the aircraft contribute, the wing dominates as it is by far the most powerful lifting surface of the aircraft. The total aerodynamic rolling moment is then expressed using the form of a derivative, that is

$$L = \mathring{L}_p \cdot p \tag{7.1}$$

where $\mathring{L}_p$ is known as the 'damping in roll' derivative.

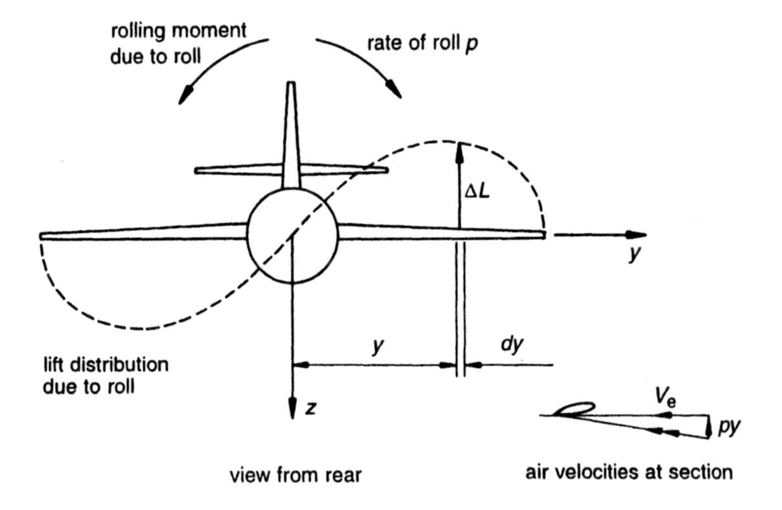

Fig. 7.1 Effect of rate of roll on an aircraft

We now obtain the equation of motion by equating the aerodynamic moments due to rate of roll and aileron deflection to the product of the roll inertia, I_x, and acceleration in roll, to give

$$\mathring{L}_p \cdot p + \mathring{L}_\xi \cdot \xi = I_x \frac{dp}{dt}$$

or rearranging

$$I_x \cdot \dot{p} - \mathring{L}_p \cdot p = \mathring{L}_\xi \cdot \xi \tag{7.2}$$

We will use this to investigate how the rate of roll varies if the pilot moves the ailerons rapidly to some angle and holds them there, i.e. we solve (7.2) subject to the conditions that at $t = 0$, $p = 0$ and $\xi = \text{constant} = \xi_1$. The general solution is

$$p = Ae^{\lambda t} - \frac{\mathring{L}_\xi \cdot \xi_1}{\mathring{L}_p} \tag{7.3}$$

where λ is $= \mathring{L}_p / I_x$. Putting in the initial condition $p = 0$ at $t = 0$ gives the constant A and the solution is finally

$$p = \frac{\mathring{L}_\xi \cdot \xi_1}{\mathring{L}_p} \left(e^{\lambda t} - 1 \right) \tag{7.4}$$

In this equation it should be remembered that $\lambda < 0$ since the damping in roll derivative is usually negative.

The rate of roll tends to be the value $p_\infty = \mathring{L}_\xi . \xi_1 / \mathring{L}_p$; in fact usually this rate is nearly achieved in a second or two as the inertia in roll is relatively small. We measure the rapidity of the response by the time taken, t_H, to reach half this rate. Then from (7.4)

$$\frac{p}{p_\infty} = \tfrac{1}{2} = e^{\lambda t_H} - 1$$

and solving for the time to half the final roll rate we find

$$t_H = \frac{I_x \ln 2}{\mathring{L}_p} \tag{7.5}$$

These results can be put in a slightly more meaningful form if we use nondimensional versions of some of the quantities involved. Let us introduce

$$i_x = \frac{I_x}{mb^2} \tag{7.6}$$

$$L_\xi = \frac{\mathring{L}_\xi}{\tfrac{1}{2}\rho V_e^2 Sb} \tag{7.7}$$

and

$$L_p = \frac{\mathring{L}_p}{\tfrac{1}{2}\rho V_e Sb^2} \tag{7.8}$$

where b = wing span. The final rate of roll can then be put in the form

$$\frac{p_\infty b}{V_e} = \frac{L_\xi}{L_p} \cdot \xi \tag{7.9}$$

For small rates of roll, the quantity on the left-hand side is twice the helix angle of the path of the wing tips of the rolling aircraft, or put another way, the final rate of roll is proportional to the speed. Using (7.6)–(7.8) the time to reach half the final roll rate becomes

$$t_H = \frac{i_x \ln 2}{L_p} \frac{m}{\tfrac{1}{2}\rho V_e S}$$

The second term on the right-hand side is the aerodynamic time unit introduced in Chapter 4, and so we can express t_H in the form

$$\frac{t_H}{\tau} = \frac{i_x \ln 2}{L_p} \tag{7.10}$$

The time to half the final rate is proportional to the aerodynamic time unit and therefore inversely proportional to the speed.

7.2.2 Pitching oscillation

If a statically stable aircraft is in steady level flight and encounters a vertical gust, or the pilot quickly moves the elevator away from the trim position and back again, it will respond in pitch, rapidly regaining its original angle of incidence. The changes in incidence in the motion following the disturbance will result in changes to the lift and hence motion normal to the flight path. The changes in the lift will cause changes to the drag and the speed will change as a result. The most important motion is that in pitch and we will idealize the situation and proceed as if there was no response in the other freedoms.

We first consider the aerodynamic terms that will appear in the equation of motion. Let the angle through which the aircraft has rotated in pitch from the initial flight direction be θ, positive in the nose-up sense; this is also the change in angle of incidence in the absence of vertical motion. The pitching moment arising from the change of incidence is

$$\tfrac{1}{2}\rho V_e^2 S\bar{\bar{c}}\frac{dC_m}{d\alpha}\theta = -\mathscr{C}V_e^2\theta, \text{ say}$$

Since $\dfrac{dC_m}{d\alpha} = \dfrac{dC_m}{dC_L}\cdot\dfrac{dC_L}{d\alpha}$ it appears that we could relate this to the cg margin, although frequency effects may also have to be taken into account. We can still expect this term to be negative for a statically stable aircraft, however.

There are also moments generated as a result of the angular velocity in pitch, particularly from the tailplane. For most aircraft the contribution from the tailplane will dominate and, from Section 5.6.1 and figure 5.10, we see that the resulting pitching moment is

$$-\tfrac{1}{2}\rho V_e^2 S_T\left(\frac{ql_T}{V_e}\right)a_1 l_T = -\tfrac{1}{2}\rho V_e S_T l_T^2 a_1\frac{d\theta}{dt} = -\mathscr{B}V_e\frac{d\theta}{dt}, \text{ say}$$

We should also make some allowance for changes in the downwash from the wing arriving at the tailplane after some time delay; however, ignoring this introduces errors of the same order as other assumptions we have made. Finally we equate the aerodynamic terms to the product of the angular acceleration in pitch and the moment of inertia (I_y), to find

$$\mathscr{A}\ddot{\theta} + \mathscr{B}V_e\dot{\theta} + \mathscr{C}V_e^2\theta = 0 \tag{7.11}$$

where $\mathscr{A} = I_y$ and the quantities $\mathscr{A}$, $\mathscr{B}$ and $\mathscr{C}$ are positive. Usually the condition $4\mathscr{A}\mathscr{C}V_e^2 > (\mathscr{B}V_e)^2$ is satisfied and the roots of the corresponding auxiliary equation are complex, so that the motion is an oscillatory one. The solution of (7.11) can then be written as

$$\theta = \theta_1 e^{-\mu t}\sin(\omega t + \varepsilon) \tag{7.12}$$

where θ_1 and ε are determined by the initial conditions. The damping coefficient is given by

$$\mu = \frac{\mathscr{B}V_e}{2A} \tag{7.13}$$

and the circular frequency by

$$\omega = V_e \sqrt{\frac{\mathscr{C}}{\mathscr{A}} - \left(\frac{\mathscr{B}}{4\mathscr{A}}\right)^2} \tag{7.14}$$

Hence the damping and frequency are both proportional to the speed to the degree of approximation used.

This oscillation is known as the 'rapid incidence adjustment' or 'short period pitching oscillation'. The incidence returns to its initial value in a second or two and the period is of the same order. Although the incidence rapidly recovers its value, the speed and flight path angle generally remain altered. These normally return to their original values as a result of a further longitudinal oscillation known as the 'phugoid' which is the subject of the next section.

7.2.3 The phugoid oscillation

We assume that we have an oscillation of an aircraft of frequency much lower than that of the rapid incidence adjustment so that the incidence is kept substantially constant throughout the motion. The lift coefficient is then also constant; also we assume that the Mach number is low and the elevator is held fixed. As the pitch angle varies the aircraft will lose and gain height. Assuming that thrust equals drag throughout the motion, the total mechanical energy of the aircraft will remain constant and we can apply the principle of conservation of energy, provided also that we ignore the energy used in the rapid incidence adjustment mode caused by changes in the drag. The lift always acts normal to the flight path and so does no work.

Assume that the aircraft is initially in horizontal flight and let V be the instantaneous speed and h the height above some datum, then we have

$$m\left(gh + \tfrac{1}{2}V^2\right) = \text{constant} \tag{7.15}$$

Since the lift coefficient is constant, we have

$$L = kV^2 \tag{7.16}$$

where

$$k = \tfrac{1}{2}\rho S C_L \tag{7.17}$$

For small angles of the flight path to the horizontal we can write the vertical equation of motion as

$$m\frac{d^2h}{dt^2} = L - mg = kV^2 - mg \tag{7.18}$$

Now let V_e be the speed of flight for equilibrium; then (7.18) gives

$$0 = kV^2 - mg \text{ or } kV_e^2 = mg \tag{7.19}$$

and substituting (7.19) into (7.18) gives

$$m\frac{d^2h}{dt^2} = k\left(V^2 - V_e^2\right) \tag{7.20}$$

Now let us define a height h_e at which the aircraft is in equilibrium and has the same total mechanical energy; then from (7.15) we have

$$gh_e + \tfrac{1}{2}V_e^2 = gh + \tfrac{1}{2}V^2 \qquad (7.21)$$

Then substituting for $(V^2 - V_e^2)$ from (7.21) into (7.20) gives

$$m\frac{d^2h}{dt^2} = 2kg(h_e - h)$$

and using k from (7.19) and writing $z = h - h_e$, we find

$$\frac{d^2z}{dt^2} + \frac{2g^2z}{V_e^2} = 0 \qquad (7.22)$$

The solution of this second-order differential equation can be written

$$z = \mathcal{A}\cos \omega t + \mathcal{B}\sin \omega t \qquad (7.23)$$

where $\omega = \sqrt{2}g/V_e$; $\mathcal{A}$ and $\mathcal{B}$ are arbitrary constants determined from the initial conditions. The solution represents simple harmonic motion with periodic time

$$t_p = \sqrt{2}\pi\frac{V_e}{g} \qquad (7.24)$$

If we substitute for g using SI units we find $t_p \simeq 0.45V_e$ seconds, where V_e is expressed in m s^{-1}.
To determine $\mathcal{A}$ and $\mathcal{B}$ we suppose that at $t = 0$, the velocity is $V = V_e + u$ inclined upwards at angle θ_i, where u and θ_i are assumed small. Then from (7.23) at $t = 0$

$$\frac{dz}{dt} = -\omega\mathcal{B} = (V_e + u)\sin\theta_i \simeq V_e\theta_i$$

to first order; then

$$\mathcal{B} = \frac{V_e^2\theta_i}{\sqrt{2}g} \qquad (7.25)$$

Now using (7.21) we see that

$$2gz = 2g(h - h_e) = V_e^2 - V^2 = V_e^2 - (V_e + u)^2 \simeq -2uV_e$$

to first order, and from (7.23) at $t = 0$,

$$z = \mathcal{A}$$

Hence we find

$$\mathcal{A} = -\frac{uV_e}{g} \qquad (7.26)$$

The complete solution is then

$$z = -\frac{uV_e}{g}\cos\omega t + \frac{V_e^2\theta_i}{\sqrt{2}g}\sin\omega t \tag{7.27}$$

The motion of the aircraft represented by the solution is illustrated in figure 7.2. Suppose that an aircraft meets a disturbance that leaves it with a small increase of speed, and the pilot takes no corrective action. In this case $\theta_i = 0$, and the flight path is a cosine wave as shown in figure 7.2(a). Note that the mechanical energy of the aircraft is increased by the disturbance and so the datum height is increased.

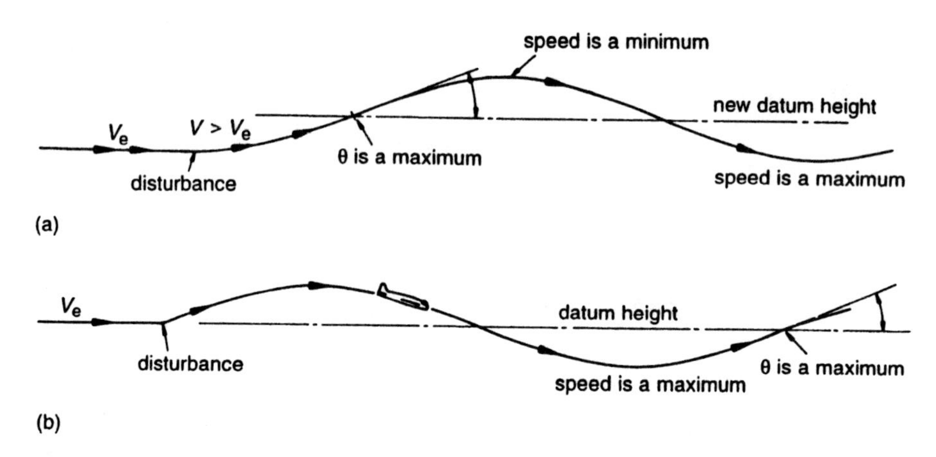

Fig. 7.2 Motion due to phugoid mode: (a) initiated by increase of forward speed, (b) initiated by increase of pitch angle

Now consider a disturbance which leaves the aircraft at the same speed but with the flight path inclined upwards. In this case $u = 0$ and the flight path is a sine wave as shown in figure 7.2(b). Note that in this case there is no change in the mechanical energy of the aircraft due to the disturbance. In both cases the speed and flight path angle (θ) vary between fixed limits and the maximum speed occurs a quarter of a wave before the maximum flight path angle. In other words the speed leads the flight path angle by a phase angle of 90°.

The analysis of the phugoid based on the full equations given in Section 9.4.1 shows that at speeds for which we can neglect compressibility the estimate of the period is fairly good. The motion is shown to be damped, mainly by the drag. It is also shown that the phugoid oscillation is affected by the cg margin which has to be larger than a certain minimum for it to exist.

7.3 'Standard' form for second-order equation

The motion of many systems can be described by a second-order differential equation. On examining systems with more than two degrees of freedom it is often found that some oscillatory modes are only weakly coupled with the remainder. An approximate treatment of them can then also be based on the simple second-order system, adding to the utility of this approach. Consider the system with the equation of motion

$$\ddot{x} + a\dot{x} + bx = f(t) \tag{7.28}$$

with a and b positive. The solution to the characteristic equation is

$$\lambda = -\frac{a}{2} \pm i\sqrt{b - \left(\frac{a}{2}\right)^2} \tag{7.29}$$

where $i = \sqrt{-1}$. Now consider the variation in the characteristics of the motion as the damping is varied. If the damping is zero there is an oscillation of circular frequency

$$\omega_n = \sqrt{b} \tag{7.30}$$

that is, ω_n is the natural frequency. In the case of critical damping, which is the minimum damping for no overswing, we have

$$b - \left(\frac{a}{2}\right)^2 = 0$$

Then the critical damping is

$$a_{crit} = 2\sqrt{b} = 2\omega_n \tag{7.31}$$

using (7.30). We now define the 'relative damping', ζ, as

$$\zeta = \text{(actual damping)/(critical damping)} \tag{7.32}$$

Then we have

$$a = 2\zeta\omega_n \text{ and } b = \omega_n^2$$

The equation of motion is then written

$$\ddot{x} + 2\zeta\omega_n\dot{x} + \omega_n^2 x = f(t) \tag{7.33}$$

and the solution of the characteristic equation is

$$\lambda = \omega_n\left(-\zeta \pm i\sqrt{1 - \zeta^2}\right) \tag{7.34}$$

For values of a greater than a_{crit} the motion is aperiodic, so this form of the equation is only relevant for lesser values. Like any method of writing equations this has advantages and disadvantages. This form of the second-order equation has been extensively investigated; for instance see reference (7.1).

7.4 Dynamics using moving axes

We have seen in Section 4.2.1 that it is necessary to use axes fixed in the aircraft, which are clearly moving axes, and some modification to the usual equations of motion

(force) = (mass) × (acceleration) and its angular equivalent must be expected. We first express the equations of motion for fixed axes using vector notation.

7.4.1 Equations of motion for a system of particles

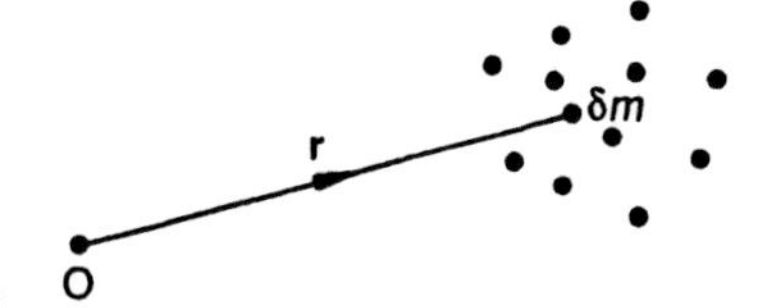

Fig. 7.3 An assemblage of small particles

Consider an assemblage of small particles, as shown in figure 7.3. Consider in particular one having a position vector **r** relative to some origin O, and mass δm. We write its velocity $\dot{\mathbf{r}}$ as **v**, its linear momentum is therefore $\delta m\mathbf{v}$ and the total linear momentum of the assemblage is

$$\mathbf{G} = \sum \delta m\mathbf{v} = \sum \delta m\dot{\mathbf{r}} = \frac{\mathrm{d}}{\mathrm{d}t}\sum \delta m\mathbf{r} \tag{7.35}$$

where $\sum$ indicates that we are forming the vector sum over all the particles. The centre of mass[1] of the particles has a position vector $\bar{\mathbf{r}}$ given by

$$m\bar{\mathbf{r}} = \sum \delta m\mathbf{r}$$

where $\mathrm{m} = \sum \delta m$ = total mass of the particles. Hence

$$\mathbf{G} \equiv \frac{\mathrm{d}}{\mathrm{d}t}(m\bar{\mathbf{r}}) = m\frac{\mathrm{d}\bar{\mathbf{r}}}{\mathrm{d}t} = m\bar{\mathbf{v}}$$

and its rate of increase is

$$\frac{\mathrm{d}\mathbf{G}}{\mathrm{d}t} = \frac{\mathrm{d}}{\mathrm{d}t}(m\bar{\mathbf{v}}) = m\bar{\mathbf{a}} \tag{7.36}$$

where $\bar{\mathbf{v}}$ and $\bar{\mathbf{a}}$ are the velocity and acceleration of the cm of the assemblage. The moment of the momentum of a particle about O is $\mathbf{r}_\wedge(m\mathbf{v})$, where '∧' indicates the vector product. We then find its rate of change to be

$$\delta m\frac{\mathrm{d}}{\mathrm{d}t}(\mathbf{r}_\wedge\mathbf{v}) = \delta m(\dot{\mathbf{r}}_\wedge\mathbf{v} + \mathbf{r}_\wedge\dot{\mathbf{v}}) = \delta m\mathbf{r}_\wedge\dot{\mathbf{v}}$$

Taking vector sums gives the total moment of the momentum. The more usual term for this is 'angular momentum', **H**, which is therefore $\mathbf{H} = \sum \mathbf{r}_\wedge \delta m\mathbf{v}$. Its rate of change is

$$\frac{\mathrm{d}\mathbf{H}}{\mathrm{d}t} = \sum \mathbf{r}_\wedge \frac{\mathrm{d}}{\mathrm{d}t}(\delta m\mathbf{v}) \tag{7.37}$$

If $\delta\mathbf{F}$ is the force on a single particle then Newton's law gives $\delta\mathbf{F} = \dfrac{\mathrm{d}}{\mathrm{d}t}(\delta m\mathbf{v})$, and we then write

$$\mathbf{F} \equiv \sum \delta\mathbf{F} = \frac{\mathrm{d}}{\mathrm{d}t}\sum \delta m\mathbf{v} = \frac{\mathrm{d}\mathbf{G}}{\mathrm{d}t} = \frac{\mathrm{d}}{\mathrm{d}t}(m\bar{\mathbf{v}}) = m\bar{\mathbf{a}} \tag{7.38}$$

using (7.35) and (7.36). This shows us the form of Newton's law that applies to the cm of an assemblage of particles.

The moment of all the forces $\delta\mathbf{F}$ about O is

$$\mathbf{Q} \equiv \sum \mathbf{r}_\wedge \delta\mathbf{F} = \sum \mathbf{r}_\wedge \frac{\mathrm{d}}{\mathrm{d}t}(\delta m\mathbf{v}) = \frac{\mathrm{d}\mathbf{H}}{\mathrm{d}t} \tag{7.39}$$

using (7.37). We can restate this in words as 'torque equals rate of change of angular momentum'. In expressing the angular momentum it is convenient to use the particle position relative to the cm. Let $\mathbf{r}'$ and $\mathbf{v}'$ be position and velocity vectors of a particle relative to the cm. Then we have

$$\mathbf{r} = \bar{\mathbf{r}} + \mathbf{r}' \text{ and } \mathbf{v} = \bar{\mathbf{v}} + \mathbf{v}'$$

The angular momentum is then

$$\mathbf{H} = \sum \delta m\mathbf{r}_\wedge \mathbf{v} = \sum (\bar{\mathbf{r}} + \mathbf{r}')_\wedge (\bar{\mathbf{v}} + \mathbf{v}')$$

and multiplying out gives

$$\mathbf{H} = \sum \delta m(\bar{\mathbf{r}}_\wedge \bar{\mathbf{v}} + \mathbf{r}'_\wedge \mathbf{v}') + \left(\sum \delta m\mathbf{r}'\right)_\wedge \bar{\mathbf{v}} + \mathbf{r}'_\wedge \sum \delta m\bar{\mathbf{v}}'$$

Now by definition $\sum \delta m\mathbf{r}' = 0$, hence also $\sum \delta m\mathbf{v}' = 0$ and writing the angular momentum about the cm as

$$\mathbf{h} = \sum \mathbf{r}'_\wedge \delta m\mathbf{v}' \tag{7.40}$$

we find finally

$$\mathbf{H} = \bar{\mathbf{r}}_\wedge \mathbf{G} + \mathbf{h} \tag{7.41}$$

This shows that the angular momentum is the sum of the moment about O of the linear momentum and the angular momentum about the cm.

7.4.2 Equations of motion for a rigid body

The equations of the previous section are easily taken over to cover this case, the small particles becoming small elements of the body. First, the parts of the forces which are internal to the body must cancel out, so that **F** and **Q** become the external applied force and moment.

With this change of definition of **F**, (7.38) applies to the rigid body. Similarly if **Q** is now the applied moment about the cm the equation of angular motion becomes

$$\mathbf{Q} = \dot{\mathbf{h}} \tag{7.42}$$

using (7.41) with $\bar{\mathbf{r}} = 0$, and (7.39).

Second, the velocity of the particles relative to the cm is only that due to the rotation of the body, i.e.

$$\mathbf{v}' = \mathbf{A}_\wedge\mathbf{r}' \tag{7.43}$$

where **A** is the angular velocity, which is a localized vector through the cm.[2] Hence using (7.40) and (7.43)

$$\mathbf{h} = \sum \mathbf{r}'_\wedge \delta m(\mathbf{A}_\wedge\mathbf{r}')$$

and using the rule for the triple vector product we find

$$\mathbf{h} = \sum \delta m r'^2\mathbf{A} - \sum \delta m(\mathbf{r}'.\mathbf{A})\mathbf{r}' \tag{7.44}$$

where r' is the magnitude of $\mathbf{r}'$.

We will need the Cartesian form of this equation and so we write

$$\mathbf{r}' = x\mathbf{i} + y\mathbf{j} + z\mathbf{k} \tag{7.45}$$

and

$$\mathbf{A} = p\mathbf{i} + q\mathbf{j} + r\mathbf{k} \tag{7.46}$$

where **i**, **j** and **k** are unit vectors along axes Ox, Oy and Oz. Then x, y and z are the coordinates of a point of the body relative to the cm and p, q and r are the angular velocities about the $Oxyz$ axes. Performing the vector operations gives us

$$\begin{aligned}\mathbf{h} = &\sum \delta m\left\{\left(y^2 + z^2\right)p - xyq - xzr\right\}\mathbf{i} \\ &+\sum \delta m\left\{\left(z^2 + x^2\right)q - yzr - yxp\right\}\mathbf{j} \\ &+\sum \delta m\left\{\left(x^2 + y^2\right)r - zxp - zyq\right\}\mathbf{k}\end{aligned} \tag{7.47}$$

The sums such as $\sum \delta m(y^2 + z^2)$ are known as the 'moments of inertia' while the sums like $\sum \delta mxy$ are the 'products of inertia'. We write the moments of inertia as I_x, I_y and I_z in the order given above and the products of inertia as $I_{xy} = \sum \delta mxy$ and so on. The angular momentum is written as the sum of its components

$$\mathbf{h} = h_1\mathbf{i} + h_2\mathbf{j} + h_3\mathbf{k} \tag{7.48}$$

where

$$\left.\begin{aligned}h_1 &= I_x p - I_{xy}q - I_{xz}r \\ h_2 &= I_y q - I_{yz}r - I_{xy}p \\ h_3 &= I_z r - I_{xz}p - I_{yz}q\end{aligned}\right\} \tag{7.49}$$

It can be shown that the axes of reference can be chosen such that the products of inertia vanish; such axes are known as 'principal axes of inertia'; the simplification obtained by choosing these is obvious.

7.4.3 Moving frames of reference

We now turn to the problem of relating quantities in a moving frame with their counterparts in a fixed frame, and are able to provide a general result. Suppose we have two frames of reference S_1 and S_2 as shown in figure 7.4

We will use the suffices 1 and 2 to denote quantities as measured in frames S_1 and S_2 respectively, *when a distinction needs to be made*. The frame S_1 is a fixed or 'inertial' frame, i.e. it is subject to much smaller accelerations than those involved in our applications, for example a frame fixed to the surface of the Earth. The frame S_2 is rotating with angular velocity $\mathbf{A}$ about a point O in frame S_1. Let $\mathbf{J} = \mathbf{J}(t)$ be any vector measured in the frame S_2. This must also be its value in the frame S_1, but its rate of change as measured in the two frames must be affected by the rotation. It is therefore required to find the relation between its rates of change in the two frames.

Figure 7.4 has been drawn from the point of view of S_1. In the figure $\overrightarrow{OP}$ represents the vector $\mathbf{J}$ at time t; in a short time δt it changes to $\overrightarrow{OR}$. Then $\overrightarrow{PR}$ is the increment $\delta\mathbf{J}_1$, i.e. the change relative to S_1. During the interval δt the point of S_2 which was at P has moved to Q, where if $\mathbf{v}_p$ is the velocity of P then

$$\overrightarrow{PQ} = \mathbf{v}_p\delta t = (\mathbf{A}_\wedge\mathbf{J})\delta t$$

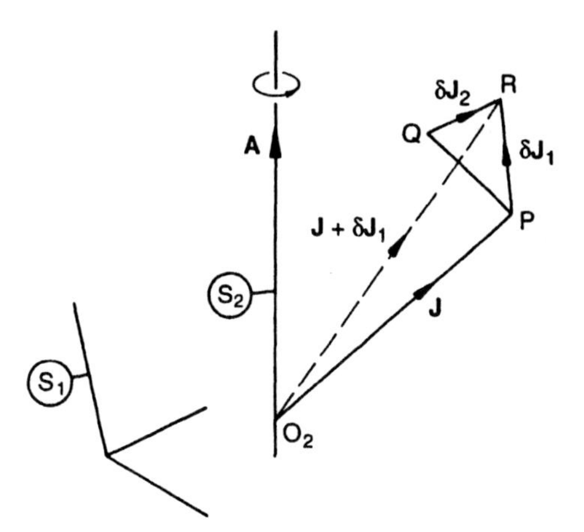

Fig. 7.4 Determination of relationship between quantities measured in fixed and moving sets of axes

Now $\overrightarrow{QR}$ is the increment of $\mathbf{J}$ relative to the frame S_2, i.e. it is $\delta\mathbf{J}_2$. Then from the triangle PQR we have

$$\overrightarrow{PR} = \overrightarrow{PQ} + \overrightarrow{QR}$$

or

$$\delta\mathbf{J}_1 = \delta\mathbf{J}_2 + (\mathbf{A}_\wedge\mathbf{J})\delta t$$

We now divide through by δt and take the limit as $\delta t \longrightarrow 0$ to find the required result:

$$\left(\frac{d\mathbf{J}}{dt}\right)_1 = \left(\frac{d\mathbf{J}}{dt}\right)_2 + \mathbf{A}_\wedge \mathbf{J} \tag{7.50}$$

We can use this equation to relate the time rate of change of any vector quantity in a moving frame to its rate of change in a fixed frame and derive appropriate equations of motion.

7.4.4 Equations of motion of a rigid body referred to body fixed axes

We now proceed to use the result of the last section to find the relationship between the linear and angular accelerations of a body relative to axes fixed in the body to the same quantities relative to fixed axes. This then leads to the equations of motion. First let us identify $\mathbf{J}$ in (7.50) with the velocity of the cm, $\bar{\mathbf{v}}$, then

$$\left(\frac{d\bar{\mathbf{v}}}{dt}\right)_1 = \left(\frac{d\bar{\mathbf{v}}}{dt}\right)_2 + \mathbf{A}_\wedge \bar{\mathbf{v}}$$

and on using (7.38) we find

$$\mathbf{F} = \dot{\bar{\mathbf{v}}}_2 + \mathbf{A}_\wedge \bar{\mathbf{v}} \tag{7.51}$$

This is the equation of linear motion of a body referred to axes fixed in the body. For the angular case we identify $\mathbf{J}$ with the angular momentum, finding

$$\left(\frac{d\mathbf{h}}{dt}\right)_1 = \left(\frac{d\mathbf{h}}{dt}\right)_2 + \mathbf{A}_\wedge \mathbf{h}$$

and on using (7.42) we find

$$\mathbf{Q} = \dot{\mathbf{h}}_2 + \mathbf{A}_\wedge \mathbf{h} \tag{7.52}$$

This is the equation of angular motion referred to axes fixed in the body.

The Cartesian forms of these equations will be needed, so we write

$$\bar{\mathbf{v}} = U\mathbf{i} + V\mathbf{j} + W\mathbf{k} \tag{7.53}$$
$$\mathbf{F} = X\mathbf{i} + Y\mathbf{j} + Z\mathbf{k} \tag{7.54}$$
$$\mathbf{Q} = L\mathbf{i} + M\mathbf{j} + N\mathbf{k} \tag{7.55}$$

Then performing the vector operations and separating the equations gives

$$\left.\begin{aligned} X &= m(\dot{U} + qW - rV \\ Y &= m(\dot{V} + rU - pW \\ Z &= m(\dot{W} + pV - qU \end{aligned}\right\} \tag{7.56}$$

and

$$\left.\begin{aligned} L &= \dot{h}_1 + qh_3 - rh_2 \\ M &= \dot{h}_2 + rh_1 - ph_3 \\ N &= \dot{h}_3 + ph_2 - qh_1 \end{aligned}\right\} \tag{7.57}$$

The last set of equations are essentially those usually known as Euler's dynamical equations; they are normally used with (7.49) substituted using principal moments of inertia.

7.4.5 Example of use of equations

Before applying these equations to the motion of an aircraft, it is instructive to demonstrate their use on a simpler problem.

Consider the bifilar pendulum AB shown in figure 7.5. It consists of a thin uniform beam of length $2l$ suspended at its ends by strings of negligible mass and length d. The beam has a mass m and radius of gyration k; attached to the end B is a cruciform vane which is suspended in an oil pot to provide damping. Provided no vertical forces are applied, the beam can be regarded as having three degrees of freedom, namely

- translation in a direction along its length
- translation normal to the length
- rotation about a vertical axis through its centre.

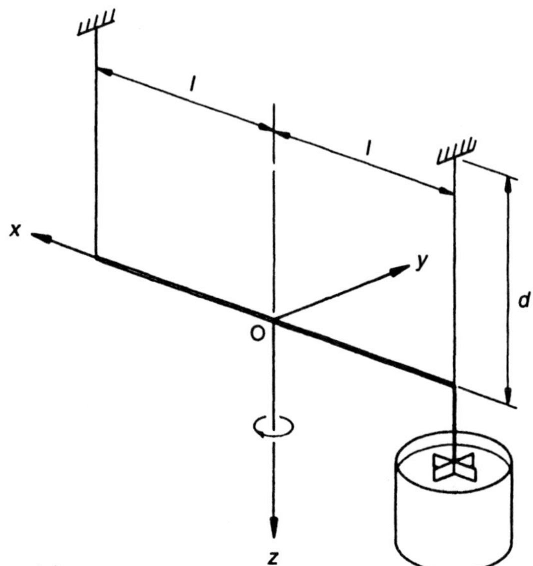

Fig. 7.5 The bifilar pendulum

We need to choose axes of reference for the analysis; any choice of axes fixed in space will result in the moment of inertia being a function of the position of the beam. Accordingly we choose axes fixed to the beam with the origin at its centre, Ox along the beam, Oy horizontal and normal to the beam and Oz vertically downwards as shown. Also let x and y be displacements of the beam from the rest position and ψ be the clockwise rotation about Oz.

With these three freedoms the equations we will need are the first two of (7.56), the third of (7.57) and (7.49). These can be simplified, as the following quantities are zero: W, q, p, I_x, I_{yz}, I_{xz} and I_{xy}; as a result h_1 and h_2 are also zero. Then we have

$$X = m(\dot{U} - rV) = m(\ddot{x} - \dot{\psi}\dot{y}) \qquad (7.58)$$

$$Y = m(\dot{V} - rU) = m(\ddot{y} - \dot{\psi}\dot{x}) \qquad (7.59)$$

$$N = \mathrm{I}_z\dot{r} = mk^2\ddot{\psi} \qquad (7.60)$$

We now consider the forces and moment applied to the beam. The total tension in the strings is mg and a small displacement x inclines the strings at an angle x/d resulting in an x-wise force of $-mgx/d$, and similarly for the y axis. A small rotation ψ gives a displacement ψl to the ends of the beam and therefore an inclination of $\psi l/d$ to the strings, giving a moment about Oz of $-mgl^2\psi/d$. Also let the damper vane give horizontal forces of $-Cx$(velocity along Ox and Oy), so that there is an x-wise force of $C\dot{x}$ The velocity of the end B in the y-direction is $\dot{y} - l\dot{\psi}$ giving a force of $-C\dot{y} + Cl\dot{\psi}$; there will also be a moment about Oz of $-Cl^2\dot{\psi} + Cl\dot{y}$. Since we have assumed the displacements to be small the corresponding velocities will also be small, and so the products $\dot{x}\dot{\psi}$ and $\dot{y}\dot{\psi}$ will be second-order small quantities and to be consistent will be dropped. The equations now become

$$m\ddot{x} + C\dot{x} + mgx/d = X(t) \qquad (7.61)$$

$$m\ddot{y} + C\dot{y} - Cl\dot{\psi} + mgy/d = Y(t) \qquad (7.62)$$

$$mk^2\ddot{\psi} - Cl\dot{y} + Cl^2\dot{\psi} + mgl^2\psi/d = N(t) \qquad (7.63)$$

where $X(t)$, $Y(t)$ and $N(t)$ are disturbance inputs. We notice that (7.61) is independent of the other two equations and so, to the degree of approximation used, the motion in the x-direction is independent of the others. We shall follow a similar approach in the next chapter when we derive the equations of motion for an aircraft; it will be seen that there are a few other similarities.

7.5 State-space description

The traditional method of writing equations of motion for a system consists of writing one equation, usually of second order, for each freedom as was done with (7.61)–(7.63) above. The more general modern method is that of the state-space description where state variables are used. A state variable can be a displacement, a velocity or whatever variable is necessary for the problem in hand. The essential point is that there is a first-order differential equation for each variable; new variables are introduced as required.

If we know all the state variables at some instant t_0, together with the inputs at t_0, then all possible outputs of the system are known at $t = t_1$. If we can determine the time behaviour of the state variables then the time behaviour of all the outputs is known for any given inputs.

In this method of analysis we

- determine the time behaviour of the state variables, so that there must be state equations relating the state variables to the inputs; and
- determine the desired output from the state variables, for which there must be output equations.

For the case of linear systems we arrange that the state equations are n simultaneous, first-order differential equations in n state variables $x_1, x_2, x_3, \ldots, x_n$ with m inputs $u_1(t), u_2(t), u_3(t), \ldots, u_m(t)$. The equations are arranged to have the form

$$\dot{x}_k = a_{k1}x_1 + a_{k2}x_2 + \ldots + a_{kn}x_n + b_{k1}u_1 + b_{k2}u_2 + \ldots + b_{km}u_m \tag{7.64}$$

where $k = 1, 2, 3, \ldots, n$. The equations for the p outputs will have the form

$$y_j = c_{j1}x_1 + c_{j2}x_2 + \ldots + c_{jn}x_n + d_{j1}u_1 + d_{j2}u_2 + \ldots + d_{jm}u_m \tag{7.65}$$

where $j = 1, 2, 3, \ldots, p$.

These equations are most conveniently written in matrix form, so the state equations are

$$\dot{\mathbf{x}} = \mathbf{Ax} + \mathbf{Bu} \tag{7.66}$$

where $\mathbf{x}$ is the $n \times 1$ 'state vector', $\mathbf{u}$ is the $m \times 1$ 'input' or 'control' vector, and $\mathbf{A}$ and $\mathbf{B}$ are coefficient matrices of dimensions $n \times n$ and $n \times m$, respectively.

The output equations are written in the form

$$\mathbf{y} = \mathbf{Cx} + \mathbf{Du} \tag{7.67}$$

where $\mathbf{y}$ is the $p \times 1$ output vector and $\mathbf{C}$ and $\mathbf{D}$ are coefficient matrices of dimensions $p \times n$ and $p \times m$, respectively. Examples of output variables in the aeronautical stability context are the normal acceleration, accelerations of particular points of the aircraft and the time rate change of displacement of the cm of the aircraft from the mean flight path. It should be noted that there may be more than one possible choice for the state variables and that their number must be equal to the number of initial conditions for a complete solution.

The advantages of the description of the problem in terms of first-order equations are as follows:

- Such equations have been extensively investigated and several methods are available for their solution.
- This form can be extended to time varying and nonlinear systems.
- It is the form from which analogue computers are set up and it is the usual form for use in step-by-step integration of equations of motion on a digital computer.
- It is suitable for optimization techniques.

7.5.1 Example of state-space description

As an example of the formation of state-space equations from the traditional form we take the bifilar pendulum discussed in Section 7.4.5. Specifically we take the coupled pair (7.62) and (7.63), divide through by m and mk^2 respectively and solve for the highest derivatives to find

$$\ddot{y} = -c\dot{y} + cl\dot{\psi} - fy - u_1(t) \tag{7.68}$$

$$\ddot{\psi} = el\dot{y} - el^2\dot{\psi} - bf\psi - u_2(t) \tag{7.69}$$

where $b = l^2/k^2$, $c = C/m$, $e = C/mk^2$, $f = g/d$, $u_1(t) = Y(t)/m$ and $u_2(t) = N(t)/mk^2$. We now write

$$y = x_1,\ \dot{y} = x_2,\ \psi = x_3,\ \dot{\psi} = x_4 \tag{7.70}$$

Our set of equations in state-space form are then

$$\dot{x}_1 = x_2 \tag{7.71}$$
$$\dot{x}_2 = -cx_2 + clx_4 - fx_1 - u_1(t) \tag{7.72}$$
$$\dot{x}_3 = x_4 \tag{7.73}$$
$$\dot{x}_4 = elx_2 - el^2x_4 - bfx_3 - u_2(t) \tag{7.74}$$

The corresponding coefficient matrix is then

$$\mathbf{A} = \begin{bmatrix} 0 & 1 & 0 & 0 \\ -f & -c & 0 & cl \\ 0 & 0 & 0 & 1 \\ 0 & el & -bl & -el^2 \end{bmatrix} \tag{7.75}$$

7.5.2 Analytical solution of state-space equations

There are two complementary approaches to solving the equations in the sense that they can be solved in either the time or the frequency domain. Both solutions depend on the idea that we can give a meaning to $e^{\mathbf{A}t}$ where $\mathbf{A}$ is a square matrix of order n. We define it by the use of the usual series for the exponential function; in this case it is

$$\begin{aligned} e^{\mathbf{A}t} &= \mathbf{I} + \mathbf{A}t + \frac{\mathbf{A}^2t^2}{2!} + \frac{\mathbf{A}^3t^3}{3!} + \ldots + \frac{\mathbf{A}^nt^n}{n!} + \ldots \\ &= \sum^{\infty} \frac{\mathbf{A}^kt^k}{k!} \end{aligned} \tag{7.76}$$

where $\mathbf{I}$ is the unit matrix. Clearly $e^{\mathbf{A}t}$ is a square matrix and we interpret $\mathbf{A}^2$ as the product of $\mathbf{A}$ with itself, and so on for higher orders. All the usual properties of the exponential function hold and can be derived from this series. In particular we derive the usual result for its derivative by differentiating term by term, as follows:

$$\begin{aligned} \frac{d}{dt}e^{\mathbf{A}t} &= \mathbf{A} + \mathbf{A}^2t + \frac{\mathbf{A}^3t^2}{2!} + \ldots \\ &= \mathbf{A}e^{\mathbf{A}t} = e^{\mathbf{A}t}\mathbf{A} \end{aligned} \tag{7.77}$$

as in the scalar case

7.5.2.1 Time domain solution

To give us a clue on the form of the solution to be expected we note that the solution of the first-order equation

$$\dot{x} = ax + bu$$

where $x = x(0)$ at $t = 0$, is

$$x = e^{at}x(0) + e^{at}\int_0^t e^{-a\tau}bu\,d\tau$$

where τ is a dummy variable. Turning to the equation

$$\dot{\mathbf{x}} = \mathbf{Ax} + \mathbf{Bu} \tag{7.78}$$

where $\mathbf{x} = \mathbf{x}(0)$ at $t = 0$, from the result

$$\frac{d}{dt}(\mathbf{AB}) = \frac{d\mathbf{A}}{dt}\mathbf{B} + \mathbf{A}\frac{d\mathbf{B}}{dt}$$

we find that

$$\begin{aligned}\frac{d}{dt}\left(e^{-\mathbf{A}t}\mathbf{x}\right) &= \frac{d}{dt}\left(e^{-\mathbf{A}t}\right)\mathbf{x} + e^{-\mathbf{A}t}\dot{\mathbf{x}} \\ &= -e^{-\mathbf{A}t}\mathbf{Ax} + e^{-\mathbf{A}t}\dot{\mathbf{x}}\end{aligned} \tag{7.79}$$

We now premultiply both sides of (7.78) by $e^{-\mathbf{A}t}$ to give

$$e^{-\mathbf{A}t}\dot{\mathbf{x}} = e^{-\mathbf{A}t}\mathbf{Ax} + e^{-\mathbf{A}t}\mathbf{Bu}$$

or

$$e^{-\mathbf{A}t}\dot{\mathbf{x}} - e^{-\mathbf{A}t}\mathbf{Ax} = e^{-\mathbf{A}t}\mathbf{Bu}$$

Then using (7.79) we find

$$\frac{d}{dt}\left(e^{-\mathbf{A}t}\mathbf{x}\right) = e^{-\mathbf{A}t}\mathbf{Bu}$$

We now integrate both sides from 0 to t to give

$$e^{-\mathbf{A}t}\mathbf{x} = \mathbf{x}(0) + \int_0^t e^{-\mathbf{A}\tau}\mathbf{Bu}\,d\tau$$

and premultiply both sides by $e^{\mathbf{A}t}$ to give the solution as

$$\mathbf{x} = e^{\mathbf{A}t}\mathbf{x}(0) + e^{\mathbf{A}t}\int_0^t e^{-\mathbf{A}\tau}\mathbf{Bu}\,d\tau \tag{7.80}$$

The result is then in the same form as the first-order solution. The matrix $e^{\mathbf{A}t}$ in this context is known as the 'state transition matrix' written as $\mathbf{\Phi}(t)$. In terms of $\mathbf{\Phi}$ the solution can now be written

$$\mathbf{x} = \mathbf{\Phi}(t)\mathbf{x}(0) + \int_0^t \mathbf{\Phi}(t-\tau)\mathbf{Bu}\,d\tau \tag{7.81}$$

7.5.2.2 *Frequency domain solution*

The phrase 'frequency domain' here implies that we are using the Laplace transform method to find the solution. We therefore write the transforms of **x** and **u** as $\bar{\mathbf{x}} = \mathcal{L}[\mathbf{x}]$ and $\bar{\mathbf{u}} = \mathcal{L}[\mathbf{u}]$. We proceed to take the Laplace transforms of both sides of (7.78) to give

$$s\bar{\mathbf{x}}(s) + \mathbf{x}(0) = \mathbf{A}\bar{\mathbf{x}}(s) + \mathbf{B}\bar{\mathbf{u}}(s)$$

or

$$s\bar{\mathbf{x}}(s) - \mathbf{A}\bar{\mathbf{x}}(s) = \mathbf{x}(0) + \mathbf{B}\bar{\mathbf{u}}(s)$$

hence

$$(s\mathbf{I} - \mathbf{A})\bar{\mathbf{x}}(s) = \mathbf{x}(0) + \mathbf{B}\bar{\mathbf{u}}(s)$$

and solving,

$$\bar{\mathbf{x}}(s) = (s\mathbf{I} - \mathbf{A})^{-1}[\mathbf{x}(0) + \mathbf{B}\bar{\mathbf{u}}(s)]$$

On taking the inverse transform we find

$$\bar{\mathbf{x}}(t) = \mathcal{L}^{-1}[(s\mathbf{I} - \mathbf{A})^{-1}\mathbf{x}(0)] + \mathcal{L}^{-1}[(s\mathbf{I} - \mathbf{A})^{-1}\mathbf{B}\bar{\mathbf{u}}(s)] \tag{7.82}$$

which must represent the same solution as (7.80). We consider the two terms separately. Taking the second and using the convolution theorem:

$$\begin{aligned} \mathcal{L}^{-1}[(s\mathbf{I} - \mathbf{A})^{-1}\mathbf{B}\bar{\mathbf{u}}(s)] &= \int_0^t \mathrm{e}^{\mathbf{A}(t-\tau)}\mathbf{B}\mathbf{u}(\tau)\mathrm{d}\tau \\ &= \mathrm{e}^{\mathbf{A}t}\int_0^t \mathrm{e}^{-\mathbf{A}\tau}\mathbf{B}\mathbf{u}(\tau)\mathrm{d}\tau \end{aligned} \tag{7.83}$$

This agrees with the last term of (7.80).

For the homogeneous case, i.e. $\mathbf{u}(t) = 0$, we have from the Laplace transform of the exponential

$$\mathbf{\Phi} = \mathrm{e}^{\mathbf{A}t} = \mathcal{L}^{-1}[(s\mathbf{I} - \mathbf{A})^{-1}]$$

so that the first term is

$$\mathrm{e}^{\mathbf{A}t}\mathbf{x}(0) \tag{7.84}$$

We now consider the form that $\mathbf{\Phi}$ must take; from matrix algebra we have

$$(s\mathbf{I} - \mathbf{A})^{-1} = \frac{\operatorname{adj}(s\mathbf{I} - \mathbf{A})}{|s\mathbf{I} - \mathbf{A}|}$$

where $\operatorname{adj}(s\mathbf{I} - \mathbf{A})$ denotes the adjoint matrix of $(s\mathbf{I} - \mathbf{A})$. The determinant $|s\mathbf{I} - \mathbf{A}|$ can be written as

$$|s\mathbf{I} - \mathbf{A}| = (s - \lambda_1)(s - \lambda_2)(s - \lambda_3) \ldots (s - \lambda_n)$$

where $\lambda_1, \lambda_2, \lambda_3, \ldots, \lambda_n$ are the eigenvalues of the system, that is the roots of $|\lambda\mathbf{I} - \mathbf{A}| = 0$. If we write a typical element of the adjoint matrix as a_{ij} a typical element of $(s\mathbf{I} - \mathbf{A})^{-1}$ takes the form

$$\alpha_{ij} = \frac{a_{ij}}{(s - \lambda_1)(s - \lambda_2)(s - \lambda_3) \ldots (s - \lambda_n)}$$

Then splitting this into partial fractions we find

$$\alpha_{ij} = \frac{c_{ij1}}{s - \lambda_1} + \frac{c_{ij2}}{s - \lambda_2} + \frac{c_{ij3}}{s - \lambda_3} + \ldots + \frac{c_{ijn}}{s - \lambda_n}$$

where c_{ij1}, c_{ij2}, c_{ij3}, ..., c_{ijn} are determined in the usual way. The case of repeated roots needs special treatment for which a textbook should be consulted. Each element of the state transition matrix will therefore be of the form

$$\Phi_{ij} = \mathcal{L}^{-1}\left[\frac{c_{ij1}}{s - \lambda_1} + \frac{c_{ij2}}{s - \lambda_2} + \frac{c_{ij3}}{s - \lambda_3} + \ldots + \frac{c_{ijn}}{s - \lambda_n}\right]$$

hence inverting the transforms we find

$$\Phi_{ij}(t) = c_{ij1}e^{\lambda_1 t} + c_{ij2}e^{\lambda_2 t} + c_{ij3}e^{\lambda_3 t} + \ldots + c_{ijn}e^{\lambda_n t} \tag{7.85}$$

7.5.2.3 Numerical example

Suppose the system equation $\dot{\mathbf{x}} = \mathbf{Ax} + \mathbf{Bu}$ is

$$\begin{bmatrix} \dot{x}_1 \\ \dot{x}_2 \end{bmatrix} = \begin{bmatrix} -4 & 1 \\ -3 & 0 \end{bmatrix} \begin{bmatrix} x_1 \\ x_2 \end{bmatrix} + \begin{bmatrix} 1 \\ 2 \end{bmatrix} \mathbf{u} \tag{7.86}$$

with u being a unit step function. The eigenvalues are given by $|\lambda\mathbf{I} - \mathbf{A}| = 0$, that is by

$$\begin{vmatrix} \lambda + 4 & -1 \\ 3 & \lambda \end{vmatrix} = 0$$

or

$$\lambda^2 + 4\lambda + 3 = 0 \text{ or } (\lambda + 1)(\lambda + 3) = 0$$

so that the eigenvalues are $\lambda_1 = -1$ and $\lambda_2 = -3$. We also have

$$\mathbf{I} = \begin{bmatrix} 1 & 0 \\ 0 & 1 \end{bmatrix} \text{ and } \mathbf{A}^2 = \begin{bmatrix} -4 & 1 \\ -3 & 0 \end{bmatrix} \begin{bmatrix} -4 & 1 \\ -3 & 0 \end{bmatrix} = \begin{bmatrix} 13 & -4 \\ 12 & -3 \end{bmatrix}$$

The state transition matrix is then

$$\Phi(t) = \begin{bmatrix} 1 - 4t + (13/2)t^2 + \ldots & 0 + t - 2t^2 + \ldots \\ 0 - 3t + 6t^2 + \ldots & 1 + 0 - (3/2)t^2 + \ldots \end{bmatrix} \tag{7.87}$$

We know for instance from (7.85) that each term of Φ has the form $c_1 e^{\lambda_1 t} + c_2 e^{\lambda_2 t}$; substituting into this the series for the exponential

$$e^{\lambda t} = 1 + \lambda t + \frac{\lambda^2 t^2}{2!} + \frac{\lambda^3 t^3}{3!} + \ldots$$

for each eigenvalue to give

$$c_1\left(1 - t + \frac{t^2}{2} + \ldots\right) + c_2\left(1 - 3t + \frac{9t^2}{2} + \ldots\right)$$

and comparing coefficients of powers of t in (7.87), element by element, we find

$$\Phi(t) = \tfrac{1}{2}\begin{bmatrix} -e^{-t} + 3e^{-3t} & e^{-t} - e^{-3t} \\ -3e^{-t} + e^{-3t} & 3e^{-t} - e^{-3t} \end{bmatrix} \tag{7.88}$$

which leads to the solution to the homogeneous equation. The rest of the solution is given by

$$\begin{aligned} \int_0^t \Phi(t-\tau)\mathbf{B}\mathbf{u}(\tau)\mathrm{d}\tau &= \tfrac{1}{2}\int_0^t \begin{bmatrix} -e^{-(t-\tau)} + 3e^{-3(t-\tau)} & e^{-(t-\tau)} - e^{-3(t-\tau)} \\ -3e^{-(t-\tau)} + e^{-3(t-\tau)} & 3e^{-(t-\tau)} - e^{-3(t-\tau)} \end{bmatrix}\begin{bmatrix} 1 \\ 2 \end{bmatrix} 1\mathrm{d}\tau \\ &= \tfrac{1}{2}\int_0^t \begin{bmatrix} e^{-(t-\tau)} - e^{-3(t-\tau)} \\ 3e^{-(t-\tau)} + e^{-3(t-\tau)} \end{bmatrix}\mathrm{d}\tau \\ &= \tfrac{1}{6}\begin{bmatrix} 2 - 3e^{-t} + e^{-3t} \\ 10 - 9e^{-t} - e^{-3t} \end{bmatrix} \end{aligned} \tag{7.89}$$

The complete solution is then

$$\mathbf{x} = \tfrac{1}{2}\begin{bmatrix} -e^{-t} + 3e^{-3t} & e^{-t} - e^{-3t} \\ -3e^{-t} + e^{-3t} & 3e^{-t} - e^{-3t} \end{bmatrix}\mathbf{x}(0) + \tfrac{1}{6}\begin{bmatrix} 2 - 3e^{-t} + e^{-3t} \\ 10 - 9c^{-t} - e^{-3t} \end{bmatrix} \tag{7.90}$$

using (7.84), (7.88) and (7.89).

7.5.3 Step-by-step solution of state-space equations

The state-space form of the equations leads to a simple method of finding the response of a system to a set of inputs. Suppose we are given a vector $\mathbf{x}_0$ of $\mathbf{x}$ values at $t = 0$, we can then calculate a vector of the $\mathbf{x}$-derivatives at $t = 0$ as

$$\dot{\mathbf{x}}_0 = \mathbf{A}\mathbf{x}_0 + \mathbf{u} \tag{7.91}$$

An estimate of the vector of values of $\mathbf{x}$ after a small time step Δt can then be calculated as

$$\mathbf{x}_1 = \mathbf{x}_0 + \dot{\mathbf{x}}_0 \Delta t \tag{7.92}$$

and the process is repeated as necessary. Better results, or similar results with larger time steps, can be obtained by using more sophisticated methods such as Runge–Kutta methods. Step-by-step methods are particularly useful in non-linear cases, for instance if **A** is a function of time or **x**, as it may be if flow is separated.

Student problems

7.1 An aircraft is flying horizontally at a speed of 250 m s^{-1} when it hits a gust which gives it an increment of forward speed of 5 m s^{-1} and inclines the flight path downward at an angle of 2°. The pilot allows the resulting phugoid oscillation to continue. Find the periodic time and the amplitudes of oscillation in altitude and forward speed. (A)

7.2 Extend the simple analysis given for the phugoid mode in Section 7.2.3 to allow for small changes of density with height to show that the periodic time found is to be factored by

$$\frac{1}{\sqrt{1 + \gamma \mathbf{M}^2 / 2n}}$$

where it is assumed that the pressure and density in the atmosphere are related by the polytropic law $p\rho^{-n}$ = constant and pressure and height are related by $\mathrm{d}p/\mathrm{d}h = -\rho g$.

7.3 Rewrite the following sets of first-order differential equations in state-space form:

(a) $\ddot{x} + a\dot{x} + bx = u(t)$
(b) $\ddot{x} + a_1\dot{x} + b_1\dot{y} + c_1x = u_1(t)$
$\ddot{y} + a_2\dot{x} + b_2\dot{y} + c_2y = u_2(t)$
(c) $\dot{x} + c_1y = u_1(t)$
$\ddot{y} + a_2\dot{x} + b_2\dot{y} + c_2x = u_2(t)$
(d) $\ddot{x} + a_1\dot{x} + b_1\dot{y} + c_1y = u_1(t)$
$\ddot{y} + a_2\dot{x} + b_2\dot{y} + c_2x = u_2(t)$
(e) $x''' + a\ddot{x} + b\dot{x} + cx = u(t)$
(f) $x''' + a_1\ddot{x} + b_1\dot{x} + c_1x + d_1\dot{y} = u_1(t)$
$\dot{x} + c_2y = u_2(t)$
(g) $\ddot{x} + a_1\ddot{y} + b_1x + c_1y = u_1(t)$
$\ddot{y} + a_2\ddot{x} + b_2x + c_2y = u_2(t)$

7.4 Solve the following set of equations in state-space form:

$$\begin{bmatrix} \dot{x}_1 \\ \dot{x}_2 \end{bmatrix} = \begin{bmatrix} -4 & 2 \\ -1 & -1 \end{bmatrix} \begin{bmatrix} x_1 \\ x_2 \end{bmatrix} + \begin{bmatrix} 1 \\ 2 \end{bmatrix} \mathbf{u}$$

with **u** being a unit step function.

Notes

1. The term 'centre of gravity' is strictly only appropriate when discussing forces due to gravity; in this chapter the term 'centre of mass' (cm) is preferred as it is more appropriate in this case; the two are, of course, very close together.
2. The reader is reminded that a localized vector is a vector through a specific point and may not, in general, be moved around in space.

Background reading

Harrison, H. R. and Nettleton, T. 1994: *Principles of engineering mechanics*. London: Edward Arnold.
Karamcheti, K. 1967: *Vector analysis and cartesian tensors with selected applications*. San Francisco: Holden-Day.
Lennox, S. C. and Chadwick, M. 1970: *Mathematics for engineers and applied scientists*. London: Heinemann Educational Books.

8
Equations of motion of a rigid aircraft

8.1 Introduction

It is perhaps a little surprising that it was only some seven years after the Wright brothers' first flight that the equations of aircraft dynamic stability were set out by G. H. Bryan in reference (8.1) in essentially the form known today. Since then many refinements have been made and a lot of research done to find values for the stability derivatives and to verify the results. In this chapter, using the results of Section 7.4.4, we will derive equations of stability for an aircraft which will be used in the remaining chapters.

8.2 Some preliminary assumptions

As before we assume that the aircraft is a perfectly rigid body. Real aircraft are of course deformable but provided that the frequencies of any structural oscillations are well away from those of oscillations we find and any steady deformations are small, we can expect reasonable agreement with experiment. We will also assume that the aircraft is initially flying steadily, i.e. it is in trimmed flight, when it is disturbed in some way.

8.2.1 Axes and notation

We have already introduced some of the notation needed in previous chapters; however, a complete statement of the basic notation will be made at this point and is shown in figure 8.1.

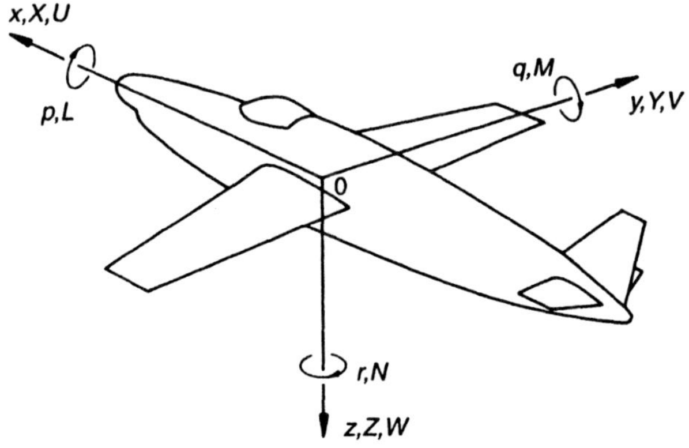

Fig. 8.1 Stability axes and velocities and forces along them, angular velocities and moments around them

We place a set of right-handed orthogonal axes with the origin at the cg. The Ox axis points roughly in the forward direction, Oy at right angles to the plane of symmetry and towards the starboard wing tip. The z-axis points roughly downwards and completes a right-handed set. For the time being the freedom to choose the precise direction of the x-axis is left to be decided to suit the problem in hand. There are at least two natural choices:

- to have the x-axis fixed initially in the direction of undisturbed flight – such axes are known as 'wind axes';
- aircraft normally have a principal axis of inertia lying roughly in the flight direction – this is sometimes a convenient direction to take as the x-axis.

Now let

- $\mathbf{V}$ be the velocity of the aircraft cg, with components U, V and W along Ox, Oy and Oz;
- $\mathbf{A}$ be the angular velocity of the aircraft, with components p, q and r about Ox, Oy and Oz;
- $\mathbf{F}$ be the force on the aircraft, with components X, Y and Z along Ox, Oy and Oz;
- $\mathbf{Q}$ be the moment on the aircraft about the cg, with components L, M and N along Ox, Oy and Oz.

The positive sense of the velocities and forces is in the direction of the axes and that of the angular velocities and moments is that of a right-hand screw advancing along the direction of the axes.

8.2.2 Plan of action

It is useful to remind ourselves at this point of the basic equations that we shall be deriving the stability equations from. These are (7.49) for the linear motion and (7.50) (with (7.42) substituted in) for the angular motion. Then the programme for the next few sections is first to express the forces and moments on the aircraft in a usable form, second to set about introducing various reasonable assumptions to simplify the equations, and then to write down the equations in various alternative forms. In fact the full equations are too cumbersome to write down in full and until we have simplified their component parts there is no compelling need to do so.

Our first step is to recognize that normally there are only two kinds of forces and moments acting on the aircraft, aerodynamic ones and the attraction due to gravity. Symbolically we write this as

$$\mathbf{F} = \mathbf{F}_a + \mathbf{F}_g \tag{8.1}$$

and

$$\mathbf{Q} = \mathbf{Q}_a \tag{8.2}$$

since by choosing to put the origin at the cg there are no moments of the weight. We shall also use these suffices, 'a' for aerodynamic and 'g' for gravitational, for the components of the respective force and moment.

To express the components of the weight at some point during a disturbance we need to be able to describe the orientation of the aircraft relative to its initial attitude, where we assume that we knew the components of the weight along the axes. Orientation is the subject of the next section.

8.3 Orientation

Three rotations about non-parallel axes will move one set of axes so as to be parallel with another set; however, there are many possible choices of combinations of rotations about the axes and the order to take them in. We choose one attributed to Euler as follows: take clockwise rotations ψ, θ and ϕ about the Oz, Oy and Ox axes *where rotation takes place about that position of the axis to which previous rotations have brought it.* We will rotate a set of axes parallel to the axes in the undisturbed state into parallelism with the position of the axes at some time t after the start of the disturbance. The angles of rotation required then define the orientation of the aircraft. Figure 8.2 shows the procedure.

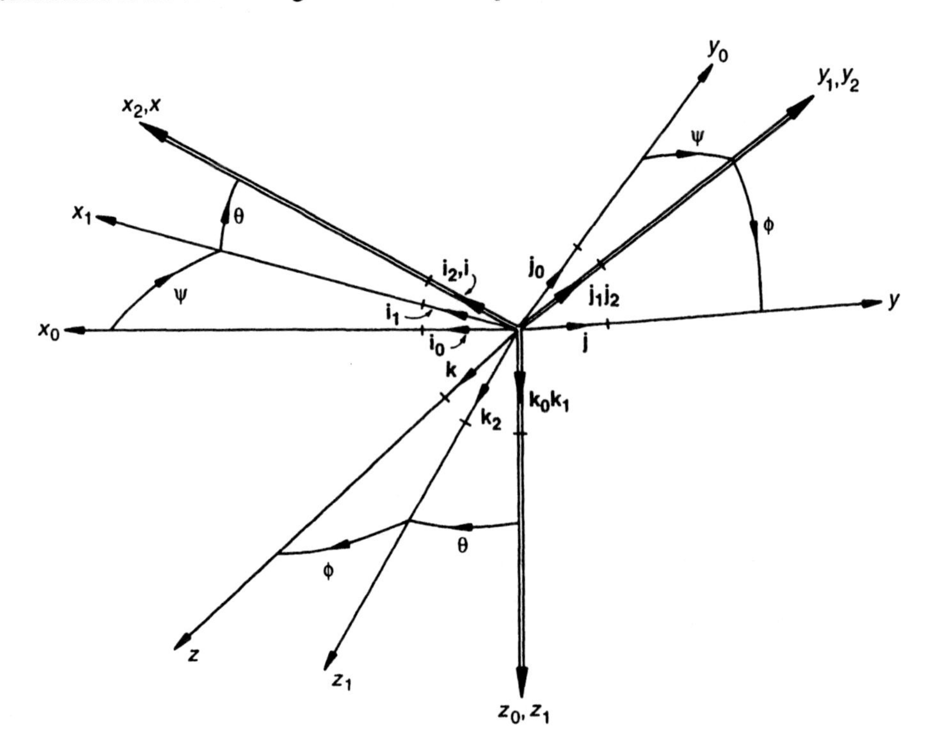

Fig. 8.2 Definition of Euler angles and unit vectors along various axes

Let us label the successive positions of the axes as $Ox_0y_0z_0$, $Ox_1y_1z_1$, $Ox_2y_2z_2$ and finally $Oxyz$. To keep track of the directions of the axes choose unit vectors $\mathbf{i}_0, \mathbf{j}_0, \mathbf{k}_0$; $\mathbf{i}_1, \mathbf{j}_1, \mathbf{k}_1$; $\mathbf{i}_2, \mathbf{j}_2, \mathbf{k}_2$ and $\mathbf{i}, \mathbf{j}, \mathbf{k}$ along the axes, respectively. We now need the relations between these unit vectors.

Consider the first rotation ψ about the Oz_0 axis; figure 8.3 shows the view looking along that axis. In the figure we have dropped perpendiculars PN and QT from the ends of the unit vectors $\mathbf{i}_0$, $\mathbf{j}_0$ onto the Ox_1 and Oy_1 axes. Then from the triangles formed we see

$$\left.\begin{aligned} \mathbf{i}_0 = \overrightarrow{OP} = \overrightarrow{ON} + \overrightarrow{NP} &= \mathbf{i}_1 \cos\psi - \mathbf{j}_1 \sin\psi \\ \mathbf{j}_0 = \overrightarrow{OQ} = \overrightarrow{OT} + \overrightarrow{TQ} &= \mathbf{j}_1 \cos\psi + \mathbf{i}_1 \sin\psi \\ \mathbf{k}_0 &= \mathbf{k}_1 \end{aligned}\right\} \tag{8.3}$$

since the perpendiculars PN and QT are parallel to $-Oy_1$ and Ox_1.

Each rotation looks like any other when viewed along its axis of rotation, allowing for the change of labels on the axes. Hence for the rotation θ about the Oy_1 axis,

$$\left.\begin{aligned} \mathbf{i}_1 &= \mathbf{k}_2 \sin\theta + \mathbf{i}_2 \cos\theta \\ \mathbf{j}_1 &= \mathbf{j}_2 \\ \mathbf{k}_1 &= \mathbf{k}_2 \cos\theta - \mathbf{i}_2 \sin\theta \end{aligned}\right\} \tag{8.4}$$

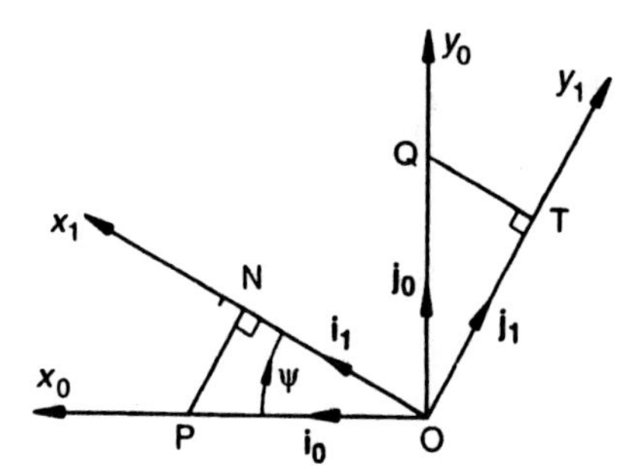

Fig. 8.3 Relations for one rotation

and for the rotation ϕ about the Ox_2 axis,

$$\left.\begin{aligned} \mathbf{i}_2 &= \mathbf{i} \\ \mathbf{j}_2 &= \mathbf{j}\cos\phi - \mathbf{k}\sin\phi \\ \mathbf{k}_2 &= \mathbf{j}\sin\phi + \mathbf{k}\cos\phi \end{aligned}\right\} \tag{8.5}$$

Then on substituting back we find

$$\left.\begin{aligned} \mathbf{i}_0 = {} & \mathbf{i}\cos\psi\cdot\cos\theta + \mathbf{j}(\cos\psi\cdot\sin\theta\cdot\sin\phi - \sin\psi\cdot\cos\phi) \\ & + \mathbf{k}(\cos\psi\cdot\sin\theta\cdot\cos\phi + \sin\psi\cdot\sin\phi) \\ \mathbf{j}_0 = {} & \mathbf{i}\sin\psi\cdot\cos\theta + \mathbf{j}(\sin\psi\cdot\sin\theta\cdot\sin\phi + \cos\psi\cdot\cos\phi) \\ & + \mathbf{k}(\sin\psi\cdot\sin\theta\cdot\cos\phi - \cos\psi\cdot\sin\phi) \\ \mathbf{k}_0 = {} & -\mathbf{i}\sin\theta + \mathbf{j}\cos\theta\cdot\sin\phi + \mathbf{k}\cos\theta\cdot\cos\phi \end{aligned}\right\} \tag{8.6}$$

which are the required relationships. The inverse relations are also needed; they are found in a similar manner and are

$$\left.\begin{aligned} \mathbf{i} = {} & \mathbf{i}_0\cos\psi\cdot\cos\theta + \mathbf{j}_0\sin\psi\cdot\cos\theta - \mathbf{k}_0\sin\theta \\ \mathbf{j} = {} & \mathbf{i}_0(\cos\psi\cdot\sin\theta\cdot\sin\phi - \sin\psi\cdot\cos\phi) \\ & + \mathbf{j}_0(\sin\psi\cdot\sin\theta\cdot\sin\phi + \cos\psi\cdot\cos\phi) + \mathbf{k}_0\cos\theta\cdot\sin\phi \\ \mathbf{k} = {} & \mathbf{i}_0(\sin\psi\cdot\sin\phi + \cos\psi\cdot\sin\theta\cdot\cos\phi) \\ & + \mathbf{j}_0(\sin\psi\cdot\sin\theta\cdot\cos\phi - \cos\psi\cdot\sin\phi) + \mathbf{k}_0\cos\theta\cdot\cos\phi \end{aligned}\right\} \tag{8.7}$$

8.3.1 Relations between the rates of change of angles

Suppose we find the orientation, or Euler, angles at time t and at time $t + \delta t$. Then by the usual process of taking differences, dividing by δt and proceeding to the limit we can define $\dot{\psi}$, $\dot{\theta}$ and $\dot{\phi}$. The vector sum of these must be the angular velocity vector of the aircraft, $\mathbf{A}$. They are, however, measured about non-orthogonal axes; because ϕ is a rotation about Ox_2, then $\dot{\phi}$ is a vector along that axis. Similarly $\dot{\theta}$ is along Oy_1 and $\dot{\psi}$ along Oz_0. Hence we can express $\mathbf{A}$ as

$$\mathbf{A} = \dot{\phi}\mathbf{i}_2 + \dot{\theta}\mathbf{j}_1 + \dot{\psi}\mathbf{k}_0$$

then the appropriate substitutions from the previous section give

$$\mathbf{A} = \dot{\phi}\mathbf{i} + \dot{\theta}(\mathbf{j}\cos\phi - \mathbf{k}\sin\phi) + \dot{\psi}(-\mathbf{i}\sin\theta + \mathbf{j}\cos\theta.\sin\phi + \mathbf{k}\cos\theta.\cos\phi)$$

The aircraft angular velocity can also be expressed as

$$\mathbf{A} = p\mathbf{i} + q\mathbf{j} + r\mathbf{k}$$

Then equating the components of **A** we find

$$\left.\begin{aligned} p &= \dot{\phi} - \dot{\psi}\sin\theta \\ q &= \dot{\theta}\cos\phi + \dot{\psi}\cos\theta\cdot\sin\phi \\ r &= -\dot{\theta}\sin\phi + \dot{\phi}\cos\theta\cdot\cos\phi \end{aligned}\right\} \tag{8.8}$$

These then are the relations between the rates of change of the orientation angles and the components of angular velocity and, as they are linear, they can be solved for $\dot{\psi}$, $\dot{\theta}$ and $\dot{\phi}$ in terms of p, q and r. When we have solved a problem these may be integrated to find the orientation of the aircraft during the disturbance.

8.4 Development of the equations

At this stage we need to define the initial condition of flight as different initial conditions lead to slightly different forms of the final equations. The standard choice is that of a straight steady unsideslipped climb at speed V_e with the x-axis at an angle Θ_e to the horizontal. In straight steady flight we have

$$\mathbf{A} = \dot{\mathbf{A}} = \dot{\bar{\mathbf{v}}} = V = 0 \tag{8.9}$$

with the result that the equations of motion reduce to

$$\mathbf{F} = \mathbf{F}_{a_e} + \mathbf{F}_{g_e} = 0 \tag{8.10}$$

and

$$\mathbf{Q}_{a_e} = 0 \tag{8.11}$$

This flight condition will be referred to as the 'datum flight condition' and the values of quantities in this condition are indicated by the suffix 'e'.

8.4.1 Components of the weight

We now consider the components of the weight along the various axes. Figure 8.4 shows the aircraft in the datum condition, with the x-axis at an angle α_e to the direction of flight. From the figure we see that the weight vector is initially

$$\mathbf{F}_{g_e} = mg\left[-\mathbf{i}_0\sin\Theta_e + \mathbf{k}_0\cos\Theta_e\right] \tag{8.12}$$

We substitute for $\mathbf{i}_0$ and $\mathbf{k}_0$ from (8.6) to obtain the weight vector in terms of **i**, **j** and **k**, that is we find its components along the disturbed aircraft axes. Then

$$\begin{aligned} \mathbf{F}_g &= mg\big[-\{\mathbf{i}\cos\psi\cdot\cos\theta + \mathbf{j}(\cos\psi\cdot\sin\theta\cdot\sin\phi - \sin\psi\cdot\cos\phi) \\ &\quad + \mathbf{k}(\cos\psi\cdot\sin\theta\cdot\cos\phi + \sin\psi\cdot\sin\phi)\}\sin\Theta_e \\ &\quad + \{-\mathbf{i}\sin\theta + \mathbf{j}\cos\theta\cdot\sin\phi + \mathbf{k}\cos\theta\cdot\cos\phi\}\cos\Theta_e\big] \\ &= X_g\mathbf{i} + Y_g\mathbf{j} + Z_g\mathbf{k} \end{aligned}$$

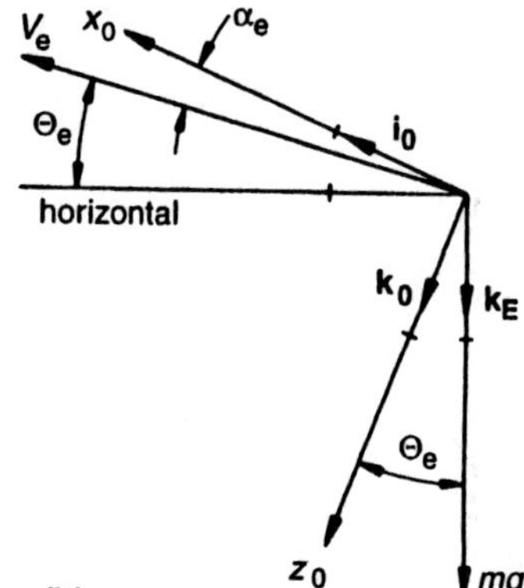

Fig. 8.4 Datum flight condition

Then equating components we have

$$\left.\begin{aligned} X_g &= -mg[\cos\psi\cdot\cos\theta\cdot\sin\Theta_e + \sin\theta\cdot\cos\Theta_e] \\ Y_g &= mg[\{\sin\psi\cdot\cos\phi - \cos\psi\cdot\sin\theta.\sin\phi\}\sin\Theta_e + \cos\theta\cdot\sin\phi\cdot\cos\Theta_e] \\ Z_g &= mg[-\{\cos\psi\cdot\sin\theta\cdot\cos\phi + \sin\psi\cdot\sin\phi\}\sin\Theta_e + \cos\theta\cdot\cos\phi.\cos\Theta_e] \end{aligned}\right\} \tag{8.13}$$

We write the aerodynamic forces in the datum condition as X_{a_e}, Y_{a_e} and Z_{a_e}. Then the equations of linear motion are

$$\left.\begin{aligned} -mg\sin\Theta_e + X_{a_e} &= 0 \\ Y_{a_e} &= 0 \\ mg\cos\Theta_e + Z_{a_e} &= 0 \end{aligned}\right\} \tag{8.14}$$

8.4.2 Small perturbations

We come now to the first and most far-reaching of the simplifying assumptions; it is that the aircraft is only disturbed by small amounts from the initial steady state. From figure 8.4 we see that the initial flight velocity, V_e, can be resolved along the initial aircraft axes into U_e and W_e where

$$U_e = V_e \cos \alpha_e \text{ and } W_e = V_e \sin \alpha_e \tag{8.15}$$

We then write

$$\left.\begin{aligned} U &= U_e + u \\ V &= v \\ W &= W_e + w \end{aligned}\right\} \tag{8.16}$$

The assumption is then specifically that

- u, v and w are much less than the speed V_e;
- p, q and r are much less than V_e/l where l is a characteristic length of the aircraft. This implies, for instance, that the helix angle of the rolling wing discussed in Section 7.2.1 is small;
- ψ, θ and ϕ are small.

The result is that squares and products of these are negligible quantities of second order and that sines of angles may be replaced by the angles and their cosines by unity. We have always to be aware that practical situations may appear where these assumptions are not justified; some cases where this happens are discussed in Chapter 12. Because only first-order terms are retained the process may also be described as one of 'linearization'. Note that we should not (and need not) assume that the angles α_e and Θ_e are small; aircraft with slender delta wings can fly at angles of incidence of the order of 30°.

8.4.2.1 *Stability derivatives*

We now assume that the aerodynamic forces and moments in a disturbance are functions of the perturbations u, v, w, p, q and r and expand them as Taylor series about the datum condition. Taking the X-force as a typical example, we write

$$X_a = X_{a_e} + \delta X_a \tag{8.17}$$

where

$$\delta X_a = \left(\frac{\partial X}{\partial u}\right)_e u + \left(\frac{\partial X}{\partial v}\right)_e v + \ldots + \left(\frac{\partial X}{\partial r}\right)_e r + X(t) \tag{8.18}$$

In writing this we have truncated the Taylor series after the linear terms in line with our assumption of small disturbances. The suffix 'e' indicates that the derivative is to be evaluated in the datum condition and $X(t)$ is the time dependent force due to, for instance, movement of the controls. For brevity we use a suffix notation, writing, for instance

$$\left(\frac{\partial X}{\partial u}\right)_e = \overset{\circ}{X}_u \tag{8.19}$$

These quantities are known as 'stability derivatives'. The derivatives used in Chapters 6 and 7 are quasi-static versions of these stability derivatives; as before the superscript ord, '$^\circ$', is to indicate that it is a dimensional, or 'ordinary', quantity.

To improve the correlation between theory and experiment, it has been found necessary to add derivatives with respect to linear accelerations, in particular the vertical acceleration $\dot{w}$. We will add only the two derivatives, $\overset{\circ}{Z}_{\dot{w}}$ and $\overset{\circ}{M}_{\dot{w}}$, but it is a simple matter to add more if required. If the aircraft were in a flight condition such that displacements from the datum flight path generated forces or moments then derivatives due to this source would be required. Determining these displacements is dealt with in Section 8.6.2.1. Examples of such cases are flight near the ground or in the flow field of another aircraft.

8.4.2.2 *Linearized equations of motion*

Bearing in mind (8.1), (8.2) and (8.16), the linearization of (7.56) and (7.57) results in

$$\left.\begin{aligned}
X_g + X_a &= m(\dot{u} + qW_e)\\
Y_g + Y_a &= m(\dot{v} + rU_e - pW_e)\\
Z_g + Z_a &= m(\dot{w} - qU_e)\\
L_a &= \dot{h}_1\\
M_a &= \dot{h}_2\\
N_a &= \dot{h}_3
\end{aligned}\right\} \tag{8.20}$$

Similarly the expressions (8.13) for the components of the weight become

$$\left.\begin{aligned} X_g &= -m(g_2 + g_1\theta) \\ Y_g &= m(g_2\psi + g_1\phi) \\ Z_g &= m(g_1 - g_2\theta) \end{aligned}\right\} \tag{8.21}$$

where for brevity we have written

$$g_1 = g \cos \Theta_e \text{ and } g_2 = g \sin \Theta_e \tag{8.22}$$

Another drastic simplification takes place with the relation between the rates of change of orientation angles and the components of angular velocity (8.8). These become

$$\left.\begin{aligned} p &= \dot{\phi} \\ q &= \dot{\theta} \\ r &= \dot{\phi} \end{aligned}\right\} \tag{8.23}$$

8.4.3 Symmetry

We will assume that the aircraft has symmetry about a vertical plane containing the longitudinal axis. This in fact amounts to two assumptions: symmetry of mass distribution and symmetry of external shape. No aircraft is ever perfectly symmetrical in either sense, but the assumption is probably as accurate as other assumptions we have made.

Symmetry of the mass distribution results in the cross-products of inertia I_{xy} and I_{yz} being zero. For instance consider $I_{xy} = \sum \delta mxy$; for every element of mass δm at a point (x, y, z) there is an equal mass at the point $(x, -y, z)$ which cancels its effect in the summation. The components of angular moments (7.49) then become

$$\left.\begin{aligned} h_1 &= I_x p - I_{xz} r \\ h_2 &= I_y q \\ h_3 &= I_z r - I_{xz} p \end{aligned}\right\} \tag{8.24}$$

representing a useful simplification.

Symmetry of external form implies that about half of the stability derivatives can be taken to be zero. If the aircraft is given a small disturbance lying in the plane of symmetry, i.e. an increment in forward or vertical velocity or in pitching velocity, then the airflow remains symmetrical. The result is that no force or moment out of the plane would appear, i.e. there is no sideforce, rolling or yawing moments. However, it is quite easy to find mechanisms by which disturbances out of the plane of symmetry, i.e. sideslip, rolling or yawing velocities, can produce forces in the plane of symmetry. In Chapter 6 we found that lift was produced on the fin in a sideslip; this must be accompanied by a drag, which is also likely to produce a pitching moment. In Chapter 7 we found that rolling the wing produces a change in the lift distribution on the wing; although theoretically the change in lift on one half wing is equal and opposite to that on the other, the drag changes will not cancel out. However, a symmetrical aircraft in an initially symmetric flight condition would produce the same drag regardless of the sense of the disturbance. This means that the drag is an even function of the disturbance and therefore is of second order.

In general asymmetric disturbances only produce second-order effects in the symmetric freedoms; however, there are some exceptions. In a sideslip the tailplane may move into a region in which the average wing downwash is different, thereby changing the pitching moment. Propellers are a fairly obvious source of effects linking the two. In a practical situation, flow separation can depend critically on such things as surface finish and so provide mechanisms for producing linking effects. For instance, a small increase in incidence can cause one wing to stall and not the other. The result is a large rolling moment and even nominally identical aircraft can behave differently.

Accepting these limitations we write the aerodynamic forces and moments in the form

$$\left.\begin{aligned}
\delta X_a &= \mathring{X}_u u + \mathring{X}_w w + \mathring{X}_q q + \mathring{X}(t) \\
\delta Y_a &= \mathring{Y}_v v + \mathring{Y}_p p + \mathring{Y}_r r + \mathring{Y}(t) \\
\delta Z_a &= \mathring{Z}_u u + \mathring{Z}_w w + \mathring{Z}_{\dot{w}} \dot{w} + \mathring{Z}_q q + \mathring{Z}(t) \\
\delta L_a &= \mathring{L}_v v + \mathring{L}_p p + \mathring{L}_r r + \mathring{L}(t) \\
\delta M_a &= \mathring{M}_u u + \mathring{M}_w w + \mathring{M}_{\dot{w}} \dot{w} + \mathring{M}_q q + \mathring{M}(t) \\
\delta N_a &= \mathring{N}_v v + \mathring{N}_p p + \mathring{N}_r r + \mathring{N}(t)
\end{aligned}\right\} \tag{8.25}$$

We also assume that the control surfaces are symmetrical about the centreline of the aircraft so that the forces and moments are written

$$\left.\begin{aligned}
\mathring{X}(t) &= \mathring{X}_\eta \eta'(t) \\
\mathring{Y}(t) &= \mathring{Y}_\xi \xi'(t) + \mathring{Y}_\zeta \zeta'(t) \\
\mathring{Z}(t) &= \mathring{Z}_\eta \eta'(t) \\
\mathring{L}(t) &= \mathring{L}_\xi \xi'(t) + \mathring{L}_\zeta \zeta'(t) \\
\mathring{M}(t) &= \mathring{M}_\eta \eta'(t) \\
\mathring{N}(t) &= \mathring{N}_\xi \xi'(t) + \mathring{N}_\zeta \zeta'(t)
\end{aligned}\right\} \tag{8.25a}$$

where $\xi'(t)$, $\eta'(t)$ and $\zeta'(t)$ are the deflections of the aileron, elevator and rudder from their trimmed positions.

8.5 Dimensional stability equations

We can now assemble the equations for the stability. Starting from (8.20) we substitute for the components of momentum from (8.24), the components of weight from (8.21) and the aerodynamic terms from (8.17) and similar expressions with (8.25) substituted. We finally subtract the equations of motion in the datum case, (8.14). The results are

$$m(\dot{u} + qW_e) = -mg_1\theta + \mathring{X}_u u + \mathring{X}_w w + \mathring{X}_q q + \mathring{X}(t) \tag{8.26}$$

$$m(\dot{v} + rU_e - pW_e) = m(g_1\phi + g_2\psi) + \mathring{Y}_v v + \mathring{Y}_p p + \mathring{Y}_r r + \mathring{Y}(t) \tag{8.27}$$

$$m(\dot{w} - qU_e) = -mg_2\theta + \mathring{Z}_u u + \mathring{Z}_w w + \mathring{Z}_{\dot{w}}\dot{w} + \mathring{Z}_q q + \mathring{Z}(t \tag{8.28}$$

$$I_x\dot{p} - I_{xz}\dot{r} = \mathring{L}_v v + \mathring{L}_p p + \mathring{L}_r r + \mathring{L}(t) \tag{8.29}$$

$$I_x\dot{q} = \mathring{M}_u u + \mathring{M}_w w + \mathring{M}_{\dot{w}}\dot{w} + \mathring{M}_q q + \mathring{M}(t) \tag{8.30}$$

$$I_z\dot{r} - I_{xz}\dot{p} = \mathring{N}_v v + \mathring{N}_p p + \mathring{N}_r r + \mathring{N}(t) \tag{8.31}$$

In these equations p, q and r are related to $\dot{\psi}$, $\dot{\theta}$ and $\dot{\phi}$ by (8.23).

There are a number of observations to be made about these equations. First, we note that (8.26), (8.28) and (8.30) are independent of (8.27), (8.29) and (8.31); this is a result of the assumptions of small perturbations and of symmetry. The first three involving u, w and q comprise motion entirely in the plane of symmetry, and are known as the 'longitudinal stability equations'. The second set involving v, p and r comprise motions out of the plane and are known as the 'lateral stability equations'. They are sets of three simultaneous linear differential equations with time as the independent variable. Whilst normally a mechanical system with three degrees of freedom would be expected to result in three second-order equations, some of these equations are only of first order.

These equations cannot be regarded as sufficient to deal with all possible circumstances. In special cases additional stability derivatives, dynamic terms or terms expressing the effects of unconventional controls may be needed. In some cases it may not be justified to assume that changes in height are small and changes in density may have to be allowed for. This will require a further equation to accommodate the extra freedom and corresponding stability derivatives. Such changes can usually be made on an *ad hoc* basis.

8.6 Concise, normalized and nondimensional stability equations

Before starting to discuss solving these equations we look first at what simplifications can be obtained by some very simple transformations. The simplest is to divide through by the mass or an inertia as appropriate and define new quantities, as we did with the bifilar pendulum in Section 7.4.1; fewer parameters are then necessary.

A step with more physical relevance is to divide each variable or parameter by a constant quantity with the same dimensions and having a significance to the problem. When a number of different physical quantities are involved confusion can arise unless a system of divisors is specified in which those for mass, length, time, velocity and so on are consistent. The process can also be described as expressing the quantities in a special system of units. Three units are taken as basic and the remainder derived from them by the use of physical laws, for instance

$$\text{force} = (\text{mass}) \times (\text{acceleration})$$

or

$$\text{moment} = (\text{force}) \times (\text{arm})$$

We describe this process as 'normalization' and the quantities produced as 'normalized'. They can alternatively be described as nondimensionalized, but this does not really imply that a consistent set of units based on physically significant quantities has been used. Equations that have been treated in this way have an important property: they take the same form in whichever set of quantities are used and so do any derived results, in particular the solutions. The usefulness of this can be enhanced if we choose the symbols used for each quantity in each set of units to resemble one another as closely as possible. As an example let us take the x-equation for the bifilar pendulum discussed in Section 7.4.5 and discuss the free motion; we write (7.61) as

$$m\ddot{x} + C\dot{x} + kx = 0 \tag{8.32}$$

where

$$k = mg/d \tag{8.33}$$

We now choose quantities to use as units. The choice for mass and length is straightforward, choosing m for the mass and d for length. Now consider C which has the units (force)/(velocity) or $MLT^{-2}.L^{-1}T$; this simplifies to MT^{-1}. A suitable quantity with the units of time is then

$$m/C = \tau, \text{ say} \tag{8.34}$$

We now introduce the variables $\hat{x}$ and $\hat{t}$ defined by

$$x = d\hat{x} \text{ and } t = \tau\hat{t} \tag{8.35}$$

Substituting into (8.32) then gives

$$m\frac{d}{\tau^2}\cdot\frac{d^2\hat{x}}{d\hat{t}^2} + C\frac{d}{\tau}\cdot\frac{d\hat{x}}{d\hat{t}} + kd\hat{x} = 0$$

Substituting for τ from (8.34) and multiplying through by m/dC^2 then gives

$$\ddot{x} + \dot{x} + \hat{k}x = 0 \tag{8.36}$$

where the mass and damping in this system are just 1's and

$$\hat{k} = \frac{m}{dC^2}\cdot kd = \frac{m}{dC^2}\cdot\frac{mg}{d}\cdot d = \frac{m^2 g}{hC^2} \tag{8.37}$$

The solution of the auxiliary equation corresponding to (8.36) is

$$\lambda_1, \lambda_2 = -1/2 \pm i\sqrt{\hat{k} - 1/4}$$

and the normalized time to halve an initial disturbance (see Section 7.2.1 and (7.5)) and periodic time are

$$\hat{t}_H = 2 \ln 2 \text{ and } \hat{t}_p = \frac{2\pi}{\sqrt{\hat{k} - 1/4}}$$

Expressing these in ordinary time using (8.35) and (8.37) we find

$$t_H = \tau(2 \ln 2) = \frac{2m}{C} \ln 2 \tag{8.38}$$

and

$$t_p = \frac{2\pi\tau}{\sqrt{\hat{k} - 1/4}} = \frac{2\pi m}{C\sqrt{\dfrac{m^2 g}{hC^2} - 1/4}} = \frac{4\pi m}{\sqrt{\dfrac{4m^2 g}{h} - C^2}} \tag{8.39}$$

The solution of the auxiliary equation corresponding to (8.32) is

$$\lambda_1, \lambda_2 = -\frac{C}{2m} \pm \mathrm{i} \cdot \frac{\sqrt{\dfrac{4m^2 g}{h} - C^2}}{2m}$$

leading directly to the results (8.38) and (8.39).

This technique has the psychological advantage that we can still think of a normalized quantity in the same way as its ordinary dimensional counterpart. Newton's third law can be restated as

(normalized force) = (normalized mass) × (normalized acceleration)

and as we become accustomed to it we can forget the word 'normalized'. A normalized force or time is really no less of a force or time.

In any given situation the systems of normalized quantities are not in general unique and for our problem we have to consider the merits of two. Suppose we are investigating the dynamic behaviour of an aircraft of given mass subject to aerodynamic and other forces and flying at a known speed. The relevant units are then those of mass, force and speed. On the other hand if our main interest is the variation of the aerodynamic forces acting on a body of certain size in an airstream of given speed, the relevant units are force, length and speed. We will call a system based on the first 'dynamic-normalized' and one on the second 'aero-normalized'. The two systems have the same units of force and speed, namely $\frac{1}{2}\rho V_e^2 S$ and V_e. If we let m be the unit of mass in the dynamic system and l_0 the unit of length in the aero system, we can then deduce the other primary units of mass, length and time according to the systems, as shown in table 8.1.

Table 8.1 Comparison of units in aerodynamic-normalized systems

	Mass	*Length*	*Time*
Dynamic	m	μl_0	τ
Aero	m/μ	l_0	$l_0/V_e = \tau/\mu$

In this table τ is the unit of time discussed in Chapter 4 and

$$\mu = \frac{m}{\frac{1}{2}\rho S l_0} \tag{8.40}$$

is a relative density parameter similar to those introduced in Chapters 4 and 5. We see that the primary units in the dynamic-normalized system are μ times those in the aero-normalized system. It is convenient to use different lengths in the longitudinal and lateral stability equations and the corresponding relative density parameters are written μ_1 and μ_2.

We have equations which can be concise, normalized or both and we want to use symbols which are as similar as possible. This means using subscripts or superscripts and the stability derivatives in particular can become rather complicated in appearance. To simplify matters we will adopt two conventions.

- Concise derivatives are written in lower case.
- Normally only the first stability derivative in any equation, set of equations, matrix or determinant will be given the superscript to indicate the normalization. Other quantities will be written in full.

8.6.1 Concise stability equations

In this case we simply divide through by the mass or inertia. We need to define concise derivatives and terms such as

$$\mathring{x}_u = -\frac{\mathring{X}_u}{m}, \quad \mathring{n}_v = -\frac{\mathring{N}_v}{m}, \quad \mathring{m}_q = -\frac{\mathring{M}_q}{I_y}, \quad \mathring{l}(t) = -\frac{\mathring{L}(t)}{I_x} \tag{8.41}$$

The definitions of all the concise derivatives include a change of sign as almost all derivatives are negative. We also need to define other quantities such as $x(t) = X(t)/m$ and so on for the other derivatives, also $e_x = -I_{xz}/I_x$ and $e_z = -I_{xz}/I_z$. Our equations (8.26)–(8.31) then become

$$\dot{u} + g_1\theta + \mathring{x}_u u + x_w w + (x_q + W_e)q = -\mathring{x}(t) \tag{8.42}$$

$$\dot{v} - g_1\phi - g_2\psi + y_v v + (y_p - W_e)p + (y_r + U_e)r = -\mathring{y}(t) \tag{8.43}$$

$$\dot{w} + g_2\theta + z_u u + z_w w + z_{\dot{w}}\dot{w} + (z_q - U_e)q = -\mathring{z}(t) \tag{8.44}$$

$$\dot{p} + e_x\dot{r} + l_v v + l_p p + l_r r = -\mathring{l}(t) \tag{8.45}$$

$$\dot{q} + m_u u + m_w w + m_{\dot{w}}\dot{w} + m_q q = -\mathring{m}(t) \tag{8.46}$$

$$\dot{r} + e_z\dot{p} + n_v v + n_p p + n_r r = -\mathring{n}(t) \tag{8.47}$$

In some investigations it is necessary to work in terms of ordinary time, for instance if a real pilot is involved such as as in a flight simulator. No further manipulation of the equations then appears possible.

8.6.2 Dynamic-normalized equations

Since the normalizing units form a consistent set, we expect the formal appearance of (8.42)–(8.47) to be unchanged. We will denote quantities which are dynamic-normalized using a superscript, cap '$\hat{}$', so that we replace u, v, w, p, q and r by $\hat{u}$, $\hat{v}$, $\hat{w}$, $\hat{p}$, $\hat{q}$ and $\hat{r}$, and g_1 and g_2 by $\hat{g}_1$ and $\hat{g}_2$ We replace differentiation with respect to time t with that with respect to normalized time $\hat{t}$ and write

$$\hat{D} = \mathrm{d}/\mathrm{d}\hat{t} \text{ where } \hat{t} = t/\tau \text{ and } \tau = \frac{m}{\frac{1}{2}\rho V_e S} \tag{8.48}$$

If density changes are to be allowed for then the datum density must be written as ρ_e and used in the divisors. The normalizing divisors for the other quantities above are given in table 8.2.[1]

Table 8.2 Divisors for dynamic normalization

Quantity	*Divisor*	*Normalized quantity*
u, v, w	V_e	$\hat{u}, \hat{v}, \hat{w}$
p, q, r	$1/\tau$	$\hat{p}, \hat{q}, \hat{r}$
g_1, g_2	$\frac{1}{2}\rho V_e^2 S/m$	$\hat{g}_1, \hat{g}_2$

It is preferable to normalize the stability derivatives according to the aero-system and adjust matters later. In aerodynamics generally it has been and still is an almost vital simplification to work in terms of nondimensional quantities for most of the time. In this system the divisor for force is the same as in the dynamic-normalized system, i.e. $\frac{1}{2}\rho V_e^2 S$, and that for moment is $\frac{1}{2}\rho V_e^2 S l_0$. The derivatives are written without a superscript and the divisors for the various kinds are given in table 8.3.

Table 8.3 Divisors for aero-normalization of stability derivatives

Force derivatives with respect to	*Example*	*Divisor*	*Moment derivatives with respect to*	*Example*	*Divisor*
Linear displacement	Z_h	$\frac{1}{2}\rho V_e^2 S/l_0$	Linear displacement	M_h	$\frac{1}{2}\rho V_e^2 S$
Linear velocity	Z_w	$\frac{1}{2}\rho V_e S$	Linear velocity	M_w	$\frac{1}{2}\rho V_e S l_0$
Linear acceleration	$Z_{\dot{w}}$	$\frac{1}{2}\rho S l_0$	Linear acceleration	$M_{\dot{w}}$	$\frac{1}{2}\rho S l_0^2$
Angular displacement	Z_η	$\frac{1}{2}\rho V_e^2 S$	Angular displacement	M_η	$\frac{1}{2}\rho V_e^2 S l_0$
Angular velocity	Z_q	$\frac{1}{2}\rho V_e S l_0$	Angular velocity	M_q	$\frac{1}{2}\rho V_e S l_0^2$

For example the aero-normalized derivative of sideforce due to sideslip velocity is

$$Y_v = \frac{\mathring{Y}_v}{\frac{1}{2}\rho V_e S}$$

Now in order to form dynamic-normalized concise derivatives, such as $\hat{m}_w$, from the corresponding aero-normalized derivatives, in this case M_w, we need to define the inertia parameters i_x, i_y and i_z where, for instance,

$$i_x = \frac{I_x}{m l_0^{\ 2}} \tag{8.49}$$

Table 8.4 gives the factors for forming the various types of these derivatives from the aero-normalized derivative.

Table 8.4 Conversion of aero-normalized derivatives to concise dynamic-normalized derivatives

Force derivatives with respect to	*Example*	*Factor*	*Moment derivatives with respect to*	*Example*	*Factor*
Linear displacement	Z_h	$-\mu$	Linear displacement	M_h	$-\mu^2/i_y$
Linear velocity	Z_w	-1	Linear velocity	M_w	$-\mu/i_y$
Linear acceleration	$Z_{\dot{w}}$	$-1/\mu$	Linear acceleration	$M_{\dot{w}}$	$-1/i_y$
Angular displacement	Z_η	-1	Angular displacement	M_η	$-\mu/i_y$
Angular velocity	Z_q	$-1/\mu$	Angular velocity	M_q	$-1/i_y$

We now write out the concise dynamic-normalized equations after some rearrangement, starting with the longitudinal set. In these equations the reference length l_0 is chosen to be the wing aerodynamic mean chord so that the relative density parameter used in the definition of these derivatives is

$$\mu_1 = \frac{m}{\frac{1}{2}\rho S\bar{\bar{c}}} \tag{8.50}$$

Then taking the pitching moment due to vertical velocity derivative as an example and using tables 8.3 and 8.4 we find

$$\hat{m}_w = -\frac{\mu_1 M_w}{i_y}$$

where

$$M_w = \frac{\overset{\circ}{M}_w}{\frac{1}{2}\rho V_e S\bar{\bar{c}}^2} \text{ and } i_y = \frac{I_y}{m\bar{\bar{c}}^2}$$

Using (8.26), (8.28), (8.30) and (8.23) we find the longitudinal set to be

$$(\hat{D} + \hat{x}_u)\hat{u} + x_w\hat{w} + (x_q + \hat{W}_e)\hat{q} + \hat{g}_1\theta = -\hat{x}(\hat{t}) \tag{8.51}$$

$$z_u\hat{u} + \left[(1 + z_{\dot{w}})\hat{D} + z_w\right]\hat{w} + (z_q - U_e)\hat{q} + \hat{g}_2\theta = -\hat{z}(\hat{t}) \tag{8.52}$$

$$m_u\hat{u} + (m_{\dot{w}}\hat{D} + m_w)\hat{w} + (\hat{D} + m_q)\hat{q} = -\hat{m}(\hat{t}) \tag{8.53}$$

$$\hat{q} - \hat{D}\theta = 0 \tag{8.54}$$

In these equations the control terms are given from (8.25a) by

$$\hat{x}(\hat{t}) = \hat{x}_\eta\eta'(\hat{t}),\ \hat{z}(\hat{t}) = z_\eta\eta'(\hat{t}) \text{ and } \hat{m}(\hat{t}) = m_\eta\eta'(\hat{t}) \tag{8.55}$$

The unit of time τ (8.48) is the same for both longitudinal and lateral sets of equations. We note that (8.51), (8.52) and (8.53) are virtually identical in form to (8.42), (8.44) and (8.46) of the ordinary concise equations and an analysis in terms of one set is rapidly rewritten in terms of the other.

In the case of the lateral stability the reference length is chosen to be the wing span so that the relative density parameter is

$$\mu_2 = \frac{m}{\frac{1}{2}\rho S b} \tag{8.56}$$

As an example of the normalized derivatives we take the rolling moment derivative due to rate of roll and using tables 8.3 and 8.4 find

$$\hat{l}_p = -L_p / i_x$$

where

$$L_p = \frac{\mathring{L}_p}{\frac{1}{2}\rho V_e S b^2} \text{ and } i_x = \frac{I_x}{mb^2}$$

Using (8.27), (8.29), (8.31) and (8.23) we find the lateral set to be

$$(\hat{D} + \hat{y}_v)\hat{v} + (y_p - \hat{W}_e)\hat{p} + (y_r + \hat{U}_e)\hat{r} - \hat{g}_1\phi - \hat{g}_2\psi = -\hat{y}(\hat{t}) \tag{8.57}$$

$$l_v\hat{v} + (\hat{D} + l_p)\hat{p} + (e_x\hat{D} + l_r)\hat{r} = -\hat{l}(\hat{t}) \tag{8.58}$$

$$n_v\hat{v} + (e_z\hat{D} + n_p)\hat{p} + (\hat{D} + n_r)\hat{r} = -\hat{n}(\hat{t}) \tag{8.59}$$

$$\hat{p} - \hat{D}\phi = 0 \tag{8.60}$$

$$\hat{r} - \hat{D}\psi = 0 \tag{8.61}$$

In these equations the control terms from (8.25a) are

$$\hat{y}(\hat{t}) = \hat{y}_\xi \xi'(\hat{t}) + \hat{y}_\zeta \zeta'(\hat{t}) \tag{8.62}$$

$$\hat{l}(\hat{t}) = \hat{l}_\xi \xi'(\hat{t}) + \hat{l}_\zeta \zeta'(\hat{t}) \tag{8.63}$$

$$\hat{n}(\hat{t}) = \hat{n}_\xi \xi'(\hat{t}) + \hat{n}_\zeta \zeta'(\hat{t}) \tag{8.64}$$

We note that (8.57), (8.58) and (8.59) are virtually identical in form to (8.43), (8.45) and (8.47) of the ordinary concise equations. However, the result of normalization is that the coefficients of the various terms in the resulting equations then become much more independent of aircraft size. Comparison between the stability characteristics of a wide range of aircraft of different sizes and speeds can then be made more easily and blunders are more easily detected.

8.6.2.1 The motion of the centre of gravity of the aircraft[2]

It may be required to find the displacement of the aircraft cg from where it would have been in the absence of the disturbance. Figure 8.5 shows the situation.

In the datum condition the aircraft velocity is $U_e\mathbf{i}_0 + W_e\mathbf{k}_0$. Whilst in disturbed flight it is, from (8.16),

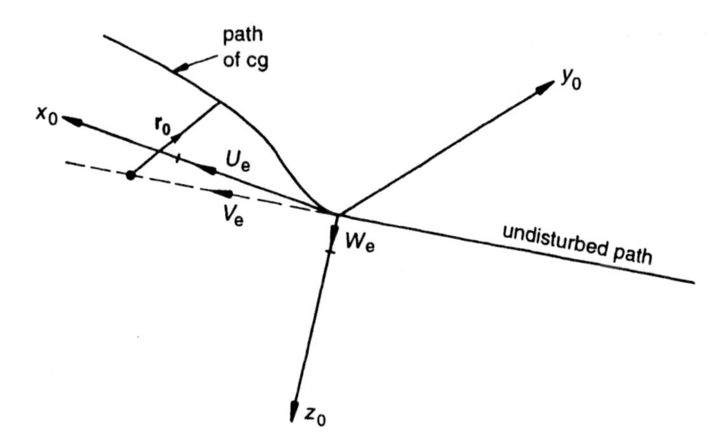

Fig. 8.5 Determination of displacement of aircraft cg due to a disturbance

$$\mathbf{V} = (U_e + u)\mathbf{i} + v\mathbf{j} + (W_e + w)\mathbf{k} \tag{8.65}$$

The rate of change of the position vector from an imaginary point P coincident with the cg at the start of the disturbance and travelling with its datum velocity to the aircraft cg in the disturbance is then

$$\dot{\mathbf{r}}_0 = (U_e + u)\mathbf{i} + v\mathbf{j} + (W_e + w)\mathbf{k} - U_e\mathbf{i}_0 - W_e\mathbf{k}_0$$

We now substitute for $\mathbf{i}$, $\mathbf{j}$ and $\mathbf{k}$ in terms of $\mathbf{i}_0$, $\mathbf{j}_0$ and $\mathbf{k}_0$ from (8.7), linearize the result and separate the components of the rate of change of displacement measured parallel to Ox_0, Oy_0 and Oz_0 to find

$$\dot{x}_0 = u + W_e\theta \tag{8.66}$$

$$\dot{y}_0 = U_e\psi + v - W_e\phi \tag{8.67}$$

$$\dot{z}_0 = U_e\theta + w \tag{8.68}$$

We see that to first order $\dot{x}_0$ and $\dot{z}_0$ are zero in asymmetric flight, as is $\dot{y}_0$ in symmetric flight.

Another requirement may be to find the change in the height, perhaps to allow for the effects of air density change. Let $\mathbf{k}_E$ be the unit vector pointing towards the Earth's centre (see figure 8.4), then the rate of gain of height is the component of velocity along $-\mathbf{k}_E$, where

$$\mathbf{k}_E = -\mathbf{i}_0 \cos \Theta_e + \mathbf{k}_0 \sin \Theta_e \tag{8.69}$$

We now substitute for $\mathbf{i}$, $\mathbf{j}$ and $\mathbf{k}$ in terms of $\mathbf{i}_0$, $\mathbf{j}_0$ and $\mathbf{k}_0$ using (8.7) into $\mathbf{V}$ from (8.65), form the dot product with $-\mathbf{k}_E$ from (8.69), and then linearize to give

$$\dot{h} = (U_e + u) \sin \Theta_e + U_e\theta \cos \Theta_e - (W_e + w) \cos \Theta_e + W_e\theta \sin \Theta_e \tag{8.70}$$

We note that the sideslip velocity has no effect to first order.

8.6.3 Stability equations in American notation

There have been some variations in the American usage, so we will describe the notation given by B. Etkin, reference (8.3). The major differences in the notation for dimensional quantities are that the amc is denoted by $\bar{c}$, moments of inertia by A, B and C, and the product of inertia I_{xz} by E. Quantities in the datum state are given the suffix zero so that the datum speed and climb angle are u_0 and θ_0. The orientation angles are denoted by the capital letters Ψ, Θ and Φ and changes in them by ψ, θ and ϕ. Control angles in the trimmed state are denoted by δ_a, δ_e and δ_r for the aileron, elevator and rudder respectively and their changes from the datum state by ξ, η and ζ. Capital letters are used for the dimensional stability derivatives; for example M_u is the counterpart of our $\mathring{M}_u$. Wind axes are referred to as stability axes.

The equations are aero-normalized so that the unit of time is $t^* = l/u_0$, where l is the reference length. Normalized quantities are indicated by the superscript hat, '^'. Differentiation with respect to normalized time is denoted by $D = d/\hat{t}$ The divisors to form force and moment coefficients are based on the instantaneous speed v_c given by

$$v_c^2 = U^2 + v^2 + w^2$$

where $U = u_0 + u$. Divisors to form the most important quantities are given in table 8.5.

Table 8.5 Divisors for normalization in the American notation

Dimensional quantity				*Divisor*	*Nondimensional quantity*			
X	Y	Z		$\frac{1}{2}\rho v_c^2 S$	C_x	C_y	C_z	
L	M	N		$\rho v_c^2 S l$	C_l	C_m	C_n	
u	v	w		u_0	$\hat{u}$	β	α	
p	q	r		$1/t^*$	$\hat{p}$	$\hat{q}$	$\hat{r}$	
	$\dot{\beta}$	$\dot{\alpha}$		$1/t^*$		$D\beta$	$D\alpha$	
m				$\rho S l$	μ			
A	B	C	E	$\rho S l^3$	i_A	i_B	i_C	i_E
t				t^*	$\hat{t}$			

The notation for nondimensional stability and control derivatives uses the appropriate force or moment coefficient plus an appropriate symbol added as a further suffix. For instance

$$C_{z_u} = \left(\frac{\partial C_z}{\partial \hat{u}}\right)_0$$

where the subscript zero indicates evaluation in the datum condition. Stability derivatives are related to the force and moment coefficients as follows, taking Z_u as an example. Thus

$$Z_u = \left(\frac{\partial Z}{\partial u}\right)_0 = C_{z_0}\rho u_0 S\left(\frac{\partial v_c}{\partial u}\right)_0 + \tfrac{1}{2}\rho u_0^2 S\left(\frac{\partial C_z}{\partial u}\right)_0$$

But $2v_c \dfrac{\partial v_c}{\partial u} = 2U\dfrac{\partial U}{\partial u}$ and hence $\left(\dfrac{\partial v_c}{\partial u}\right)_0 = 1$. Also

$$\left(\frac{\partial C_z}{\partial u}\right)_0 = \frac{1}{u_0}\left(\frac{\partial C_z}{\partial \hat{u}}\right)_0 = \frac{1}{u_0}C_{z_u}$$

Since wind axes are used, we have $C_{z_0} = -C_{L_0}$ and hence

$$Z_u = -\rho u_0 S C_{L_0} + \tfrac{1}{2}\rho u_0 S C_{z_u} \tag{8.71}$$

For most derivatives the term corresponding to the first is zero as the aircraft is in trim in the datum state.

The normalized equations of longitudinal stability are then

$$\left(2\mu D - 2C_{L_0}\tan\theta_0 - C_{x_u}\right)\hat{u} \quad - C_{x_\alpha}\alpha \quad + C_{L_\theta}\theta = 0 \tag{8.72}$$

$$\left(2C_{L_0} - C_{z_u}\right)\hat{u} + \left(2\mu D - C_{z_{\dot\alpha}}D - C_{z_\alpha}\right)\alpha \left[\left(2\mu + C_{z_q}\right)D - C_{L_0}\tan\theta_0\right]\theta - C_{z_\eta}\eta = 0 \tag{8.73}$$

$$-C_{m_u}\hat{u} \quad -\left(C_{m_{\dot\alpha}}D + C_{m_\alpha}\right)\alpha \quad +\left(i_B D^2 - C_{m_q}D\right)\theta - C_{m_\eta}\eta = 0 \tag{8.74}$$

In these equations the reference length l is taken as half the wing mean chord and hence

$$\mu = \frac{2m}{\rho S\bar{c}} \text{ and } t^* = \frac{\bar{c}}{2\mu_0} \tag{8.75}$$

For the lateral stability the equations are

$$\left(2\mu D - C_{y_\beta}\right)\beta \quad -C_{y_p}\hat{p} + \left(2\mu - C_{y_r}\right)\hat{r} \quad - C_{y_\zeta}\zeta = 0 \tag{8.76}$$

$$-C_{l_\beta}\beta + \left(i_A D - C_{l_p}\right)\hat{p} - \left(i_E D + C_{l_r}\right)\hat{r} \quad - C_{l_\xi}\xi - C_{l_\zeta}\zeta = 0 \tag{8.77}$$

$$-C_{n_\beta}\beta - \left(i_E D + C_{n_p}\right)\hat{p} + \left(i_C D - C_{n_r}\right)\hat{r} \quad - C_{n_\xi}\xi - C_{n_\zeta}\zeta = 0 \tag{8.78}$$

$$\hat{p} + \hat{r}\tan\theta_0 \quad - D\phi = 0 \tag{8.79}$$

$$\hat{r}\sec\theta_0 \quad - D\psi = 0 \tag{8.80}$$

In these equations the reference length is taken as the wing semispan and hence

$$\mu = \frac{2m}{\rho Sb} \text{ and } t^* = \frac{b}{2u_0} \tag{8.81}$$

Etkin also gives equations for the motion of free controls and the effects on the main equations. Expressions for converting between these derivatives and those used in the rest of this book are given in Sections 9.2.8 and 11.2.7.

Student problems

8.1 Derive the six linearized stability equations corresponding to (8.26) to (8.31) for the case of an aircraft initially in a steady level correctly banked turn. Are the longitudinal and lateral motions still uncoupled?

8.2 Verify that all the nondimensional quantities defined in this chapter are, in fact, nondimensional.

8.3 Rework the derivation of the equations for aircraft stability adding terms to account for the angular momentum of the rotating parts of an engine. To do this rewrite (7.48) as

$$\mathbf{h} = (h_1 + h_e)\mathbf{i} + h_2\mathbf{j} + h_3\mathbf{k}$$

where $h_e = I_e\omega$ is the angular momentum of the engine and it is assumed that the axis of rotation of the engine is parallel to the x-axis. How can it be judged whether the neglect of engine gyroscopic effects is important?

8.4 Consider the case of a completely axisymmetric missile and show that only 11 stability derivatives are needed. Rewrite equations (8.43), (8.44), (8.46) and (8.47) assuming that the effects of forward speed change, roll rate and gravity can be neglected; also add the term $n_{\dot{v}}\dot{v}$ to the yawing moment equation. Add i times the vertical force equation to the sideforce equation and i times the yawing moment equation to the pitching moment equation, where $i = \sqrt{-1}$, to show that the equations of free motion can be reduced to two, as follows

$$\dot{\bar{v}} + \mathring{y}_v\bar{v} + y_r\bar{r} - U_e\bar{r}^* = 0$$

and

$$\dot{\bar{r}} + n_v\bar{v} + n_r\bar{r} + n_{\dot{v}}\dot{\bar{v}} = 0$$

where $\bar{v} = v + iw$, $\bar{r} = q + ir$ and $\bar{r}^* = q - ir$.

Notes

1. The notation is based on that proposed by Hopkin, reference (8.2), and is consistent with that used in the Data Items of the Engineering Sciences Data Unit.
2. This section can be omitted when first studying this chapter.

9
Longitudinal dynamic stability

9.1 Introduction

The main aim of this chapter is to discuss the stability of aircraft in longitudinal motion. After some general remarks on stability derivatives we deal with the estimation of the longitudinal derivatives. The longitudinal stability equations for the case of free motion are then solved and we discuss the modes obtained.

Throughout this chapter we shall be using wind axes as defined in Section 8.2.1.

9.2 General remarks on stability derivatives[1]

Like all other aerodynamic coefficients, stability derivatives are functions of Reynolds and Mach numbers, the geometry of the body and its orientation to the airflow. However, we cannot define a derivative if it depends on time in some manner. A lifting body such as a wing moving through a fluid leaves behind it a vortex wake. In subsonic flow these vortices affect the pressure on all points of the body. This means that the forces on the body are theoretically a function of all the past history of the motion and the concept of a stability derivative is not in general tenable. The derivatives are independent of time in only two cases: first, where the motion is quasi-steady, and second, where the disturbance has the form exp (λt), where λ is a real, imaginary or complex parameter, defined in (9.1) below, and the motion has existed for infinite time. The resulting derivatives are known as 'quasi-steady' and 'exponential' derivatives, respectively. It must be emphasized that stability derivatives are restricted to small disturbances.

In the case of estimation of the quasi-steady derivatives it is assumed that flow conditions at each point of the body are the same as for exactly steady conditions at linear or angular velocities equal to the instantaneous values. If motion is sufficiently slow the shape of the vortex wake and the distribution of vortex strength differ from the exactly steady ones by only a negligible amount. The great majority of stability calculations are made using quasi-steady derivatives, apparently with adequate results.

The results of theoretical investigations into exponential derivatives shows that they can be expressed as functions of a nondimensional frequency parameter

$$\nu = \frac{\omega l}{V}$$

and a nondimensional damping parameter

$$\sigma = \frac{\kappa l}{V}$$

where l is a characteristic length and

$$\lambda = \kappa + i\omega \tag{9.1}$$

Since the advent of powerful digital computers it has become possible to model the flow around an aircraft in a disturbance, including its vortex wake. This has generated new information on the validity of stability derivatives.

Experimental determinations of stability derivatives can be made using windtunnels, on full scale aircraft or in specialized experimental facilities such as whirling arms. There are several ways in which windtunnels can be used. A tunnel equipped with a six-component balance can be used to find quasi-steady values for derivatives due to the linear velocity disturbances u, v and w. For instance if the model is mounted on the balance and rotated about the z-axis to some yaw angle ψ there is a component of the tunnel wind speed along the y-axis of $V \sin \psi$. In a sideslip of velocity v on the full size aircraft the air has a relative velocity $-v$ along the y-axis. Plotting the windtunnel results against ψ and reversing the sign of the slope then gives the derivatives.

Derivatives due to angular velocity in roll can be measured by mounting the model on a strain-gauge balance which is itself mounted on a bearing with its axis aligned with the wind direction. The model and balance are then rotated by a suitable motor. A simple measurement of the damping in roll derivative L_p can be made by applying a known constant rolling moment and measuring the rate of roll. Oscillatory tests can be made by mounting the model on a bearing aligned with any of the three axes, and using springs to restrain the model. The model can be given a forced oscillation in which case the frequency and amplitude of oscillation need to be measured as well as the force or phase lag. Alternatively experimental conditions can be adjusted to produce resonance, and the results deduced from relationships resulting from the mathematical conditions for resonance. Another alternative is to have the model free to oscillate and then to disturb it; values of the stiffness and damping derivatives are then deduced from the frequency and logarithmic decrement of the oscillation. In oscillatory testing the frequency parameter ν is varied by varying the tunnel speed or spring stiffnesses. Bearings with very low friction, such as air bearings or magnetic suspension, give the best results. Helical springs can give rise to problems due to parasitic oscillations; torsion springs are usually better behaved. Complex 'derivative engines' have been built which combine motions in more than one freedom. Instead of measuring forces an alternative is to measure pressures using systems with good high frequency response.

As an example of the effect that frequency can have on a stability derivative, figure 9.1 shows the effect on the derivative Z_w for a wing of aspect ratio 4 at a Mach number of 0.5. In this figure the parameter ν is based on the mean chord. The curve shown has been estimated from theory but it is a good fit to some experimental results at full scale which deduced the derivative from the frequency response of the aircraft while flying through turbulent air. It is evident from the graph that Z_w decreases steadily with frequency.

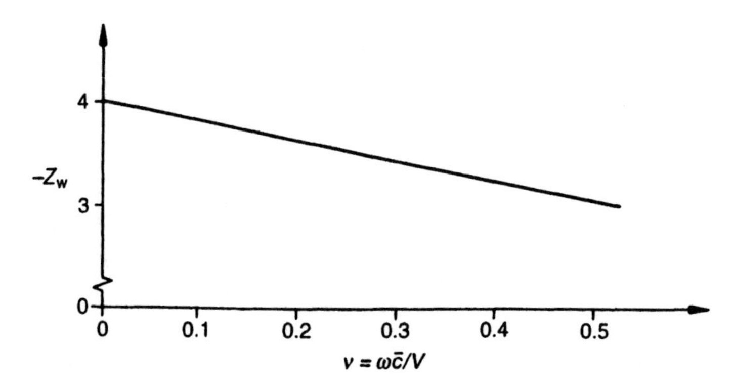

Fig. 9.1 Variation of Z_w with frequency

The result of all the experimentation is that in general the concept of stability derivatives is a valid one which can be relied on well away from incidences at which flow separations may occur and in some cases can be validated at incidences at which flow separations are liable to occur. The expressions for the derivatives which we will derive in the next few sections will all be for the quasi-static approximation.

9.2.1 Derivatives due to change in forward velocity

The derivatives that we are concerned with in this section are

- X_u, the change in forward force due to change in forward velocity;
- Z_u, the change in downward force due to change in forward velocity;
- M_u, the change in pitching moment due to change in forward velocity.

We consider an aircraft initially in a steady straight climb at angle Θ_e and speed V_e, which is given a small increment u in forward velocity as shown in figure 9.2. Since we are using wind axes the x-axis has zero incidence to the flight direction, then $U_e = V_e$ and the forward velocity in the disturbance becomes $U = V_e + u$; see (8.15) and (8.16).

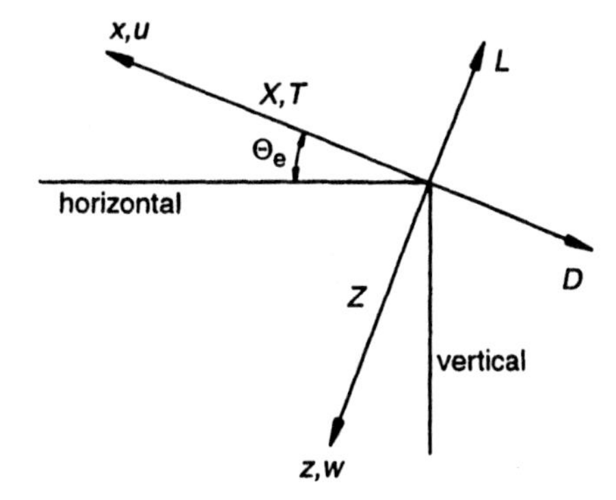

Fig. 9.2 Determination of forward velocity derivatives

Writing down the forces in the disturbance gives

$$\begin{aligned} X_a &= T - D = T - \tfrac{1}{2}\rho U^2 S C_D \\ Z_a &= \quad\; - L = \;\; - \tfrac{1}{2}\rho U^2 S C_L \\ M_a &= \qquad\qquad\quad \tfrac{1}{2}\rho U^2 S\bar{\bar{c}} C_m \end{aligned}$$

where T, C_D, C_L and C_m are functions of U, and it has been assumed that the thrust is along the x-axis. We now differentiate partially with respect to U giving

$$\begin{aligned} \frac{\partial X_a}{\partial U} &= \frac{\partial T}{\partial U} - \rho U S C_D - \tfrac{1}{2}\rho U^2 S \frac{\partial C_D}{\partial U} \\ \frac{\partial Z_a}{\partial U} &= \quad - \rho U S C_L - \tfrac{1}{2}\rho U^2 S \frac{\partial C_L}{\partial U} \\ \frac{\partial M_a}{\partial U} &= \quad - \rho U S \bar{\bar{c}} C_m - \tfrac{1}{2}\rho U^2 S \bar{\bar{c}} \frac{\partial C_m}{\partial U} \end{aligned}$$

We now evaluate these in the initial condition, as required by the definition of a stability derivative, Section 8.4.2.1, by putting $U_e = V_e$, $C_D = C_{De}$, $C_L = C_{Le}$ and $C_m = C_{me} = 0$. This last equality occurs because the aircraft was initially in trim. These now give

$$X_u = \frac{\overset{\circ}{X}_u}{\frac{1}{2}\rho V_e S} = \frac{1}{\frac{1}{2}\rho V_e^2 S}\frac{\partial T}{\partial U}\bigg|_e - 2C_{De} - V_e\frac{\partial C_D}{\partial U}\bigg|_e \tag{9.2}$$

$$Z_u = \frac{\overset{\circ}{Z}_u}{\frac{1}{2}\rho V_e S} = - 2C_{L_e} - V_e\frac{\partial C_L}{\partial U}\bigg|_e \tag{9.3}$$

$$M_u = \frac{\overset{\circ}{M}_u}{\frac{1}{2}\rho V_e S\bar{\bar{c}}} = + V_e\frac{\partial C_m}{\partial U}\bigg|_e \tag{9.4}$$

where $|_e$ indicates evaluation in the datum condition. Note that in the absence of speed effects the derivative M_u is zero. The sources of the last terms in each of these equations are effects such as viscosity and compressibility and are expressed by the Reynolds and Mach numbers. Taking Mach number, the usual case, as an example we can write

$$V_e\frac{\partial}{\partial U}\bigg|_e = \frac{V_e}{c_s}\frac{\partial}{\partial U/c_s}\bigg|_e = \mathrm{M}_e\frac{\partial}{\partial \mathrm{M}}\bigg|_e \tag{9.5}$$

where c_s is the speed of sound. Derivatives with respect to Mach number may be evaluated from empirical data (70011) or from experimental or computational fluid dynamic data.

As an example we take the estimation of the contribution of the wing to the final term of (9.3) for Z_u, using empirical data. The lift curve slope, a, is given as a function of the form

$$\frac{a}{A} = \mathrm{f}(A\beta', A\tan\Lambda, \lambda)$$

where $\beta' = \sqrt{1 - \mathrm{M}^2}$, λ is the taper ratio and A is the aspect ratio. Now $C_L = a\alpha$ and, as the incidence is constant, we find

$$\mathrm{M}_e\frac{\partial C_L}{\partial \mathrm{M}}\bigg|_e = A\mathrm{M}_e\alpha\frac{\partial \mathrm{f}}{\partial \mathrm{M}}\bigg|_e = A\mathrm{M}_e\alpha\left[\frac{\partial \mathrm{f}}{\partial(A\beta')}\cdot\frac{\partial(A\beta')}{\partial \mathrm{M}}\right]_e$$
$$= A^2\mathrm{M}_e\alpha\frac{\partial \mathrm{f}}{\partial(A\beta')}\bigg|_e\cdot\frac{\partial\beta'}{\partial \mathrm{M}}\bigg|_e$$

Now

$$\frac{\partial\beta'}{\partial \mathrm{M}} = -\frac{\mathrm{M}}{\sqrt{1 - \mathrm{M}^2}}$$

hence

$$\mathrm{M}_e\frac{\partial C_L}{\partial \mathrm{M}}\bigg|_e = -\frac{A^2\mathrm{M}_e^2\alpha}{\sqrt{1 - \mathrm{M}_e^2}}\cdot\frac{\partial \mathrm{f}}{\partial(A\beta')}\bigg|_e \tag{9.6}$$

9.2.2 Derivatives due to downward velocity

The derivatives that we are concerned with in this section are

- X_w, the change in forward force due to change in downward velocity;
- Z_w, the change in downward force due to change in downward velocity;
- M_w, the change in pitching moment due to change in downward velocity.

We again consider an aircraft initially in a steady straight climb at angle Θ_e and speed V_e as in figure 9.3(a). The situation after a downward velocity increment is shown in figure 9.3(b). In the initial condition we have

$$X_{a_e} = T - D, Z_{a_e} = -L, M_{a_e} = 0 \tag{9.7}$$

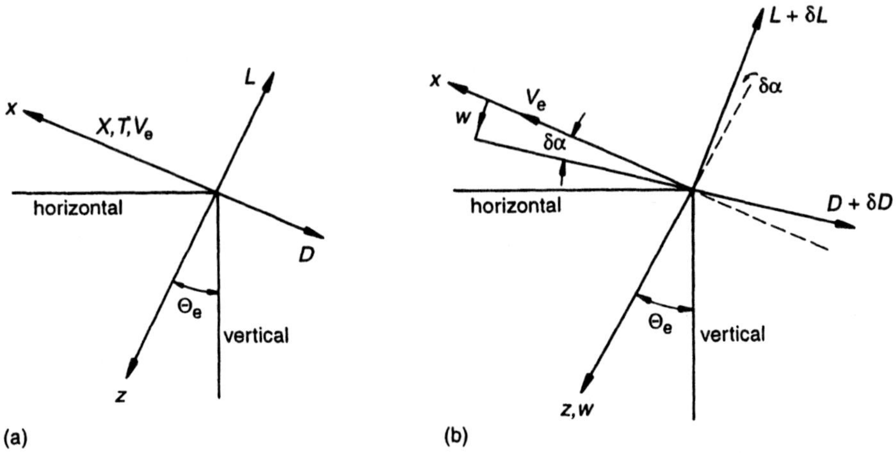

Fig. 9.3 Determination of vertical velocity derivatives: (a) initially, (b) after a downward velocity increment

The effect of the downward velocity is to increase the incidence by

$$\delta\alpha \simeq w/V_e \tag{9.8}$$

while the lift and drag are increased by δL and δD and rotated through $\delta\alpha$. Then resolving we find

$$X_{a_e} + \delta X_a = (L + \delta L)\sin\delta\alpha - (D + \delta D)\cos\delta\alpha + T \tag{9.9}$$

$$Z_{a_e} + \delta Z_a = -(L + \delta L)\cos\delta\alpha - (D + \delta D)\sin\delta\alpha \tag{9.10}$$

where we have again assumed that the thrust is along the x-axis and is independent of incidence. We also write

$$\delta L = \frac{\partial L}{\partial\alpha}\delta\alpha,\ \delta D = \frac{\partial D}{\partial\alpha}\delta\alpha \text{ and } \delta M_a = \frac{\partial M}{\partial\alpha}\delta\alpha$$

We now substitute for δL and δD into (9.9) and (9.10) and linearize by writing $\sin\delta\alpha = \delta\alpha$, $\cos\delta\alpha = 1$, neglecting second-order terms like $\frac{\partial L}{\partial\alpha}(\delta\alpha)^2$. Then

$$X_{a_e} + \delta X_a = L\delta\alpha - D - \frac{\partial D}{\partial \alpha}\delta\alpha + T \tag{9.11}$$

$$Z_{a_e} + \delta Z_a = -L - \frac{\partial L}{\partial \alpha}\delta\alpha - D\delta\alpha \tag{9.12}$$

After substituting for X_{a_e} and Z_{a_e} from (9.7), for $\delta\alpha$ from (9.8), and cancelling, (9.11) and (9.12) become

$$\delta X_a = \left(L - \frac{\partial D}{\partial \alpha}\right)\frac{w}{V_e}$$

$$\delta Z_a = -\left(\frac{\partial L}{\partial \alpha} + D\right)\frac{w}{V_e}$$

also

$$\delta M_a = \frac{\partial M}{\partial \alpha}\cdot\frac{w}{V_e}$$

We differentiate with respect to w and normalize to give

$$X_w = \frac{\overset{\circ}{X}_w}{\frac{1}{2}\rho V_e S} = \frac{1}{\frac{1}{2}\rho V_e^2 S}\left(L - \frac{\partial D}{\partial \alpha}\right) = C_{Le} - \frac{\partial C_{De}}{\partial \alpha} \tag{9.13}$$

$$Z_w = \frac{\overset{\circ}{Z}_w}{\frac{1}{2}\rho V_e S} = \frac{-1}{\frac{1}{2}\rho V_e^2 S}\left(\frac{\partial L}{\partial \alpha} + D\right) = -\frac{\partial C_{Le}}{\partial \alpha} - C_{De} \tag{9.14}$$

$$M_w = \frac{\overset{\circ}{M}_w}{\frac{1}{2}\rho V_e S\bar{\bar{c}}} = \frac{1}{\frac{1}{2}\rho V_e^2 S\bar{\bar{c}}}\frac{\partial M}{\partial \alpha} = \frac{\partial C_{me}}{\partial \alpha} \tag{9.15}$$

Some interpretation of these equations is required in order to obtain usable results. The lift and drag involved are those of the complete aircraft. From (5.17) and (5.18) the tailplane lift coefficient is

$$C_L^T = a_1(\alpha + \eta_T - \varepsilon) + a_2\eta + a_3\tau$$

and the overall lift coefficient using (5.16) is

$$C_L = a\alpha + \frac{S_T}{S}[a_1(\alpha + \eta_T - \varepsilon) + a_2\eta + a_3\tau]$$

Hence we find

$$\frac{\partial C_{\mathrm{Le}}}{\partial \alpha} = \frac{\partial C_{\mathrm{L}}}{\partial \alpha} = a\alpha + \frac{S_{\mathrm{T}}}{S} a_1 \left(1 - \frac{\partial \varepsilon}{\partial \alpha}\right) \tag{9.16}$$

assuming that the elevator and tab are fixed. There is also the likelihood of the propulsion unit(s) producing a small lift force due to incidence, something that in particular propellers do (89047).

If the drag coefficient can be expressed in the form of the parabolic drag law $C_{\mathrm{De}} = a + bC_{\mathrm{Le}}^2$ then

$$\frac{\partial C_{\mathrm{De}}}{\partial \alpha} = 2bC_{\mathrm{Le}} \frac{\partial C_{\mathrm{Le}}}{\partial \alpha} \tag{9.17}$$

where a and b are assumed to be constant.

The derivative $\dfrac{\partial C_{\mathrm{me}}}{\partial \alpha}$ can be written

$$\frac{\partial C_{\mathrm{me}}}{\partial \alpha} = \frac{\partial C_{\mathrm{me}}}{\partial C_{\mathrm{Le}}} \cdot \frac{\partial C_{\mathrm{Le}}}{\partial \alpha} = -\frac{\partial C_{\mathrm{Le}}}{\partial \alpha} \cdot H_{\mathrm{n}} \tag{9.18}$$

where H_{n} is the cg margin, stick fixed, defined in Section 5.4.1.

9.2.3 Derivatives due to angular velocity in pitch

The derivatives that we are concerned with in this section are

- X_{q}, the change in forward force due to angular velocity in pitch;
- Z_{q}, the change in downward force due to angular velocity in pitch;
- M_{q}, the change in pitching moment due to angular velocity in pitch, or 'damping in pitch'.

Unlike the previous six derivatives we have to estimate the contributions from the various components of the aircraft separately, adding together later. We start by estimating the tailplane contribution, which is usually the most significant. We consider an aircraft flying with a small rate of pitch, as shown in figure 9.4.

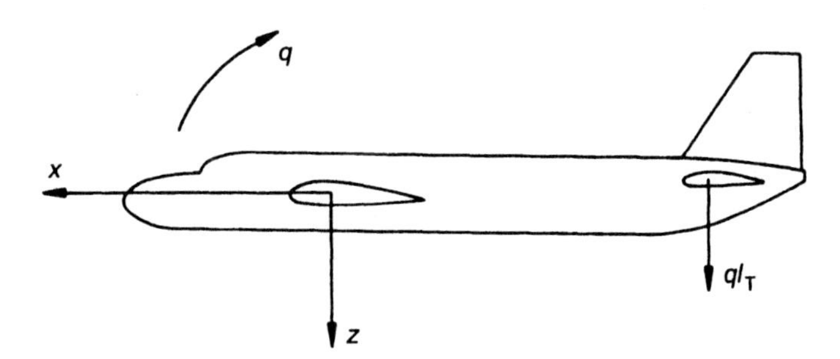

Fig. 9.4 Determination of pitching velocity derivatives

Pitching the aircraft at rate q gives a downward velocity to the tailplane of $w_{\mathrm{T}} = ql_{\mathrm{T}}$ approximately, where l_{T} is the tailplane arm, which is usually a very close approximation to the distance the tailplane aerodynamic centre is behind the cg. This downward velocity has the same effect on forces on the tailplane as downward velocity had on forces on the whole aircraft in the previous section. Using those results we find from (9.13) and (9.14)

$$\delta\overset{\circ}{X}_{\mathrm{T}} = \overset{\circ}{X}{}_{\mathrm{w}}^{\mathrm{T}} \cdot w_{\mathrm{T}} = \tfrac{1}{2}\rho V_{\mathrm{e}} S_{\mathrm{T}}\left(C_{\mathrm{L}}^{\mathrm{T}} - \frac{\partial C_{\mathrm{D}}^{\mathrm{T}}}{\partial \alpha}\right) \cdot q l_{\mathrm{T}} = \delta\overset{\circ}{X}_{\mathrm{a}}$$

$$\delta\overset{\circ}{Z}_{\mathrm{T}} = \overset{\circ}{Z}{}_{\mathrm{w}}^{\mathrm{T}} \cdot w_{\mathrm{T}} = -\tfrac{1}{2}\rho V_{\mathrm{e}} S_{\mathrm{T}}\left(\frac{\partial C_{\mathrm{L}}^{\mathrm{T}}}{\partial \alpha} + C_{\mathrm{D}}^{\mathrm{T}}\right) \cdot q l_{\mathrm{T}} = \delta\overset{\circ}{Z}_{\mathrm{a}}$$

where subscript or superscript 'T' indicates quantities for the tailplane. Differentiating with respect to q and normalizing we find

$$X_{\mathrm{q}} = \frac{\overset{\circ}{X}_{\mathrm{q}}}{\tfrac{1}{2}\rho V_{\mathrm{e}} S\bar{\bar{c}}} = \frac{S l_{\mathrm{T}}}{S\bar{\bar{c}}}\left(C_{\mathrm{L}}^{\mathrm{T}} - \frac{\partial C_{\mathrm{D}}^{\mathrm{T}}}{\partial \alpha}\right) = \bar{V}_{\mathrm{T}}\left(C_{\mathrm{L}}^{\mathrm{T}} - \frac{\partial C_{\mathrm{D}}^{\mathrm{T}}}{\partial \alpha}\right) \tag{9.19}$$

$$Z_{\mathrm{q}} = \frac{\overset{\circ}{Z}_{\mathrm{q}}}{\tfrac{1}{2}\rho V_{\mathrm{e}} S\bar{\bar{c}}} = -\frac{S l_{\mathrm{T}}}{S\bar{\bar{c}}}\left(\frac{\partial C_{\mathrm{L}}^{\mathrm{T}}}{\partial \alpha} - C_{\mathrm{D}}^{\mathrm{T}}\right) \simeq -\bar{V}_{\mathrm{T}} a_1 \tag{9.20}$$

where it has been assumed that the tailplane lift curve slope is much larger than the drag coefficient. The first of these derivatives is usually neglected.

The downward force $\delta\overset{\circ}{Z}_{\mathrm{T}}$ on the tailplane gives a pitching moment about the cg of

$$\delta\overset{\circ}{M} = \delta\overset{\circ}{Z}_{\mathrm{T}} \cdot l_{\mathrm{T}} = -\tfrac{1}{2}\rho V_{\mathrm{e}} S_{\mathrm{T}}\left(\frac{\partial C_{\mathrm{L}}^{\mathrm{T}}}{\partial \alpha} + C_{\mathrm{D}}^{\mathrm{T}}\right) \cdot q l_{\mathrm{T}}^2$$

giving

$$M_{\mathrm{q}} = \frac{\overset{\circ}{M}_{\mathrm{q}}}{\tfrac{1}{2}\rho V_{\mathrm{e}} S\bar{\bar{c}}^2} = -\frac{S l_{\mathrm{T}}^2}{S\bar{\bar{c}}^2}\left(\frac{\partial C_{\mathrm{L}}^{\mathrm{T}}}{\partial \alpha} + C_{\mathrm{D}}^{\mathrm{T}}\right) \simeq -\bar{V}_{\mathrm{T}} \frac{l_{\mathrm{T}}}{\bar{\bar{c}}} a_1 \tag{9.21}$$

If the wing is highly swept or of low aspect ratio, or if the aircraft is tailless, then the derivatives due to the wing are appreciable and have to be estimated. On a wing with high aspect ratio and sweep the tips of the wing will have a significant downward velocity due to nose-up pitching velocity, in a similar manner to that of a tailplane. Additional lift is generated by the tips giving a damping moment. In the case of a wing of low aspect ratio, parts of the wing well ahead of the cg have a significant upward velocity due to nose-up pitching velocity and parts well behind a significant downward one. The effect is similar to camber on the wing so that the no lift line is rotated giving a lift and a pitching moment. These effects are best estimated using a lifting surface theory or from semi-empirical data (90010).

9.2.4 Derivatives due to vertical acceleration

The derivatives that we are concerned with in this section are

- $Z_{\dot{w}}$, the change in downward force due to vertical acceleration;
- $M_{\dot{w}}$, the change in pitching moment due to vertical acceleration.

The derivative $X_{\dot{w}}$ is usually neglected; the others are also known as the 'downwash lag' derivatives. There is no simple satisfactory theory for these derivatives. It is normal to estimate these derivatives on the assumption that the downwash at the tailplane corresponds to the wing incidence at an instant earlier by the time taken for the aircraft to fly the distance between the wing and the tailplane. We take the time difference to be l_T/V_e and hence the relevant wing incidence to be

$$\alpha - \frac{l_T}{V_e} \cdot \frac{d\alpha}{dt} \tag{9.22}$$

where α is here the incidence at the current instant. From (5.18) the tailplane incidence is given in general by

$$\alpha^T = \alpha - \varepsilon + \eta_T$$

and hence at the current instant it is, using (9.22) and $\varepsilon = \alpha.(d\varepsilon/d\alpha)$

$$\alpha^T = \alpha - \left(\alpha - \frac{l_T}{V_e} \cdot \frac{d\alpha}{dt}\right)\frac{d\varepsilon}{d\alpha} + \eta_T = \alpha\left(1 - \frac{d\varepsilon}{d\alpha}\right) + \frac{l_T}{V_e} \cdot \frac{d\alpha}{dt} \cdot \frac{d\varepsilon}{d\alpha} + \eta_T$$

From (9.8) we have on differentiating with respect to time

$$\frac{d(\delta\alpha)}{dt} = \frac{d\alpha}{dt} = \frac{\dot{w}}{V_e}$$

then the downward force on the tailplane and hence on the aircraft is

$$\mathring{Z}_a = -\tfrac{1}{2}\rho V_e^2 S_T a_1 \alpha^T = -\tfrac{1}{2}\rho V_e^2 S_T a_1 \left[\alpha\left(1 - \frac{d\varepsilon}{d\alpha}\right) + \frac{l_T}{V_e} \cdot \frac{d\varepsilon}{d\alpha} \cdot \frac{\dot{w}}{V_e} + \eta_T\right]$$

Differentiating partially with respect to $\dot{w}$ and normalizing we find

$$Z_{\dot{w}} = \frac{\mathring{Z}_{\dot{w}}}{\frac{1}{2}\rho S\bar{\bar{c}}} = -\bar{V}_T a_1 \frac{d\varepsilon}{d\alpha} \tag{9.23}$$

As in the previous section $\mathring{Z}_a$ gives a pitching moment $\mathring{M}_a = \mathring{Z}_a . l_T$, and hence the derivative

$$M_{\dot{w}} = \frac{\mathring{M}_{\dot{w}}}{\frac{1}{2}\rho S\bar{\bar{c}}^2} = -\bar{V}_T \frac{l_T}{\bar{\bar{c}}} a_1 \frac{d\varepsilon}{d\alpha} \tag{9.24}$$

Values for the derivative $M_{\dot{w}}$ using a computational model which replaced the wing and its vortex wake by a system of discrete vortices have been obtained by Hancock and Lam in reference (9.1). They investigated two particular configurations and found that the tailplane contribution, as given above, was 65–75 per cent of their more accurate value. In addition they found a wing contribution of 20–30 per cent of the estimate of the tailplane contribution given above. Overall the theory above underestimated the derivative by about 40 per cent.

9.2.5 Derivatives due to elevator angle

The derivatives that we are concerned with in this section are

- Z_η, the change in downward force due to elevator angle;
- M_η, the change in pitching moment due to elevator angle.

The derivative X_η is usually negligible. A change of elevator angle η' gives a downward force of

$$\delta \mathring{Z}_a = -\tfrac{1}{2}\rho V_e^2 S_T a_2 \eta'$$

and hence the derivative

$$Z_\eta = \frac{\mathring{Z}_\eta}{\tfrac{1}{2}\rho V_e^2 S} = -\frac{S_T}{S} a_2 \tag{9.25}$$

and the pitching moment derivative is

$$M_\eta = \frac{\mathring{M}_\eta}{\tfrac{1}{2}\rho V_e^2 S\bar{\bar{c}}} = -\bar{V}_T a_2 \tag{9.26}$$

where it has been assumed that the moment arm about the cg of the lift produced by elevator deflection can be also approximated by l_T.

9.2.6 Derivatives relative to other axes

If derivatives are known referred to one set of axes but are required in another set that are rotated or translated from the first, then conversion formulae must be used (86041).

9.2.7 Conversion of derivatives to concise forms

Table 9.1 summarizes the required conversions and is based on Table 8.4.

Table 9.1 Conversion of derivatives to concise forms

X-force	*Z-force*	*Pitching moment*
$\hat{x}_u = -X_u$	$\hat{z}_u = -Z_u$	$\hat{m}_u = -\mu_1 M_u / i_y$
$\hat{x}_w = -X_w$	$\hat{z}_w = -Z_w$	$\hat{m}_w = -\mu_1 M_w / i_y$
$\hat{x}_{\dot{w}} = -X_{\dot{w}} / \mu_1$	$\hat{z}_{\dot{w}} = -Z_{\dot{w}} / \mu_1$	$\hat{m}_{\dot{w}} = -M_{\dot{w}} / i_y$
$\hat{x}_q = -X_q / \mu_1$	$\hat{z}_q = -Z_q / \mu_1$	$\hat{m}_q = -M_q / i_y$
$\hat{x}_\eta = -X_\eta$	$\hat{z}_\eta = -Z_\eta$	$\hat{m}_\eta = -\mu_1 M_\eta / i_y$

9.2.8 Conversions to derivatives in American notation

The conversions between these derivatives and those expressed in the American notation have been shown by Babister, reference (9.2), to be as follows.

The stability derivatives C_{m_u}, $C_{x\alpha}$, $C_{z\alpha}$ and $C_{m\alpha}$ in the American notation are equal to their counterparts M_u, X_w, Z_w and M_w in the current notation.

The stability derivatives C_{x_q}, C_{z_q}, C_{m_q}, $C_{x_{\dot{\alpha}}}$, $C_{z_{\dot{\alpha}}}$ and $C_{m_{\dot{\alpha}}}$ in the American notation are equal to twice their counterparts X_q, Z_q, M_q, $X_{\dot{w}}$, $Z_{\dot{w}}$ and $M_{\dot{w}}$ in the current notation.
Two derivatives have more complicated relations, namely

$$\left.\begin{aligned} C_{x_u} &= X_u - 2C_x \\ C_{z_u} &= Z_u - 2C_z \end{aligned}\right\} \tag{9.27}$$

The control derivatives C_{x_η}, C_{z_η} and C_{m_η} are equal to their counterparts X_η, Z_η and M_η.

In these equations $C_x = X/\frac{1}{2}\rho V_e^2 S$ and $C_z = Z/\frac{1}{2}\rho V_e^2 S$, where X and Z are the components of aerodynamic force, including the thrust, along the Ox and Oz wind axes.

9.3 Solution of the longitudinal equations

The concise dynamic-normalized longitudinal equations are (8.51)–(8.54); since we are using wind axes we put $W_e = \widehat{W}_e = 0$ and $U_e = V_e$, so that $\widehat{U}_e = 1$. The equations are then

$$(\widehat{D} + \widehat{x}_u)\widehat{u} \quad + x_w\widehat{w} \quad + x_q\widehat{q} \quad + \widehat{g}_1\theta = -\widehat{x}(\widehat{t}) \tag{9.28}$$

$$z_u\widehat{u} \quad + \left[(1 + z_{\dot{w}})\widehat{D} + z_w\right]\widehat{w} \quad + (z_q - 1)\widehat{q} \quad + \widehat{g}_2\theta = -\widehat{z}(\widehat{t}) \tag{9.29}$$

$$m_u\widehat{u} \quad + (m_{\dot{w}}\widehat{D} + m_w)\widehat{w} + (\widehat{D} + m_q)\widehat{q} \quad = -\widehat{m}(\widehat{t}) \tag{9.30}$$

$$\widehat{q} - \widehat{D}\theta = 0 \tag{9.31}$$

The reader is reminded of the convention used in writing these equations that only the first stability derivatives carries the superscript $\widehat{\ }$, and that these equations can be converted to ordinary concise equations by change of the superscript; see Section 8.6. In these equations the control terms are, from (8.55),

$$\widehat{x}(\widehat{t}) = \widehat{x}_\eta\eta'(\widehat{t}),\ \widehat{z}(\widehat{t}) = z_\eta\eta'(\widehat{t}) \text{ and } \widehat{m}(\widehat{t}) = m_\eta\eta'(\widehat{t}) \tag{9.32}$$

The weight component terms g_1 and g_2 are given by (8.22) and in their normalized form $\widehat{g}_1$ and $\widehat{g}_2$ are obtained from table 8.2. Since we are using wind axes Θ_e is the climb angle and we find

$$\widehat{g}_1 = \frac{mg\cos\Theta_e}{\frac{1}{2}\rho V_e^2 S} = C_{Le} \tag{9.33}$$

and

$$\widehat{g}_2 = \frac{mg\sin\Theta_e}{\frac{1}{2}\rho V_e^2 S} = C_{Le}\tan\Theta_e \tag{9.34}$$

A schematic representation of these equations is shown in figure 9.5 which indicates the various interactions between the three freedoms. In this figure we have omitted the small derivatives $\widehat{x}_q$, $\widehat{z}_q$, $\widehat{z}_{\dot{w}}$ and $\widehat{x}_\eta$. The derivative $\widehat{m}_u$ has been labelled 'compressibility' because it is

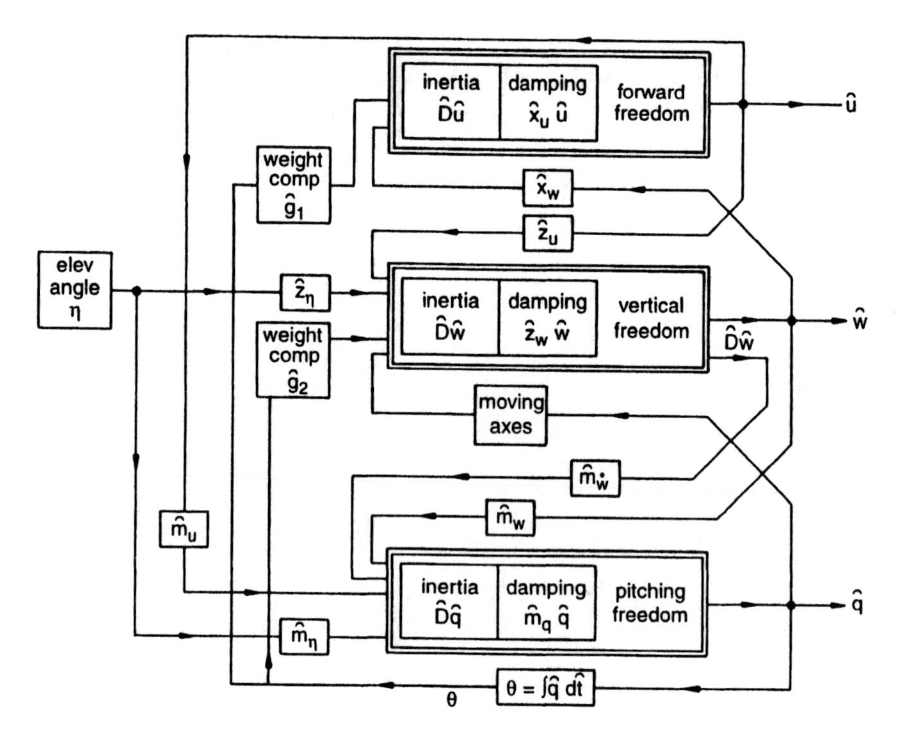

Fig. 9.5 Schematic of longitudinal stability equations

normally zero at low Mach number; however, other derivatives may also have contributions due to compressibility. It should be noted that there are no direct spring terms in the three freedoms.

There are a number of useful quantities we can find using these equations depending on the values given to the right-hand sides. We can

- make them zero and hence study the free motion, i.e. the 'homogeneous' case, giving information on the stability;
- give them constant values to study the response resulting from a disturbance;
- insert expressions for the elevator angle and find the response to pilot actions or study automatic control problems;
- insert expressions to express the effects of encountering a gust or flying through a distribution of random gusts and so examine the ride quality.

The first two topics are the subject of the rest of this chapter whilst the other two are left to Chapters 10 and 13.

9.3.1 Solution of the equations of free motion

To find the solution of the equations for the case of free motion with the controls fixed we put the right-hand sides of the equations to zero and assume a solution of the form

$$\left.\begin{aligned} \hat{u} &= k_1 e^{\lambda \hat{t}} \\ \hat{w} &= k_2 e^{\lambda \hat{t}} \\ \theta &= k_3 e^{\lambda \hat{t}} \end{aligned}\right\} \tag{9.35}$$

where k_1, k_2, k_3 and λ are real or complex quantities. We now substitute for $\hat{q}$ in (9.28) to (9.30) from (9.31), and substitute the assumed solutions (9.35). Each term then contains a factor exp $(\lambda\hat{t})$ which can be cancelled out, leaving the equations

$$\left.\begin{aligned}(\lambda + \hat{x}_u)k_1 \quad & + x_w k_2 \quad && + (x_q\lambda + \hat{g}_1)k_3 = 0\\ z_u k_1 \quad & + \left[(1 + z_{\dot{w}})\lambda + z_w\right]k_2 && + \left[(z_q - 1)\lambda + \hat{g}_2\right]k_3 = 0\\ m_u k_1 \quad & + (m_{\dot{w}}\lambda + m_w)k_2 && + (\lambda^2 + m_q\lambda)k_3 = 0\end{aligned}\right\} \tag{9.36}$$

Apart from the trivial solution $k_1 = k_2 = k_3 = 0$, these equations are incompatible unless their determinant vanishes, since there are four unknowns and only three equations. Forming the determinant and equating it to zero gives

$$F(\lambda) = \begin{vmatrix}\lambda + \hat{x}_u & x_w & x_q\lambda + \hat{g}_1\\ z_u & (1 + z_{\dot{w}})\lambda + z_w & (z_q - 1)\lambda + \hat{g}_2\\ m_u & m_{\dot{w}}\lambda + m_w & (\lambda^2 + m_q\lambda)\end{vmatrix} = 0 \tag{9.37}$$

Expanding out the determinant gives a quartic in λ which we write

$$F(\lambda) = A_1\lambda^4 + B_1\lambda^3 + C_1\lambda^2 + D_1\lambda + E_1 = 0 \tag{9.38}$$

The coefficients A_1, B_1, C_1, D_1 and E_1 are functions of the derivatives and $\hat{g}_1$ and $\hat{g}_2$ as follows:

$$A_1 = 1 + \hat{z}_{\dot{w}} \tag{9.39}$$

$$B_1 = z_w + (x_u + m_q)(1 + z_{\dot{w}}) + (1 - z_q)m_{\dot{w}} \tag{9.40}$$

$$\begin{aligned}C_1 = {} & x_u z_w - x_w z_u + \left[x_u(1 + z_{\dot{w}}) + z_w\right]m_q + \left[x_u(1 - z_q) + x_q z_u - \hat{g}_2\right]m_{\dot{w}}\\ & + (1 - z_q)m_w - x_q m_u(1 + z_{\dot{w}})\end{aligned} \tag{9.41}$$

$$\begin{aligned}D_1 = {} & \left[x_u z_w - x_w z_u\right]m_q + (\hat{g}_1 z_u - \hat{g}_2 x_u)m_{\dot{w}} + \left[x_u(1 - z_q) + x_q z_u - \hat{g}_2\right]m_w\\ & - \left[x_w(1 - z_q) + x_q z_w + \hat{g}_1(1 + z_{\dot{w}})\right]m_u\end{aligned} \tag{9.42}$$

$$E_1 = (\hat{g}_1 z_u - \hat{g}_2 x_u)m_w - (\hat{g}_1 z_w - \hat{g}_2 x_w)m_u \tag{9.43}$$

It is common practice to divide through by the coefficient A_1 if $\hat{z}_{\dot{w}}$ has not been taken to be zero.

The polynomial such as (9.38) derived in this way from the equations describing a system is known as the 'characteristic' equation and its roots, in this case four in number, are the 'eigenvalues'. The eigenvalues once found can be substituted in turn back into (9.34) to obtain values for ratios of the k's. Since the right-hand sides of (9.34) are zeros it is not possible to find k_1, k_2 and k_3 absolutely, only the ratios k_2/k_1 and k_3/k_1, for instance. These are the 'eigenvectors'. If one of the k's for a given eigenvalue is given an arbitrary value then a set of values can be found and plotted on an Argand diagram to give a 'shape of the mode'.

If we call the roots of (9.38) λ_1, λ_2, λ_3 and λ_4, then the general solution is

$$\left.\begin{aligned}\hat{u} &= k_{11}e^{\lambda_1\hat{t}} + k_{12}e^{\lambda_2\hat{t}} + k_{13}e^{\lambda_3\hat{t}} + k_{14}e^{\lambda_4\hat{t}}\\ \hat{w} &= k_{21}e^{\lambda_1\hat{t}} + k_{22}e^{\lambda_2\hat{t}} + k_{23}e^{\lambda_3\hat{t}} + k_{24}e^{\lambda_4\hat{t}}\\ \theta &= k_{31}e^{\lambda_1\hat{t}} + k_{32}e^{\lambda_2\hat{t}} + k_{33}e^{\lambda_3\hat{t}} + k_{34}e^{\lambda_4\hat{t}}\end{aligned}\right\} \tag{9.44}$$

If we are given sufficient initial information on $\hat{u}$, $\hat{w}$ and θ we can determine these constants. Notice that the ratios k_{1i}:k_{2i}:k_{3i} will equal the ratios of the eigenvector components k_1:k_2:k_3 determined for the eigenvalue λ_i. In the case of a complex eigenvalue $\mu \pm i\omega$, two of the terms in each of (9.44) are replaced by terms of the form

$$k_{j1}e^{\mu\hat{t}}\cos\omega\hat{t} + k_{j2}e^{\mu\hat{t}}\sin\omega\hat{t},\ j = 1,\ 2 \text{ and } 3$$

9.3.2 Stability of the motion

The stability of the aircraft is determined solely by the eigenvalues as can be seen from (9.44). We can identify three cases.

1. **λ is real and positive.** This represents a divergence, which doubles its initial amplitude exponentially in a normalized time (see Section 7.2.1 and (7.5)) of

$$\hat{t}_D = \frac{\ln 2}{\lambda} \text{ or, in ordinary time, } t_D = \frac{0.693\tau}{\lambda} \text{ seconds} \tag{9.45}$$

   Figure 9.6(a) illustrates the variation of the disturbance with time, which is obviously unstable.

2. **λ is real and negative.** This represents a convergence, which halves its initial amplitude exponentially in a normalized time of

$$\hat{t}_H = \frac{\ln 2}{(-\lambda)} \text{ or, in ordinary time, } t_H = \frac{0.693\tau}{(-\lambda)} \text{ seconds} \tag{9.46}$$

   Figure 9.6(b) illustrates the variation of the disturbance with time, which is stable as the aircraft returns to a steady state.

3. **λ is complex.** In this case we write $\lambda = \mu + i\omega$. This represents an oscillation, divergent or convergent, depending whether $\mu > 0$ or $\mu < 0$. The time to double or halve the initial amplitude can be found from (9.45) or (9.46) respectively, replacing λ by μ. The normalized period time is

$$\hat{t}_p = \frac{2\pi}{\omega} \text{ or, in ordinary time, } t_p = \frac{2\pi\tau}{\omega} \text{ seconds} \tag{9.47}$$

   Figure 9.6(c) illustrates the motion for $\mu > 0$ and figure 9.6(d) for $\mu < 0$; in this latter case the aircraft is regarded as stable. The case $\mu = 0$, giving a steady oscillation, would be a quite undesirable characteristic in an aircraft, unless the frequency was very low.

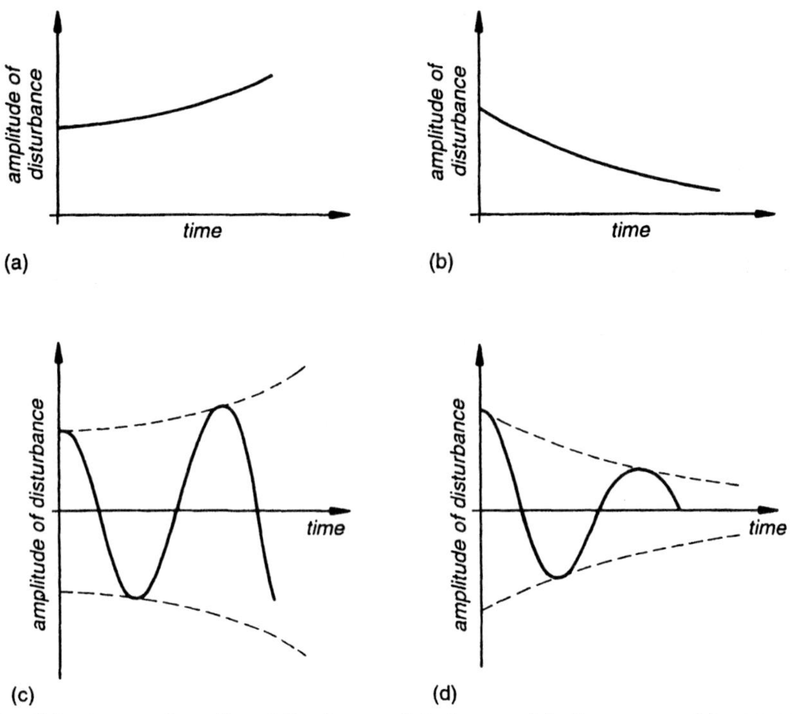

Fig. 9.6 Possible forms of motion following a disturbance: (a) divergence, (b) convergence, (c) divergent oscillation, (d) convergent oscillation

We summarize the requirements on λ for an aircraft to be stable as:

- λ to be negative, if real;
- λ to have a negative real part, if complex.

There are many other parameters that could be found to characterize the damping, such as ζ (see (7.32)), but the time to halve (or double) the initial amplitude has the merit that it can be compared to the pilot's reaction time. Another method of expressing the damping is by the 'characteristic time', which is defined as the time for a disturbance to fall to 1/e of its initial value. This can be written as

$$\hat{t}_c = \tau/(-\lambda)$$

9.3.3 Test functions

Before we solve the characteristic equation (9.38) we ask if the stability of the aircraft can be determined by a simple test applied directly to coefficients A_1, B_1, C_1, D_1 and E_1. The possibility is illustrated by considering the quadratic equation

$$\lambda^2 + a\lambda + b = 0$$

where a and b are real. The solution is

$$\lambda = \frac{-a \pm \sqrt{a^2 - 4b}}{2}$$

Consider the various cases:

1. If $a > 0$ and $b > 0$ and if $a^2 > 4b$ there are two negative real roots and the system is stable, or if $a^2 < 4b$ there is a complex pair giving a stable oscillation.
2. If $a < 0$ and $b > 0$ and if $a^2 > 4b$, then one root is

$$\lambda = \frac{+|a| + \sqrt{a^2 - 4b}}{2} > 0$$

so that this combination is not allowable. If $a^2 < 4b$ there is a complex pair with a positive real part, giving an unstable oscillation.
3. If $a > 0$ and $b < 0$ then $a^2 > 4b$, both roots are real and one is

$$\lambda = \frac{-a + \sqrt{a^2 + 4|b|}}{2} > 0$$

so that this combination is not allowable.
4. If $a < 0$ and $b < 0$ then $a^2 > 4b$; both roots are real and one is positive.

The overall result is that for a stable second-order system we require both a and b to be positive. The general condition for stability is stated for the characteristic polynomial $F(\lambda)$ of order n:

$$F(\lambda) = p_n\lambda^n + p_{n-1}\lambda^{n-1} + \ldots + p_1\lambda + p_0 = 0 \tag{9.48}$$

where $p_n > 0$. A necessary and sufficient condition for stability is that the test functions below are all positive:

$$T_1 = p_{n-1}$$

$$T_2 = \begin{vmatrix} p_{n-1} & p_n \\ p_{n-3} & p_{n-2} \end{vmatrix} \tag{9.49}$$

$$T_3 = \begin{vmatrix} p_{n-1} & p_n & 0 \\ p_{n-3} & p_{n-2} & p_{n-1} \\ p_{n-5} & p_{n-4} & p_{n-3} \end{vmatrix} \tag{9.50}$$

and so on, up to T_n. It can also be shown that, if a system is unstable, the number of roots with positive real parts is given by the number of sign changes in the sequence

$$T_1,\ T_j/T_{j-1},\ \text{for } 2 \le j \le n \tag{9.51}$$

In the case of a quadratic the conditions become

$$T_1 = p_{n-1} = a > 0$$

and

$$T_2 = p_1 p_0 = ab > 0, \text{ so } b > 0$$

agreeing with the result obtained above.

For a quartic the conditions become

$$T_1 = p_3 > 0 \tag{9.52}$$

$$T_2 = p_3 p_2 - p_4 p_1 > 0 \tag{9.53}$$

$$T_3 = p_3 p_2 p_1 - p_4 p_1^2 - p_3^2 p_0 = T_2 p_1 - p_3^2 p_0 > 0 \tag{9.54}$$

$$T_4 = T_3 p_0 > 0 \tag{9.55}$$

From (9.52) we require $p_3 > 0$, from (9.54) $T_3 > 0$ and from (9.55) $p_0 > 0$. Choosing $p_1 > 0$ then implies from (9.54) that $T_2 > 0$. Since p_1, p_2, $p_4 > 0$ then (9.53) shows that $p_3 > 0$ also. The four conditions required to ensure the stability of a fourth-order system are then, in the notation used in (9.38),

$$B_1 > 0,\ D_1 > 0,\ E_1 > 0 \text{ and } R = B_1 C_1 D_1 - A_1 D_1^2 - B_1^2 E_1 > 0 \tag{9.56}$$

In these conditions $A_1 > 0$, and if satisfied they imply that $C_1 > 0$ also. The quantity R is known as 'Routh's discriminant' and we will show that it has an important property.

Suppose that the characteristic equation (9.38) has a pair of equal and opposite roots of the form $\lambda = \pm\beta$, where β may be real or imaginary. Substituting these roots into the characteristic equation in turn gives

$$A_1\beta^4 + B_1\beta^3 + C_1\beta^2 + D_1\beta + E_1 = 0$$

and

$$A_1\beta^4 - B_1\beta^3 + C_1\beta^2 - D_1\beta + E_1 = 0$$

Adding and subtracting these successively gives

$$A_1\beta^4 + C_1\beta^2 + E_1 = 0 \tag{9.57}$$

and

$$B_1\beta^3 + D_1\beta = 0$$

or

$$B_1\beta^2 + D_1 = 0 \tag{9.58}$$

as $\beta \neq 0$ in general. From (9.58) $D_1 = -B_1\beta^2$, then substituting into R from (9.56) gives

$$R = -B_1^2(A_1\beta^4 + C_1\beta^2 + E_1)$$

hence $R = 0$ from (9.57). When β is real the roots represent a divergence and a convergence with equal time constants. If it is imaginary then there is an oscillation of constant amplitude. Considering (9.58) again, if B_1 and D_1 are of opposite signs then β is purely real, whilst if they are of the same sign β is purely imaginary. In cases of aeronautical interest they are almost always both positive, so that the vanishing of R indicates an oscillation with neutral stability.

9.3.4 Iterative solution of the characteristic quartic

In Sections 7.2.2 and 7.2.3 we discussed the phugoid and short period pitching oscillations; the latter we shall refer to as the SPPO from now on. The phugoid we found to be of long period and weakly damped, whilst the SPPO is of short period and heavily damped. We therefore expect the eigenvalues of the characteristic equation to consist of a complex pair of small modulus corresponding to the phugoid, and a complex pair of large modulus corresponding to the SPPO. These facts lead us to an iterative method of solution of the quartic. There are of course other types of methods of solution which can be used, but iterative methods are preferred for their rapidity and control over accuracy. The existence of this normal pattern of roots leads to a particular pattern for the coefficients, which is that the coefficients B_1 and C_1 are much larger than D_1 and E_1. We assume that the quartic can be factorized into two quadratics, so that taking $A_1 = 1$, we write

$$\begin{aligned} F(\lambda) &= \lambda^4 + B_1\lambda^3 + C_1\lambda^2 + D_1\lambda + E_1 \\ &= (\lambda^2 + a_1\lambda + b_1).(\lambda^2 + a_2\lambda + b_2) = 0 \end{aligned}$$

Then multiplying out and equating coefficients of like powers we have

$$B_1 = a_1 + a_2 \tag{9.59}$$

$$C_1 = a_1a_2 + b_1 + b_2 \tag{9.60}$$

$$D_1 = a_1b_2 + a_2b_1 \tag{9.61}$$

$$E_1 = b_1b_2 \tag{9.62}$$

Following the argument above we assume that $a_1 \gg a_2$ and $b_1 \gg a_1a_2 + b_2$. Then denoting approximate values by a prime:

$$\left.\begin{aligned} &\text{from (9.59)}: \ a_1' = B_1 \\ &\text{from (9.60)}: \ b_1' = C_1 \\ &\text{from (9.62)}: \ b_2' = E_1 / b_1 = E_1 / C_1 \\ &\text{from (9.61)}: \ a_2' = \frac{D_1 - a_1b_2}{b_1} = \frac{D_1 - B_1E_1/C_1}{C_1} = \frac{C_1D_1 - B_1E_1}{C_1^2} \end{aligned}\right\} \tag{9.63}$$

Then the approximate factorization is

$$F(\lambda) \simeq (\lambda^2 + B_1\lambda + C_1)\cdot\left(\lambda^2 + \frac{C_1D_1 - B_1E_1}{C_1^2}\lambda + \frac{E_1}{C_1}\right) \tag{9.64}$$

Of course it is only valid to draw conclusions from this relation if the frequencies of the modes are well separated, so that the modes are only loosely coupled. The larger coefficients B_1 and C_1 appear in the first factor which therefore applies to the SPPO, and the second factor applies to the phugoid mode. This enables us to see which coefficients have the major influence on each mode. In particular the conditions $C_1 > 0$ and $E_1 > 0$ from (9.56) refer to the SPPO and phugoid modes respectively.

We can improve the values of the coefficients of the quadratics using the values we have as first approximations to find second approximations, denoted by two primes:

$$\left.\begin{aligned} a_1'' &= B_1 - a_2' \\ b_1'' &= C_1 - a_1''a_2' - b_2' \\ b_2'' &= E_1 / b_1'' \\ a_2'' &= \frac{D_1 - a_1'' b_2''}{b_1''} \end{aligned}\right\} \tag{9.65}$$

To achieve slightly quicker convergence second approximations are used wherever possible. The process can be repeated until the required accuracy is achieved. A spreadsheet setup for solving longitudinal quartics is given in the appendix at the end of this chapter.

Worked example 9.1

The non-dimensional equations of disturbed longitudinal free motion of a certain aircraft are

$$(\hat{D} + 0.085)\hat{u} - 0.088\hat{w} + 0.16\theta = 0$$

$$0.32\hat{u} + (1.018\hat{D} + 2.42)\hat{w} - 0.96\hat{D}\theta = 0$$

$$1.14\hat{u} + (0.81\hat{D} + 29.7)\hat{w} + (\hat{D} + 3.58)\hat{D}\theta = 0$$

and lead to the quartic

$$\lambda^4 + 6.807\lambda^3 + 37.147\lambda^2 + 3.16\lambda + 1.06 = 0$$

Solve the quartic and find the ratio $\hat{w}/\theta$ (i.e. its eigenvector) for the mode with the shorter period.

Solution

From (9.63) we find the first approximations as follows:

$$\begin{aligned} a_1' &= B_1 = 6.807 \\ b_1' &= C_1 = 37.147 \\ b_2' &= E_1 / C_1 = 1.06/37.147 = 0.0285 \\ a_2' &= \frac{C_1D_1 - B_1E_1}{C_1^2} = \frac{37.147 \times 3.16 - 6.807 \times 1.06}{37.147^2} = 0.0798 \end{aligned}$$

The second approximations are then calculated from (9.65) as follows:

$$\begin{aligned} a_1'' &= B_1 - a_2' = 6.807 - 0.0798 = 6.7272 \\ b_1'' &= C_1 - a_1'' a_2' - b_2' = 37.147 - 6.7272 \times 0.0798 - 0.0285 = 36.588 \\ b_2'' &= E_1 / b_1'' = 1.06/36.588 = 0.028\,98 \\ a_2'' &= \frac{D_1 - a_1'' b_2''}{b_1''} = \frac{3.16 - 6.7272 \times 0.028\,98}{36.588} = 0.081\,06 \end{aligned}$$

Further iterations give the results $a_1 = 6.7259$, $b_1 = 36.57$, $a_2 = 0.080\,95$ and $b_2 = 0.028\,94$, or

$$(\lambda^2 + 6.7259\lambda + 36.57)(\lambda^2 + 0.080\,95\lambda + 0.028\,94) = 0$$

Then solving the quadratics the roots are $\lambda = -3.3629 \pm i5.0263$ and $\lambda = -0.0405 \pm i0.1654$.
We require the eigenvector for $\hat{w}/\theta$ for $\lambda = -3.3629 + i5.0263$, taking only the upper sign; taking the lower sign would result in the complex conjugate. Then we make the substitutions (9.35) to get equations of the form (9.36), that is

$$\left.\begin{aligned} (\lambda + 0.32)k_1 - 0.088k_2 + 0.16k_3 &= 0 \\ 0.32k_1 + (1.018\lambda + 2.42)k_2 - 0.96k_3 &= 0 \\ 1.14k_1 + (0.81\lambda + 29.7)k_2 + (\lambda^2 + 3.58\lambda)k_3 &= 0 \end{aligned}\right\}$$

We need to solve for k_2/k_3, so taking the last two equations, dividing through by k_3 and substituting for λ and taking the third terms onto the right, we have

$$\begin{aligned} 0.32k_1/k_3 + (-1.003 + i5.117)k_2/k_3 &= -3.2317 + i4.803 \\ 1.14k_1/k_3 + (26.976 + i4.071)k_2/k_3 &= 25.9936 + i15.812 \end{aligned}$$

We then multiply the first equation by $-1.14/0.32$ to give

$$-1.14k_1/k_3 + (3.573 - i18.224)k_2/k_3 = 11.513 - i17.208$$

Adding this to the second gives

$$(30.549 - i14.158)k_2/k_3 = 37.5066 - i1.396$$

and solving, $k_2/k_3 = 1.0282 + i0.4308$.

9.3.5 Relation between the coefficient E_1 and the static stability

In level flight we have $\hat{g}_2 = 0$ from (9.34) and if we neglect compressibility effects then $\hat{m}_u = 0$; see (9.4) and (9.5). Then from (9.43) we can write

$$E_1 = \hat{g}_1\hat{z}_u m_w \tag{9.66}$$

From (9.33) we have $\hat{g}_1 = C_{Le}$ and from table 9.1, $\hat{z}_u = -Z_u$, then using (9.3) we find $\hat{z}_u = 2C_{Le}$.
Again from table 9.1 $\hat{m}_w = -\dfrac{\mu_1 M_w}{i_y}$, and using (9.15) and (9.18) we find

$$\hat{m}_w = \frac{\mu_1}{i_y}\frac{\partial C_{Le}}{\partial \alpha} \cdot H_n \tag{9.67}$$

Substituting these results into (9.66) gives

$$E_1 = \frac{2C_{Le}^2\mu_1}{i_y}\frac{\partial C_{Le}}{\partial \alpha} \cdot H_n \tag{9.68}$$

The result is that $E_1 \propto H_n$, and so the stability criterion $E_1 > 0$ from (9.56) is equivalent to the criterion $H_n > 0$ from (5.41). From the approximate factorization (9.64) we also see that $E_1 > 0$ is one condition for a stable phugoid, assuming $C_1 > 0$.

When the cg margin is zero the phugoid factor becomes

$$\lambda^2 + \frac{D_1}{C_1}\lambda = 0$$

i.e. there is a pair of real roots, $\lambda = 0$ and $\lambda \simeq -D_1/C_1$. This means that as E_1 goes from positive through zero to negative, the complex phugoid pair become a real negative pair, then one goes through zero and becomes positive making the aircraft unstable.

If the assumption of level flight is dropped and all the characteristics of the aircraft are allowed to depend on speed through compressibility or otherwise, then it has been shown that a quantity K_n defined as

$$K_n = -\frac{dC_m}{dC_R} \tag{9.69}$$

is related to E_1 by

$$E_1 = \frac{2C_R^2\mu_1}{i_y}\frac{\partial C_R}{\partial \alpha}\cdot K_n \tag{9.70}$$

where $C_R^2 = C_{Le}^2 + C_{De}^2$. The quantity K_n is known as the 'static margin' and is a generalized form of the cg margin. The definitions (9.69) and (5.39) should be compared, as should (9.68) and (9.70) above. The theory is sufficiently general to enable the inclusion of the effects of quasi-steady structural distortion.

9.3.6 Relation between the coefficient C_1 and the manoeuvre stability

When the magnitudes of the various terms in the expression (9.41) for C_1 are examined for most aircraft it is found that the dominant terms are

$$C_1 \simeq \hat{z}_w m_q + m_w \tag{9.71}$$

From table 8.4 we see that $\hat{z}_w = -Z_w$ and $\hat{m}_q = -M_q/i_y$. Then using (9.14) and (9.21) we find

$$\hat{z}_w = \frac{\partial C_{Le}}{\partial \alpha} \text{ and } \hat{m}_q = \frac{\overline{V}_T l_T}{i_y \bar{\bar{c}}} a_1$$

where it has been assumed that $\hat{m}_q$ is derived solely from the tailplane and neglecting C_{De} in comparison with $\partial C_{Le}/\partial\alpha$. Then substituting these and (9.67) into (9.71) we have

$$C_1 \simeq \frac{\mu_1}{i_y}\frac{\partial C_{Le}}{\partial \alpha}\left(H_n + \frac{\overline{V}_T l_T}{\mu_1 \bar{\bar{c}}} a_1\right)$$

Substituting for H_n from (5.39) gives

$$C_1 \approx \frac{\mu_1}{i_y}\frac{\partial C_{Le}}{\partial \alpha}\left((h_0 - h) + \bar{V}_T\frac{a_1}{a}\left\{1 - \frac{d\varepsilon}{d\alpha}\right\} + \frac{\bar{V}_T l_T}{\mu_1 \bar{\bar{c}}}a_1\right)$$

and substituting for the second μ_1 from (8.50) and comparing with (5.61) and (5.64) we have

$$C_1 \approx \frac{\mu_1}{i_y}\frac{\partial C_{Le}}{\partial \alpha}H_m \tag{9.72}$$

where H_m is the manoeuvre margin. The condition $C_1 > 0$ implied by (9.56) is then equivalent to the condition $H_m > 0$ from (5.71).

9.4 Discussion of the longitudinal modes

If, as is usually the case, the two longitudinal modes are well separated in frequency and effectively uncoupled, quite accurate results can be obtained by introducing some simplifying assumptions into the stability equations (9.28) to (9.31). An alternative approach would be to use the approximate factorization (9.64) and to keep only the dominant terms in the quartic coefficients.

9.4.1 The phugoid mode

In Chapter 7 we learnt that the phugoid is a long period oscillation, and it was assumed that the SPPO has the effect of keeping the incidence constant.

Using a full solution of the stability equations and finding the eigenvectors corresponding to this mode we can plot them, to some arbitrary scale, on an Argand diagram giving the shape of the mode. A typical shape of the mode is shown in figure 9.7.

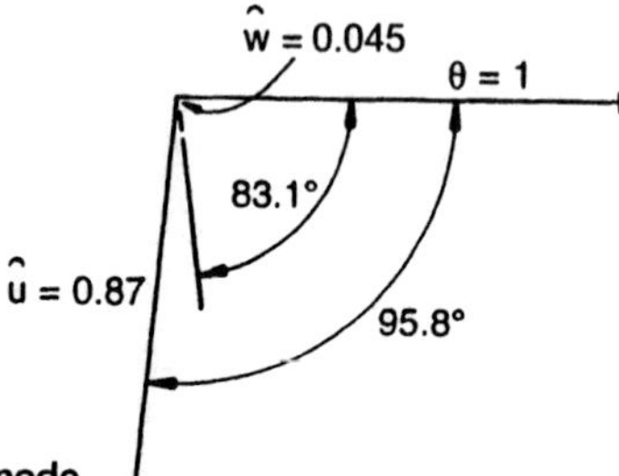

Fig. 9.7 Shape of the phugoid mode

The diagram shows the relative magnitudes and phase angles of the disturbance quantities at any instant.[2] We make two remarks on the diagram: the small size of the vertical velocity vector shows that the motion is at almost constant incidence; also the forward velocity change vector, $\hat{u}$, is approximately 90° in phase ahead of the pitch angle θ. The approximate treatment of the phugoid in Section 7.2.3 found that the velocity increment leads the pitch angle by exactly 90°.

We now adopt the assumptions of Section 7.2.3, namely that the incidence is constant and that the pitching moment equation is always satisfied. We also assume level flight and neglect some small derivatives. These assumptions amount to

$$\hat{w} = \hat{x}_q = \hat{z}_q = \hat{g}_2 = 0$$

We omit (9.30); then (9.28) and (9.29) become

$$\left.\begin{aligned}(\hat{D} + \hat{x}_u)\hat{u} + \hat{g}_1\theta &= 0\\ z_u\hat{u} - \hat{D}\theta &= 0\end{aligned}\right\}$$

As before we assume the solutions to be of exponential form and make the substitutions (9.35), which leads to the characteristic equation

$$\lambda^2 + \hat{x}_u\lambda + \hat{g}_1 z_u = 0 \tag{9.73}$$

This shows that the damping of the phugoid mode depends on the drag at low speeds – see (9.2) – and if the damping term is ignored it leads to the same periodic time as found in Section 7.2.3.

Such drastic assumptions are not necessary and we can make some physically reasonable assumptions with the aim of just reducing the stability equations to a second-order set.[3] The main assumptions are that the vertical and pitching accelerations are negligible, which we justify on the basis that motion in this mode is very slow. We also assume that we can neglect the derivative $\hat{x}_q$. These assumptions amount to putting

$$\hat{D}\hat{w} = \hat{D}\hat{q} = \hat{x}_q = 0 \tag{9.74}$$

Making these changes to (9.28) to (9.30) gives

$$\left.\begin{aligned}(\hat{D} + \hat{x}_u)\hat{u} + x_w\hat{w} \qquad\qquad\qquad + \hat{g}_1\theta &= 0\\ z_u\hat{u} + z_w\hat{w} + (z_q - 1)\hat{D}\theta + \hat{g}_2\theta &= 0\\ m_u\hat{u} + m_w\hat{w} \qquad + m_q\hat{D}\theta \qquad &= 0\end{aligned}\right\} \tag{9.75}$$

where (9.31) has been used. As before we now assume the solutions to be of exponential form and make the substitutions (9.35), leading to

$$\begin{vmatrix}\lambda + \hat{x}_u & x_w & \hat{g}_1\\ z_u & z_w & (z_q - 1)\lambda + \hat{g}_2\\ m_u & m_w & m_q\lambda\end{vmatrix} = 0 \tag{9.76}$$

Multiplying out the determinant and dividing through by the coefficient of λ^2 obtained gives

$$\lambda^2 + \left[\hat{x}_u - \left\{\frac{x_w[z_u m_q + m_u(1 - z_q)] + \hat{g}_2 m_w}{z_w m_q + m_w(1 - z_q)}\right\}\right]\lambda + \left\{\frac{E_1}{z_w m_q + m_w(1 - z_q)}\right\} = 0 \tag{9.77}$$

where E_1 is given by (9.43). We can make one deduction from this equation. From (9.4) and (9.5) we see that the derivative M_u is proportional to $\partial C_m/\partial \mathbf{M}$. Now suppose that the speed of an aircraft is increased slightly and that the increase of Mach number **M** causes a nose-down pitching moment to appear. The aircraft will respond with a nose-down inclination which will give a forward component of the weight which will tend to increase the speed; this effect is therefore destabilizing. The sign of M_u is negative in this case and from table 9.1 we see that $\hat{m}_u > 0$. As $\hat{z}_w$ is positive the effect on the coefficient E_1, given by (9.43), is to decrease it, confirming the destabilizing effect.

If we now assume that $\hat{m}_u = 0$, neglect $\hat{z}_q$ compared to unity and assume level flight then

$$\lambda^2 + \left[\hat{x}_u - x_w\left\{\frac{z_u m_q}{z_w m_q + m_w}\right\}\right]\lambda + \hat{g}_1\left\{\frac{z_u m_w}{z_w m_q + m_w}\right\} = 0 \tag{9.78}$$

The denominator of the two fractions in the above was shown in Section 9.3.6 to be proportional to the manoeuvre margin and $\hat{m}_w$ is proportional to the cg margin; see (9.15) and (9.18). If damping is ignored this equation then leads to a periodic time for the phugoid that is equal to the value found in Section 7.2.3 multiplied by the square root of the ratio of the manoeuvre margin to the cg margin. If the cg margin is large enough to satisfy the condition $\hat{z}_w m_q + m_w \gg z_u m_q$, then the first term in the square brackets in (9.78) dominates and the phugoid is primarily damped out by the drag in this case also.

We can compare the estimates of periodic time and time to half-amplitude given by (7.24), (9.73) and (9.78) and the exact result for a specific numerical case. The results are given in table 9.2.

Table 9.2 Comparison of estimates of times for phugoid

	7.24	*9.73*	*9.78*	*Exact*
t_p	87.9 s	87.95 s	109.7 s	124.7 s
t_H	—	206.7 s	164.6 s	232 s

It can be seen that none of the approximations is particularly good although that of (9.78) is a clear improvement for the periodic time over the other two.

9.4.2 The short period pitching oscillation

In Chapter 7 we learnt that the SPPO is a rapid motion, often an oscillation with heavy damping. We can again plot the eigenvectors giving the shape of the mode. A typical result is shown in figure 9.8.

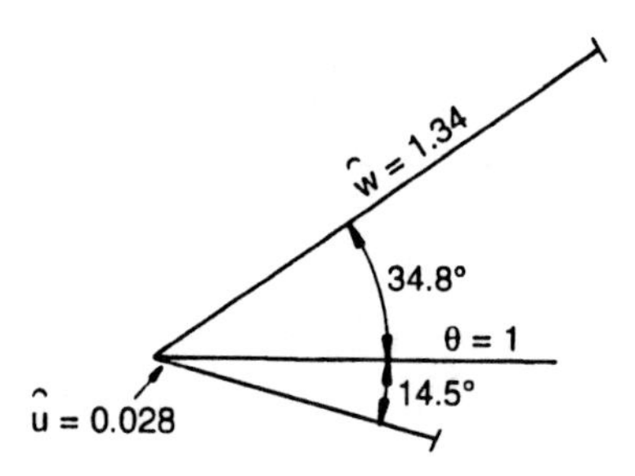

Fig. 9.8 Shape of the SPPO mode

We note the relatively small response in forward speed; in general the motion is too rapid for much change in forward speed to take place and motion is predominantly in pitch and in heave, i.e. vertically. This mode is rather more important than the phugoid as the latter is more easily controlled by the pilot and the SPPO gives rise to much higher vertical accelerations due to elevator movement or vertical gusts. The motion differs from that considered in Section 7.2.2 in that considerable vertical motion takes place as well as the pitching motion.

The observation that speed changes are small leads to a convenient approximate treatment. By neglecting the equation for the forward freedom and its derivatives we can reduce (9.29) and (9.30) to

$$\left.\begin{aligned}(\hat{D} + \hat{z}_w)\hat{w} \qquad\qquad\qquad - \hat{q} &= 0\\ (m_{\dot{w}}\hat{D} + m_w)\hat{w} + (\hat{D} + m_q)\hat{q} &= 0\end{aligned}\right\} \tag{9.79}$$

where level flight has been assumed and the small derivatives $\hat{z}_{\dot{w}}$ and $\hat{z}_q$ have been neglected, although they could have been included. This leads in the usual way to the characteristic equation

$$F'(\lambda) = \lambda^2 + (\hat{m}_q + z_w + m_{\dot{w}})\lambda + (z_w m_q + m_w) = 0 \tag{9.80}$$

This shows that the damping derivatives $\hat{m}_q$ and $\hat{z}_w$ and the derivative $\hat{m}_{\dot{w}}$ all contribute to the damping, rather than just the tailplane damping term $\hat{m}_q$ as suggested in the approximate analysis of Section 7.2.2. The fact that $\hat{m}_{\dot{w}}$ is usually underestimated, see Section 9.2.4, is not of great importance as it is usually much less than $\hat{m}_q + z_w$. We noted in Section 9.2 that Z_w decreases with frequency; the good agreement between theory and experiment for the SPPO may then be the fortuitous result of underestimation of $M_{\dot{w}}$ and the overestimation of Z_w by quasi-static theory. The periodic time is given by

$$t_p = \frac{2\pi\tau}{\sqrt{(\hat{z}_w m_q + m_w) - \left(\dfrac{m_q + z_w + m_{\dot{w}}}{2}\right)^2}} \tag{9.81}$$

This is usually dominated by the terms in the first bracket under the square root sign which are proportional to the manoeuvre margin, see Section 9.3.7, rather than the cg margin as suggested by the approximate analysis in Section 7.2.2. We also note that the normalized circular frequency in pitch is

$$\hat{\omega}_\theta = \omega_\theta \tau \simeq \sqrt{\hat{z}_w m_q + m_w} \tag{9.82}$$

which will be used in Chapter 12.

9.4.3 The effects of forward speed and cg position

The effect of variation of speed on the longitudinal stability has been calculated by solving the quartic equations for a medium range airliner in the 300-seat class at an altitude of 20 000 ft. This aircraft normally has an automatic control system to allow the use of small values of the cg margin; however, in this case an appropriate value has been assumed for the cg margin and no control system is included. The results are shown plotted in figure 9.9.

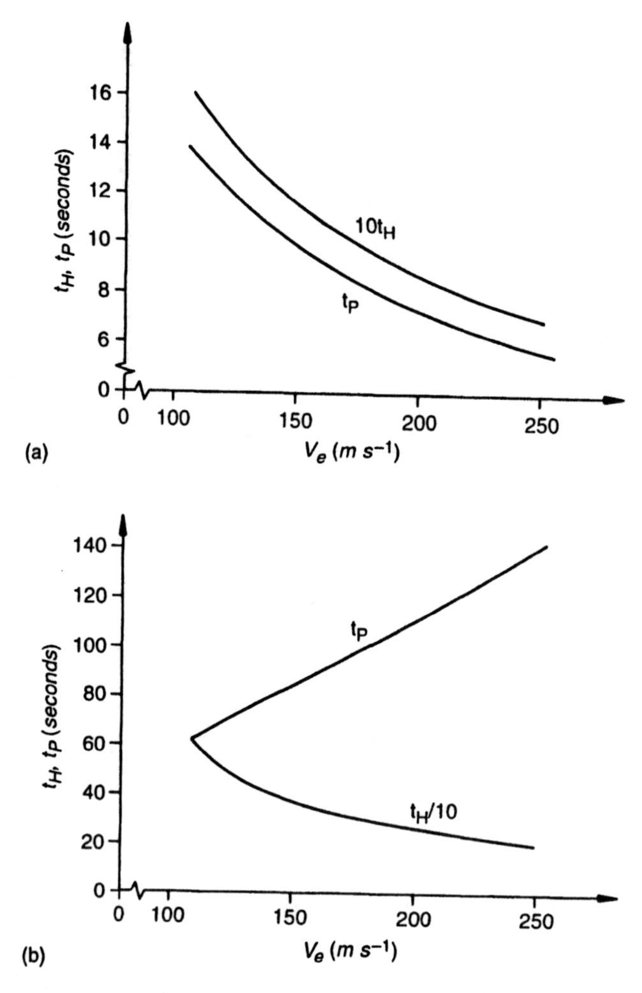

Fig. 9.9 Variation of a typical aircraft characteristics with speed: (a) SPPO periodic time and time to half-amplitude, (b) phugoid periodic time and time to half-amplitude

Figure 9.9(a) shows the effect of speed on the time to half-amplitude and periodic time for the SPPO mode. Both speeds are seen to fall with increasing speed. Figure 9.9(b) shows the effect of speed on the time to half-amplitude and periodic time for the phugoid mode. The periodic time is seen to increase almost linearly with speed in line with the approximate theories discussed, whilst the time to half-amplitude falls with speed.

To discuss the effect of cg variation we use the root locus plot which is a useful general technique in which the roots of the characteristic equation are plotted against some relevant parameter on an Argand diagram. In this case the parameter we use is the cg margin H_n to which the derivative $\hat{m}_w$ is proportional. Figure 9.10 shows a typical result of this technique, for the same aircraft; the speed is 194 m s^{-1} and the height is 20 000 ft.

Since complex roots appear in conjugate pairs the diagram is symmetrical and only half needs to be plotted. However, the SPPO and phugoid roots are so well separated that it is necessary to plot the SPPO mode in the upper half and use the lower half for the phugoid mode to a different scale. In the diagram the cg margin has been varied from 0.25 to -0.1, that is a very stable value to an unstable one. Considering first the SPPO mode we see that as cg

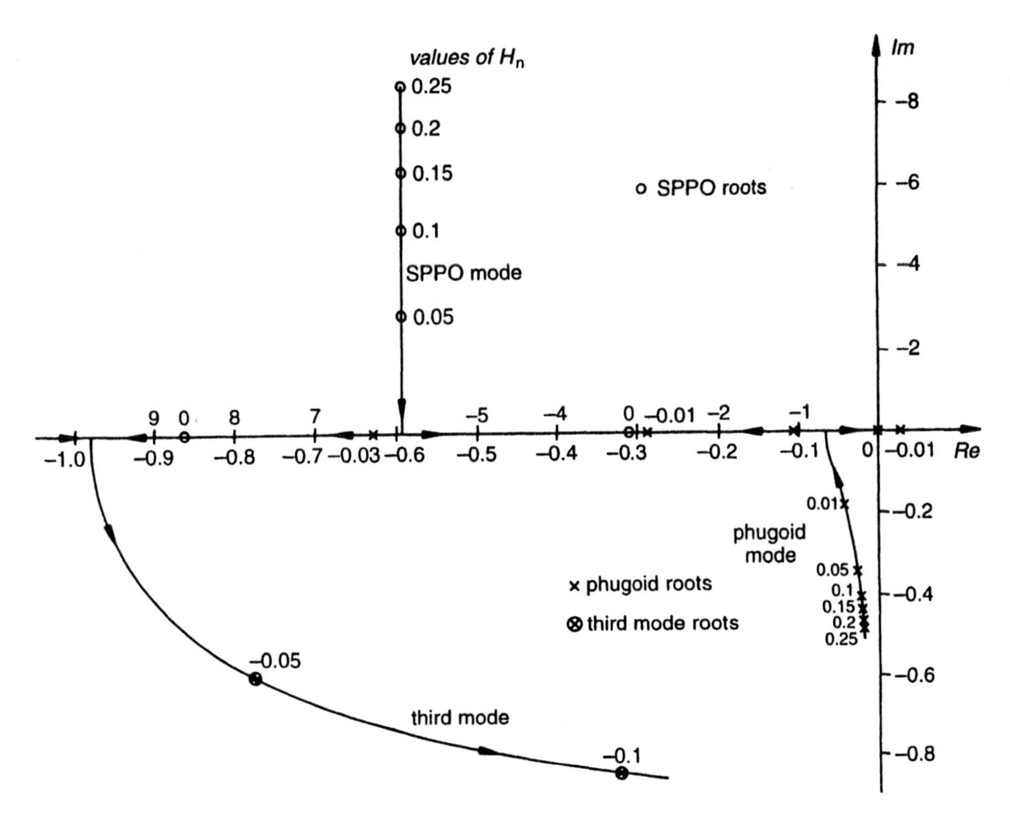

Fig. 9.10 Root locus plot for variation of cg margin. Upper part: SPPO mode; lower part: phugoid and third mode (different scales)

margin is reduced, i.e. the cg is moved backwards, the real part is almost constant whilst the imaginary part decreases steadily. This means that the damping is almost constant and the periodic time increases. At a cg margin between 0.05 and 0.01 the curve meets the axis and the complex pair becomes a real pair; after this, one root moves to the right and the other to the left. For the sake of clarity values of H_n less than zero are not plotted. In the lower half of the diagram we see that the phugoid mode behaves in a similar manner with the curve meeting the axis at a value of H_n between 0.01 and zero, and the two real roots again move in opposite directions. At a value of cg margin of exactly zero one root is zero as foreseen in Section 9.3.5; thereafter the solution always gives one positive, i.e. unstable, real root. As the cg margin is reduced still further the left-moving real root of the phugoid pair meets the right-moving root of the SPPO pair at a cg margin between -0.035 and -0.05. At this point another mode, known as the 'third mode', appears. This is an oscillatory mode involving all three freedoms; as the cg margin is reduced still further it rapidly becomes unstable. The method of solution of the quartic described in Section 9.3.4 fails for this mode.

Appendix: Solution of longitudinal quartic using a spreadsheet

This is a fairly simple routine for finding the roots of the longitudinal quartic and is based on the results of Section 9.3.4. It was developed using the Quatro®-Pro spreadsheet and has been tested on other popular spreadsheets. It should run on any spreadsheet that has the facility to turn off automatic calculation and allows the use of IF statements. It is assumed that the quartic has been divided through by the coefficient of the fourth power term, A_1.

The first step is to turn off automatic calculation; this is necessary as there are 'circular' cell references because this is an iterative process. We then enter some headings as follows:

into D1: Solution of the longitudinal quartic
into D2: Insert coefficients before copying formulae into block to right
into D3: Coefficients
into H3: First approximations
into M3: Final values
into Q3: SPPO roots
into U3: Phugoid roots

We leave some columns on the left for any preliminary calculations that may be needed and proceed to label the individual columns. Assuming that the character '^' is used to centre text in a cell, we label cells D4 to K4 as follows:

^B, ^C, ^D, ^E, ^a1, ^b1, ^b2, ^a2

where we have used B for B_1 and a1 for a_1. Then label cells M4 to X4 with:

^a1, ^b1, ^b2, ^a2, DISCRIM, TYPE, ROOT1, ROOT2, DISCRIM, TYPE, ROOT1, ROOT2

Then in cells D5 to G5 we insert the following test values: 5, 7, 0.2, 0.14. In cells H5 to K5 we put formulae to calculate the first approximations as follows:

+D5, +E5, +G5/E5, (E5*F5–D5*G5)/E5/E5

Column L is left clear, and the following iteration formulae put into cells M5 to P5:

+D5−P5, +E5−M5*P5−O5, +G5/N5, (F5–M5*O5)/N5

These formulae are the equivalent of (9.65). The function key to cause a calculation to take place (F9 in Quatro-Pro) should be pressed a few times and the final values of a1, b1, b2 and a2 should rapidly converge onto the values 4.985 675, 6.908 314, 0.020 265, 0.014 325.

The formulae to solve the quadratics are now inserted into the cells Q4 to X4 as follows:

into Q5: +M5*M5−4*N5 which calculates the discriminant for the SPPO
into R5: @IF(Q5>0,"REAL","COMPLEX") which indicates whether the roots are a real or complex pair
into S5: @IF(Q5>0,(−M5+@SQRT(Q5))/2,−M5/2) which gives the first root if there is a real pair, or gives the real part if there is a complex pair
into T5: @IF(Q5>0,(−M5−@SQRT(Q5))/2,@SQRT(−Q5)/2) which gives the second root if there is a real pair, or gives the imaginary part if there is a complex pair

This sequence is now repeated for the phugoid in cells U5 to X5:

into U5: +P5*P5-4*O5
into V5: @IF(U5>0,"REAL","COMPLEX")
into W5: @IF(U5>0,(−P5+@SQRT(U5))/2,−P5/2)
into X5: @IF(U5>0,(−P5−@SQRT(U5))/2,−@SQRT(−U5)/2)

The values in the cells Q5 to X5 should now read:

−2.7763,COMPLEX,−2.492 84,0.833 112,−0.080 86,COMPLEX,−0.007 16,0.142 176

A further check on the results is required to cover the case of quartics which have real roots; a trial quartic should be produced by multiplying together four linear factors and using this process to obtain the roots. The factors chosen should satisfy the condition that there are two small roots and two large ones.

To use the setup insert the values of the coefficients B, C, D and E into the appropriate columns, and then copy the formulae of cells H5 to K5 to the block below as far as needed. This calculates the first approximations. Next copy cells M5 to P5 to the block below and press F9 a few times to iterate the calculation until the values do not change. Finally copy cells Q5 to X5 to the block below to solve the quadratics. The values of the coefficients can be changed afterwards, if required; the F9 key will need to be used a few times to correct the results.

Student problems

9.1 An aircraft fitted with straight tapered wings has the following characteristics: wing area = 25 m^2, aspect ratio = 6, taper ratio = 0.5, the cg is 0.6 m aft of the leading edge of the wing amc, the aerodynamic centre is at 0.2 $\bar{\bar{c}}$, tail arm = 6 m, tailplane area = 3.7 m^2, $d\varepsilon/d\alpha = 0.35$, $a = 4.6$, $a_1 = 3.1$, $a_2 = 1.6$, $C_D = 0.014 + 0.052C_L^2$. Find the following longitudinal stability derivatives neglecting engine and Mach number effects if C_L = 0.21: X_u, X_w, Z_u, Z_w, $Z_{\dot{w}}$, Z_q, Z_η, M_u, M_w, $M_{\dot{w}}$, M_q, M_η. (A)

9.2 Using the approximations to the SPPO and phugoid given by (9.73), (9.78) and (9.80) find expressions for the relative damping and the natural frequency. Find the values given the stability derivatives: $\hat{x}_u = 0.0607$, $x_w = -0.355$, $z_u = 0.778$, $z_w = 5.465$, $m_w = 49.67$, $m_{\dot{w}} = 1.41$, $m_q = 5.02$. The other derivatives may be taken as zero and $g_1 = 0.53$.

9.3 Using the derivatives given in the previous problem suggest approximate expressions for the quartic coefficients B_1 and C_1 given by (9.40) and (9.41). Use these to compare the characteristic equations for the SPPO mode given in (9.63) and (9.80).

9.4 Using the approximation for the phugoid given by (9.73) calculate the periodic time and time to half-amplitude for a turboprop aircraft flying at low altitude at a speed of 130 m s^{-1}. The wing loading is 3.1 kN m^{-2} and $C_D = 0.015 + 0.05C_L^2$. (A)

9.5 Using the approximation for the phugoid given by (9.73) find the roots of the characteristic equation and the eigenvector $\hat{u}/\theta$. You are given $\hat{x}_u = 0.029$, $\hat{z}_u = 0.16$, $\hat{g}_1 = 0.081$. (A)

9.6 The equations of disturbed longitudinal motion lead to the following characteristic equation:

$$\lambda^4 + 15.7\lambda^3 + 81.7\lambda^2 + 4.16\lambda + 0.1 = 0$$

Find the roots of this equation. Identify and describe the modes with which they are associated. Find the times to half-amplitude and periodic times, given that $\tau = 5.72$ s. Find also the eigenvectors for the mode with the shorter period. You are given the following values of the concise stability derivatives for this aircraft: $\hat{x}_u = 0.052$, $x_w = -0.046$, $z_u = 0.1$, $z_w = 1.48$, $m_u = 4.5$, $m_w = 79$, $m_{\dot{w}} = 12.9$, $m_q = 1.27$, $\hat{g}_1 = 0.08$, $x_q = z_{\dot{w}} = z_q = 0$. (A)

9.7 Rework the theory leading to the approximation for the phugoid characteristic equation given in (9.78) for an aircraft in a climb. Show that if engine and Mach number effects on the derivatives can be neglected the phugoid mode becomes unstable at an angle of climb of $\tan^{-1}(2C_D/C_L)$, assuming that the cg margin is large.

Notes

1. This section may be omitted when first studying this chapter.
2. A phasor diagram in electrical engineering also shows relative phase angles although the magnitudes of the quantities are constant. To represent oscillatory quantities it is often pictured as rotating at the circular frequency. In our cases of divergent or convergent oscillatory modes, if the shape of the mode were pictured as rotating, the end points of the vectors would trace out logarithmic spirals.
3. The rest of this section can be omitted when first studying this chapter.

10
Longitudinal response

10.1 Introduction

In Section 9.3 four applications for the longitudinal stability equations were outlined; two of them were dealt with subsequently and the rest left for this chapter. These were the response to elevator angle and the response to a gust or a distribution of random gusts. We are particularly interested in finding the normal acceleration of both the cg and various points of the aircraft; the latter also requires a knowledge of the pitching acceleration of the aircraft. The ultimate purpose is to be able to design an aircraft which is safe to fly and fully capable of achieving its missions.

10.2 Response to elevator movement

We study the elevator responses to determine how quickly manoeuvres can be performed and what loads and hence stress levels may be imposed on the aircraft. Two methods are worthy of consideration; the first uses the Laplace transform and the second uses numerical integration of the equations in state-space form. The corresponding frequency responses are also of great interest.

10.2.1 Response using Laplace transform

The relevant equations are (9.28), (9.29) and (9.30). In these we substitute for $\hat{q}$ from (9.31) and for the right-hand sides from (9.32) and neglect the small elevator derivative $\hat{x}_\eta$ to obtain

$$\left.\begin{aligned}
(\hat{D} + \hat{x}_u)\hat{u} \quad & + x_w\hat{w} && +(x_q\hat{D} + \hat{g}_1)\theta = 0 \\
z_u\hat{u} \quad + & \left[(1 + z_{\dot{w}})\hat{D} + z_w\right]\hat{w} && + \left[(z_q - 1)\hat{D} + \hat{g}_2\right]\theta = -z_\eta\eta'(\hat{t}) \\
m_u\hat{u} \quad + & (m_{\dot{w}}\hat{D} + m_w)\hat{w} && + (\hat{D}^2 + m_q\hat{D})\theta = -m_\eta\eta'(\hat{t})
\end{aligned}\right\} \tag{10.1}$$

We now multiply through each of these equations by $e^{-s\hat{t}}$ and integrate with respect to $\hat{t}$ from zero to infinity, remembering that $\hat{D} = d/d\hat{t}$. The result is

$$\left.\begin{aligned}
(s + \hat{x}_u)\bar{u} \quad & + x_w\bar{w} && +(x_q s + \hat{g}_1)\bar{\theta} = 0 \\
z_u\bar{u} \quad + & \left[(1 + z_{\dot{w}})s + z_w\right]\bar{w} && + \left[(z_q - 1)s + \hat{g}_2\right]\bar{\theta} = -z_\eta\bar{\eta} \\
m_u\bar{u} \quad + & (m_{\dot{w}}s + m_w)\bar{w} && + (s^2 + m_q s)\bar{\theta} = -m_\eta\bar{\eta}
\end{aligned}\right\} \tag{10.2}$$

where we have written $\bar{u}$, $\bar{w}$, $\bar{\theta}$ and $\bar{\eta}$ for the Laplace transforms of $\hat{u}$, $\hat{w}$, θ and η' and assumed that their initial values were zero. These are now a set of simultaneous algebraic equations with the solution given by Cramer's rule as

$$\frac{\bar{u}}{G_1(s)} = \frac{\bar{w}}{G_2(s)} = \frac{\bar{\theta}}{G_3(s)} = \frac{\bar{\eta}}{F(s)} \tag{10.3}$$

In these equations $F(s)$ is the determinant of the left-hand sides of (10.2). Now the left-hand sides of these equations are the same as those of (9.36) with s replaced by λ, hence $F(s)$ is given by (9.38). It is therefore the characteristic equation and has the same roots. The functions G_i $(i = 1, 2, 3)$ are the determinants of the left-hand side with the ith column replaced by the right-hand sides of (10.2), for instance

$$G_1 = \begin{vmatrix} 0 & \hat{x}_w & x_q s + \hat{g}_1 \\ -z_\eta & (1 + z_{\dot{w}})s + z_w & (z_q - 1)s + \hat{g}_2 \\ -m_\eta & m_{\dot{w}} s + m_w & s^2 + m_q s \end{vmatrix}$$

Expanding the determinants we find

$$\begin{aligned} G_1 &= s^2\left[\hat{z}_\eta(x_w - x_q m_{\dot{w}}) - m_\eta x_q(1 + z_{\dot{w}})\right] \\ &\quad + s\left[z_\eta(x_w m_q - x_q m_w - \hat{g}_1 m_{\dot{w}}) + m_\eta\left\{x_w(1 - z_q) + x_q z_w + \hat{g}_1(1 + z_{\dot{w}})\right\}\right] \\ &\quad + \left[-z_\eta \hat{g}_1 m_w + m_\eta(z_w \hat{g}_1 - x_w \hat{g}_2)\right] \end{aligned} \tag{10.4}$$

$$\begin{aligned} G_2 &= -s^3 \hat{z}_\eta \\ &\quad + s^2\left[-z_\eta(x_u + m_q) - m_\eta(1 - z_q)\right] \\ &\quad + s\left[-z_\eta(x_u m_q - x_q m_u) - m_\eta\left\{x_u(1 - z_q) + x_q z_u + \hat{g}_2\right\}\right] \\ &\quad + \left[z_\eta m_u \hat{g}_1 - m_\eta(z_u \hat{g}_1 - x_u \hat{g}_2)\right] \end{aligned} \tag{10.5}$$

$$\begin{aligned} G_3 &= s^2\left[\hat{z}_\eta m_{\dot{w}} - m_\eta(1 + z_{\dot{w}})\right] \\ &\quad + s\left[z_\eta(m_w + m_{\dot{w}} x_u) - m_\eta\left\{(1 + z_{\dot{w}})x_u + z_w\right\}\right] \\ &\quad + \left[z_\eta(x_u m_w - x_w m_u) - m_\eta(x_u z_w - x_w z_u)\right] \end{aligned} \tag{10.6}$$

Then we write

$$\bar{u} = \frac{G_1(s)}{F(s)}\bar{\eta} = H_{u\eta}(s)\cdot\bar{\eta}(s) \tag{10.7}$$

where $H_{u\eta}$ is the 'transfer function' connecting the forward velocity change with elevator angle change; there are similar expressions for $\bar{w}$ and $\bar{\theta}$ Note that transfer functions can only be defined for linear systems. Once the elevator angle has been specified as a function of time and its transform found, the expression for $\bar{u}$ will be in the form of the ratio of two polynomials

$$\bar{u} = N(s)/D(s)$$

where $D(s)$ will usually be closely related to $F(s)$. Using the technique of partial fractions this expression can be written in terms of the roots of $D(s)$ as

$$\bar{u} = \sum_{j=1}^{n} \frac{U_j}{s - \lambda_j} \tag{10.8}$$

where the roots of $D(s)$ have been written, $\lambda_1, \lambda_2, \ldots, \lambda_j, \ldots, \lambda_n$ and $D(s)$ of degree n. The quantities U_j can be found from

$$U_j = \left.\frac{(s - \lambda_j)N(s)}{D(s)}\right|_{s=\lambda_j} \tag{10.9}$$

The factor $(s - \lambda_j)$ cancels the same factor in the denominator when factorized so that the result of the expression is finite. We now make inverse use of the transform pair

$$\mathcal{L}\left[e^{\lambda t}\right] = \frac{1}{(s - \lambda)} \tag{10.10}$$

to obtain the result

$$\hat{u} = \sum_{j=1}^{n} U_j e^{\lambda_j \hat{t}} \tag{10.11}$$

and similarly for the other disturbance variables.

10.2.2 Frequency response

An alternative view of the response of an aircraft is obtained if we consider the frequency response which can be found from the transfer function as follows. Suppose a system with transfer function $H(s)$ is given an input

$$X_i(t) = A_j e^{i\omega t} \tag{10.12}$$

Then the Laplace transform of the output is

$$\bar{X}_0 = H(s)\bar{X}_i = A_i \frac{H(s)}{s - i\omega}$$

using (10.10). Now we suppose that the transfer function is in the form of the ratio of two polynomials

$$H(s) = G(s)/F(s) \tag{10.13}$$

then

$$\bar{X}_0 = A_i \frac{G(s)}{(s - i\omega)F(s)}$$

Noting that the roots of the denominator are $i\omega$, λ_1, λ_2, ..., λ_j, ..., λ_n, and using (10.8) and (10.9) we find

$$X_0(t) = A_i \left\{ \frac{G(i\omega)}{F(i\omega)} e^{i\omega t} + \sum_{j=1}^{n} e^{\lambda_j t} \cdot \left[\frac{(s - \lambda_j)G(s)}{(s - i\omega)F(s)} \right]_{s=\lambda_j} \right\}$$

If we restrict interest to stable systems, only the first term is appreciable after a long enough time and the steady state output is

$$X_0(t) = A_i \frac{G(i\omega)}{F(i\omega)} e^{i\omega t} \tag{10.14}$$

Then from (10.12), (10.13) and (10.14) we find

$$\frac{X_0(t)}{X_i(t)} = H(i\omega), \text{ say} \tag{10.15}$$

which is the frequency response. In general it is a complex quantity and can be written in the form

$$H(i\omega) = Me^{-i\varphi} \tag{10.16}$$

Here $M = |H(i\omega)|$ is the gain and $\varphi = -\arg[H(i\omega)]$ is the amount by which the output lags the input. The frequency response can be represented as a vector plot with frequency as parameter, or Nyquist diagram. Alternatively logarithmic plots of gain and phase angle against frequency can be made, known as a Bode diagram. In this case, for the purpose of calculating the response, the denominator also can be factorized and the response can then be represented as that of a set of first and second order systems in series.

10.2.3 Response using numerical integration of state-space equations

The particular technique we choose to implement is known as the Runge–Kutta method in its classical form: see reference (10.1). The relevant equations are (9.28) to (9.32). In order to rewrite them in the state-space form, we must first eliminate $\hat{D}\hat{w}$ from the third of these. Solving the second for $\hat{D}\hat{w}$ and substituting into the third we can write the equations in the form of (7.64) as

$$\left.\begin{aligned} \hat{D}\hat{u} &= -\hat{x}_u\hat{u} - x_w\hat{w} - x_q\hat{q} - \hat{g}_1\theta \\ \hat{D}\hat{w} &= -z_u^*\hat{u} - z_w^*\hat{w} - z_q^*\hat{q} - z_\eta^*\eta' \\ \hat{D}\hat{q} &= -m_u^*\hat{u} - m_w^*\hat{w} - m_q^*\hat{q} - m_\eta^*\eta' \\ \hat{D}\theta &= \hat{q} \end{aligned}\right\} \tag{10.17}$$

where

$$\left.\begin{aligned}
\hat{z}_u^* &= \frac{z_u}{1 + z_{\dot{w}}} \\
z_w^* &= \frac{z_w}{1 + z_{\dot{w}}} \\
z_q^* &= \frac{z_q - 1}{1 + z_{\dot{w}}} \\
z_\eta^* &= \frac{z_\eta}{1 + z_{\dot{w}}} \\
m_u^* &= m_u - \frac{m_{\dot{w}} z_u}{1 + z_{\dot{w}}} \\
m_w^* &= m_w - \frac{m_{\dot{w}} z_w}{1 + z_{\dot{w}}} \\
m_q^* &= m_q - \frac{m_{\dot{w}} (z_q - 1)}{1 + z_{\dot{w}}} \\
m_\eta^* &= m_\eta - \frac{m_{\dot{w}} z_\eta}{1 + z_{\dot{w}}}
\end{aligned}\right\} \tag{10.18}$$

We let the initial values of the dependent variables at the beginning of a time step be $\hat{u}_0$, $\hat{w}_0$, $\hat{q}_0$ and θ_0 the length of the step be Δt. The following quantities are then calculated:

$$\left.\begin{aligned}
k_{1u} &= -\hat{x}_u \hat{u}_0 - x_w \hat{w}_0 - x_q \hat{q}_0 - \hat{g}_1 \theta_0 \\
k_{1w} &= -z_u^* \hat{u}_0 - z_w^* \hat{w}_0 - z_q^* \hat{q}_0 - z_\eta^* \eta'(0) \\
k_{1q} &= -m_u^* \hat{u}_0 - m_w^* \hat{w}_0 - m_q^* \hat{q}_0 - m_\eta^* \eta'(0) \\
k_{1\theta} &= \hat{q}_0
\end{aligned}\right\}$$

where $\eta'(0)$ is the elevator angle increment at the beginning of the step. Then from these we calculate the quantities

$$\left.\begin{aligned}
k_{2u} &= -\hat{x}_u(\hat{u}_0+\Delta t_1 k_{1u}) - x_w(\hat{w}_0+\Delta t_1 k_{1w}) - x_q(\hat{q}_0+\Delta t_1 k_{1q}) - \hat{g}_1(\theta_0+\Delta t_1 k_{1\theta}) \\
k_{2w} &= -z_u^*(\hat{u}_0+\Delta t_1 k_{1u}) - z_w^*(\hat{w}_0+\Delta t_1 k_{1w}) - z_q^*(\hat{q}_0+\Delta t_1 k_{1q}) - z_\eta^* \eta'(\Delta t_1) \\
k_{2q} &= -m_u^*(\hat{u}_0+\Delta t_1 k_{1u}) - m_w^*(\hat{w}_0+\Delta t_1 k_{1w}) - m_q^*(\hat{q}_0+\Delta t_1 k_{1q}) - m_\eta^* \eta'(\Delta t_1) \\
k_{2\theta} &= \hat{q}_0+\Delta t_1 k_{1q}
\end{aligned}\right\}$$

where $\Delta t_1 = \Delta t/2$. Then we calculate the quantities

$$\left.\begin{aligned}
k_{3u} &= -\hat{x}_u(\hat{u}_0+\Delta t_1 k_{2u}) - x_w(\hat{w}_0+\Delta t_1 k_{2w}) - x_q(\hat{q}_0+\Delta t_1 k_{2q}) - \hat{g}_1(\theta_0+\Delta t_1 k_{2\theta}) \\
k_{3w} &= \text{etc.}
\end{aligned}\right\}$$

and then

$$\left.\begin{aligned} k_{4u} &= -\hat{x}_u(\hat{u}_0+\Delta t k_{3u}) - x_w(\hat{w}_0+\Delta t k_{3w}) - x_q(\hat{q}_0+\Delta t k_{3q}) - \hat{g}_1(\theta_0+\Delta t k_{3\theta}) \\ k_{4w} &= \text{etc.} \end{aligned}\right\}$$

The dependent variables at the end of the step are then calculated from

$$\left.\begin{aligned} \hat{u}_1 &= \hat{u}_0 + \Delta t(k_{1u} + 2k_{2u} + 2k_{3u} + k_{4u})/6 \\ \hat{w}_1 &= \hat{w}_0 + \Delta t(k_{1w} + 2k_{2w} + 2k_{3w} + k_{4w})/6 \\ \hat{q}_1 &= \text{etc.} \end{aligned}\right\}$$

In calculations of this type checks must be made on the results before they can be accepted. If the aircraft is stable, then it is an easy matter to calculate the final steady values of the dependent variables which should be achieved if the calculation is allowed to proceed for long enough. Checks can often be made on the frequency of any oscillation. It is also most important to test the effect of step size; this should be set at a value for which a reduction causes no appreciable change in the results. The accuracy of the results is finally limited by the number of figures carried in the computer used.

10.2.4 Typical response characteristics of an aircraft

Using the method outlined in the previous section calculations have been made of the response to a step change in elevator angle of 0.04 rad ($\simeq$ 2.3°) for the aircraft used in the calculations in Section 9.4.3 at the same speed and height.

Figure 10.1(a) shows the variation of the forward speed change $\hat{u}$, the vertical speed change $\hat{w}$, the pitching velocity $\hat{q}$ and the pitch angle θ with time. The major part of the figure is the response in the phugoid mode. It can be seen that the pitch angle leads the forward velocity by rather more than 90° as discussed earlier. The response in vertical velocity is rather smaller than the other response; the aircraft is stiff in the vertical sense due to the powerful effect of wing lift changes. The small damping in the phugoid mode can be appreciated by comparing the two negative peaks shown in pitch angle. Also shown are the final values of the quantities; downward movement of the elevator produces a downward inclination of the flight path (negative θ) and this increases the forward speed. The initial response is in the SPPO mode and is shown to an enlarged scale in figure 10.1(b) where to aid clarity the sign of $\hat{q}$ has been reversed. The initial response is strongest in pitching velocity due to the powerful pitching moment produced by the elevator. The damping can be seen to be very heavy with barely a complete cycle visible; the forward speed can be seen to have only a small change during this time.

Calculations have also been made of the frequency response using the theory of Section 10.2.2 for the same conditions and are shown in figure 10.2.

Figure 10.2(a) shows the magnitude of the frequency response in pitch angle as a function of normalized circular frequency $\hat{\omega} = \omega\tau$. A very strong peak can be seen at the frequency of the phugoid mode due to the low damping, whilst there is no obvious peak at the frequency of the SPPO mode due to the large damping. Figure 10.2(b) shows the corresponding variation in the phase angle φ. It can be seen that there is a phase angle change of nearly 180° occurring at the frequency of the phugoid mode, typical of that normally occurring as the applied frequency passes through the natural frequency of a second-order system. No such change is obvious for the SPPO, again due to the damping. The response characteristics of the other quantities show very similar effects. The response in forward velocity magnitude is much stronger at the frequency of the phugoid mode than the SPPO mode, whilst that of the vertical

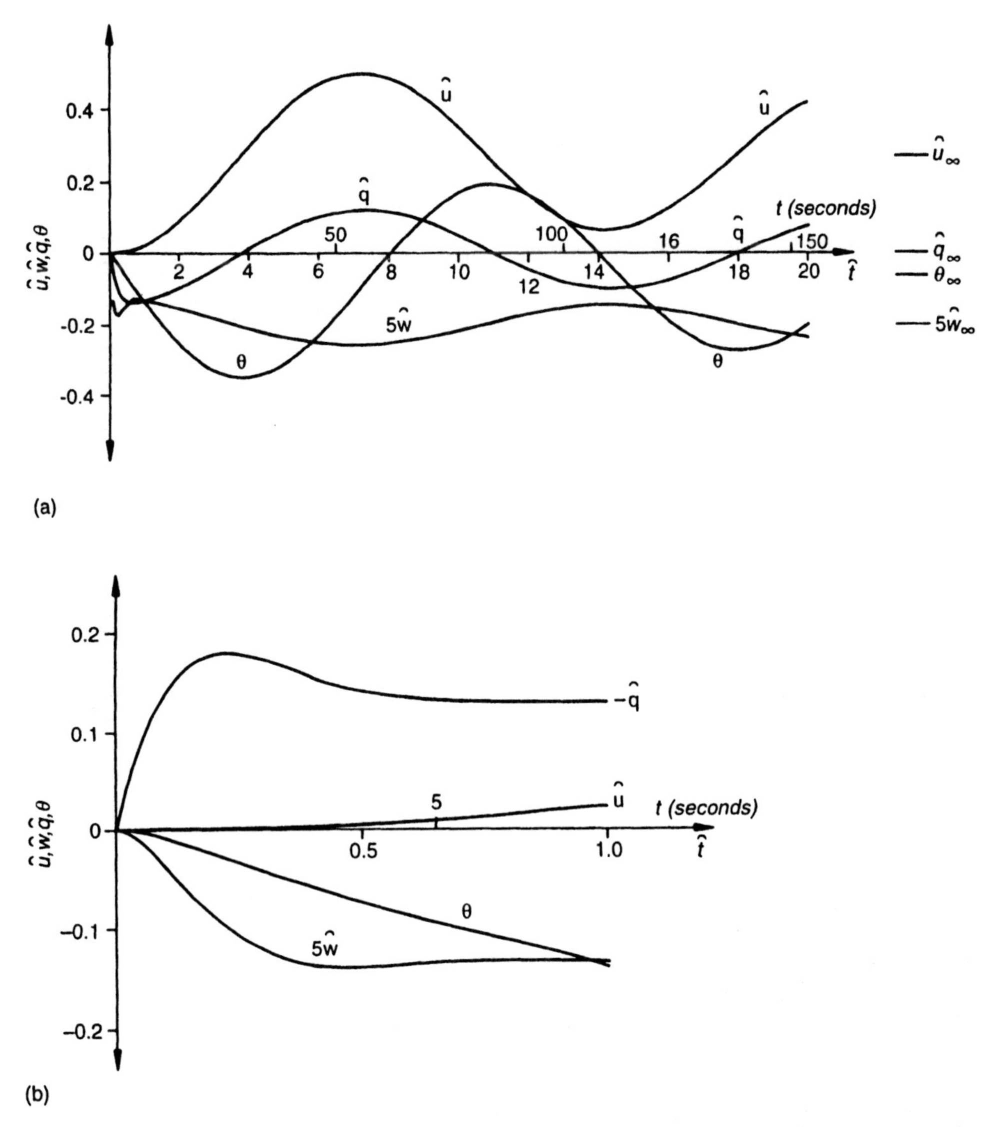

Fig. 10.1 **Longitudinal response of a typical aircraft to a step in elevator angle: (a) response showing mainly the phugoid mode, (b) first part of curves in detail showing response in SPPO mode**

velocity is very similar for both modes. The phase angle characteristics both show the near 180° change at the frequency of the phugoid mode.

10.2.5 Normal acceleration response to elevator angle

The primary effect of the elevator is to give a pitching moment which excites the SPPO mode initially and only later does the response in the phugoid mode become apparent. The initial motion imposes the greater normal acceleration and so is normally of the most interest. We can therefore analyze this using the simplified treatment of the SPPO in Section 9.4.2. Taking equations (9.79) and adding elevator angle terms from (9.32) we find

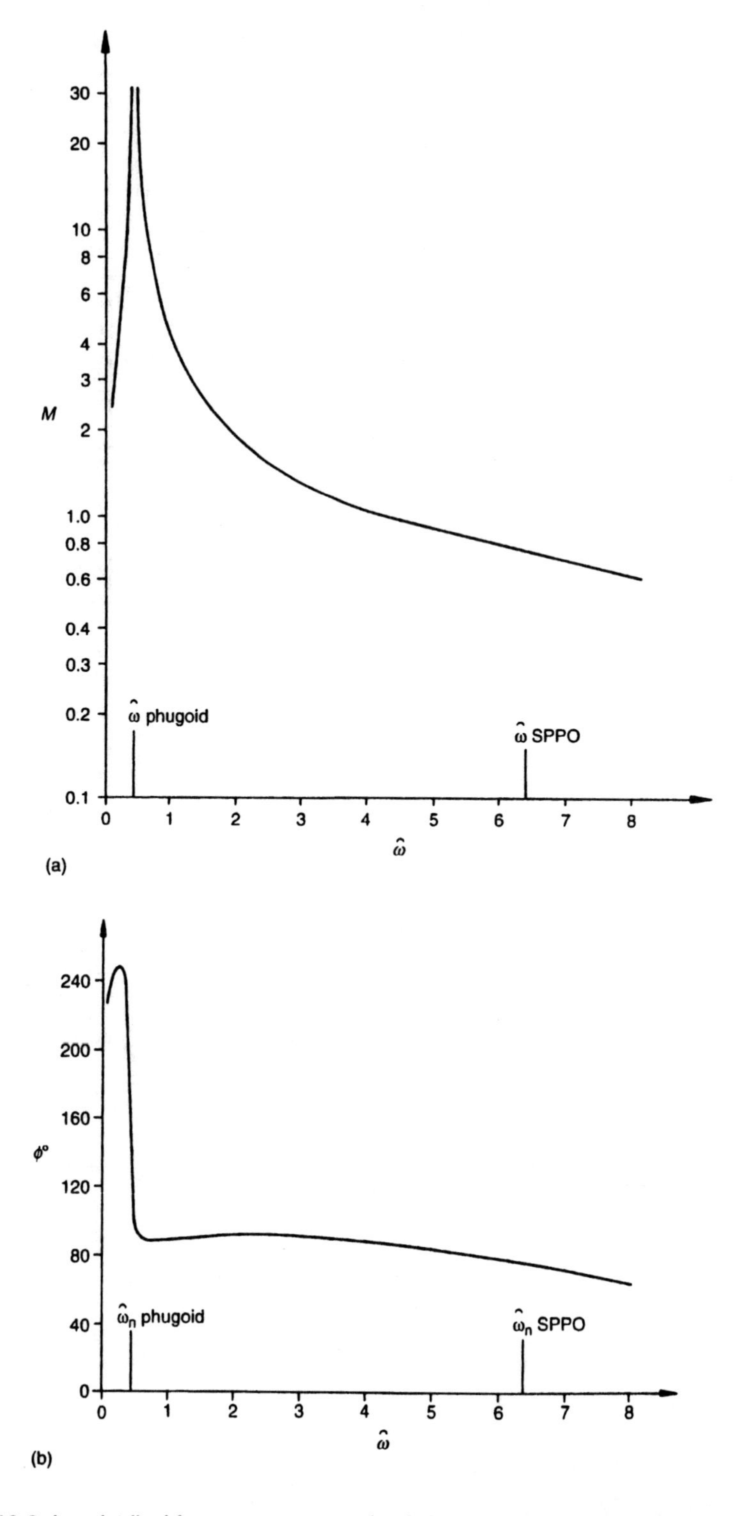

Fig. 10.2 **Longitudinal frequency response in pitch angle: (a) modulus of response, (b) phase angle**

$$(\hat{D} + \hat{z}_w)\hat{w} \qquad\qquad -\hat{q} = -z_\eta \eta'(\hat{t}) \tag{10.19}$$

$$(m_{\dot{w}}\hat{D} + m_w)\hat{w} + (\hat{D} + m_q)\hat{q} = -m_\eta \eta'(\hat{t}) \tag{10.20}$$

Since we are using axes fixed in the aircraft the normal acceleration is not simply dw/dt, and we need to find an appropriate relation. One method of calculating the acceleration is to use (force)/(mass) and then from the third equation of (8.22) we see that the acceleration along Oz is $\dot{w} - qU_e$. Then writing the upward acceleration as a factor n times the acceleration due to gravity g we have

$$ng = -\dot{w} + qV_e$$

where we have put $U_e = V_e$, since we are using wind axes. Using table 8.2 we can substitute $w = \hat{w}V_e$, $q = \hat{q}\tau$, and $\hat{D} = d/d\hat{t}$ where τ is given by (8.48) to find

$$n = -\frac{1}{g}\left\{\frac{d\hat{w}}{d\hat{t}}\frac{V_e}{\tau} - \frac{\hat{q}V_e}{\tau}\right\} = -\frac{\frac{1}{2}\rho V_e^2 S}{mg}\left(\hat{D}\hat{w} - \hat{q}\right)$$

and hence

$$n = -\frac{1}{C_{L_e}}(\hat{D}\hat{w} - \hat{q}) \tag{10.21}$$

This can be simplified using (10.19) which can be rearranged as

$$\hat{D}\hat{w} - \hat{q} = -(\hat{z}_w\hat{w} + z_\eta\eta')$$

then substituting into (10.21)

$$n = \frac{1}{C_{L_e}}\left(\hat{z}_w\hat{w} + z_\eta\eta'\right) \tag{10.22}$$

From this we see that the response in normal acceleration can be found from that in vertical velocity.[1] The derivative z_η is often small, and if it is neglected the normal acceleration becomes proportional to the vertical velocity or equivalently to the incidence change.

We proceed to transform (10.19) and (10.20) as before giving

$$\left.\begin{aligned}(s + \hat{z}_w)\bar{w} \qquad\qquad & - \bar{q} = -z_\eta\bar{\eta}(s) \\ (m_{\dot{w}}s + m_w)\bar{w} + (s + m_q)\bar{q} & = -m_\eta\bar{\eta}(s)\end{aligned}\right\}$$

assuming that w, q and η' were zero initially. Then using Cramer's rule we can solve these equations for the transformed vertical velocity in the form

$$\bar{w} = \frac{G_1'(s)}{F'(s)}\bar{\eta}(s) = H_{w\eta}(s)\bar{\eta}(s) \tag{10.23}$$

where $H_{w\eta}$ is the transfer function connecting the vertical velocity change to the elevator angle change. In this equation

$$G_1'(s) = -\left[\hat{z}_\eta s + (z_\eta m_q + m_\eta)\right] \tag{10.24}$$

and $F'(s)$ is the characteristic equation corresponding to (10.19) and (10.20) and is given by (9.80), with s replacing λ. The response in pitching velocity can be found in the same way, if required. We also take the Laplace transform of (10.22) to give

$$\bar{n} = \frac{1}{C_{L_e}}\left(\hat{z}_w \bar{w} + z_\eta \bar{\eta}\right) \tag{10.25}$$

Then substituting for $\bar{w}$ from (10.23) we find

$$\bar{n} = \frac{J}{C_{L_e} F'}\bar{\eta} = H_{n\eta}(s)\bar{\eta}(s) \tag{10.26}$$

where $J(s) = \hat{z}_w G_1' + z_\eta F'$, and $H_{n\eta}$ is the transfer function connecting normal acceleration to elevator angle change. Substituting for G_1' from (10.24) and for F' from (9.80), $J(s)$ is given by

$$J(s) = \hat{z}_\eta s^2 + (m_q + m_{\dot{w}})z_\eta s + m_w z_\eta - m_\eta z_w \simeq - m_\eta z_w \tag{10.27}$$

if, as usual, $\hat{z}_\eta$ is small. The response in normal acceleration as a function of time or frequency can be found from (10.26) using the methods already described.

10.3 Response to gusts

In this section we now drop the first of our basic assumptions in Chapter 1 and improve on our approximate treatment of gusts in Section 4.2.3. We study the response to gusts in order to determine the effect on passenger comfort or the suitability of the aircraft as a weapons platform and the effects on the stress levels and fatigue life of the airframe. Up to this point we have assumed that the aircraft has been flying through a stationary atmosphere; in fact the air is almost never at rest but is in a state of continuous random fluctuation. We obtain our first ideas of the effects of the unsteady motion of the atmosphere by considering the aircraft to fly into a region where the air has a parallel motion in some direction, known as a discrete gust. We later outline statistical methods to represent the apparently random motion of the air and the resulting motion of the aircraft.

10.3.1 Response to discrete gusts

The type of gust we will concentrate our attention on and which is generally the most important in practice is the vertical gust; horizontal gusts are as important near the ground. However, the methods we will discuss can be applied to all other forms of gust. In its simplest form the gust is pictured as a vertical jet of air fixed in position and having constant velocity. The gust is taken to be very large in extent in the direction of flight compared with the length of the aircraft, as shown in figure 10.3; this means that we can treat the aircraft as if it were a point.

Fig. 10.3 Aircraft entering a step gust

This assumption that the gust velocity changes instantaneously is of course impossible and the dotted line is a more reasonable variation. We also assume that the gust extends for a large distance in the spanwise direction. A further assumption that we make is that the aerodynamic forces and moments can change instantaneously. In fact the circulation around a wing takes an infinite time to adjust to the gust, but is close to the final value after a time corresponding to about 10 chord-lengths. For a wing of infinite span the variation of lift due to an upgust with time is shown in figure 10.4 and is known as the Küssner function.

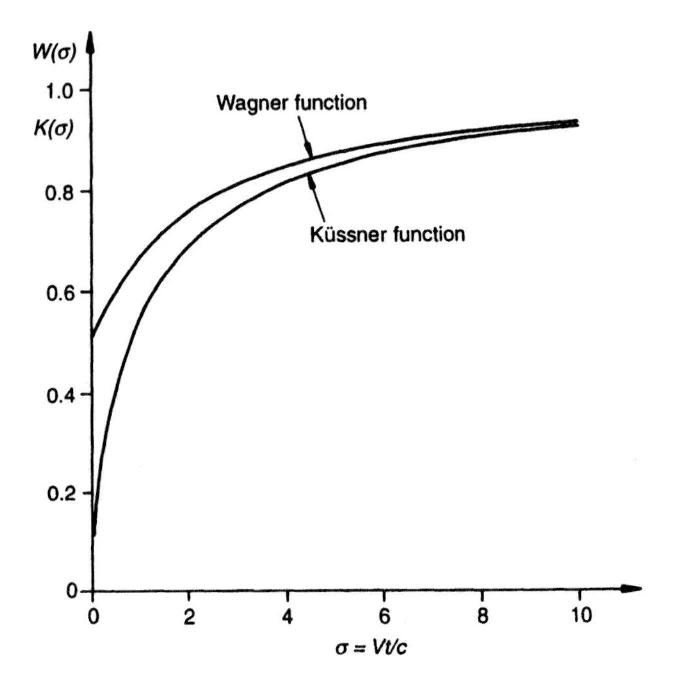

Fig. 10.4 Wagner and Küssner functions

A similar effect takes place if the incidence of a wing is changed in a stepwise fashion; this is also shown in figure 10.4 and is known as the Wagner function. It can be seen that there is little difference between them except initially. Both functions can be approximated using exponential functions, which enables them to be incorporated into more sophisticated analyses than we shall be describing. It has been shown that quasi-static derivatives give reasonable results for gust response but of course for accurate results it is necessary to use unsteady derivatives particularly at higher frequencies. The effect of the lag in build-up of lift is to reduce the maximum normal acceleration produced by a gust. The agreement with measurements for the crude theory for gusts in Section 4.2.3 can be improved by applying a gust alleviation factor to the results of (4.7).

Returning to the problem in hand the situation is shown in figure 10.5(a), where we assume that the gust velocity is taken to be a function of distance along the flight path.

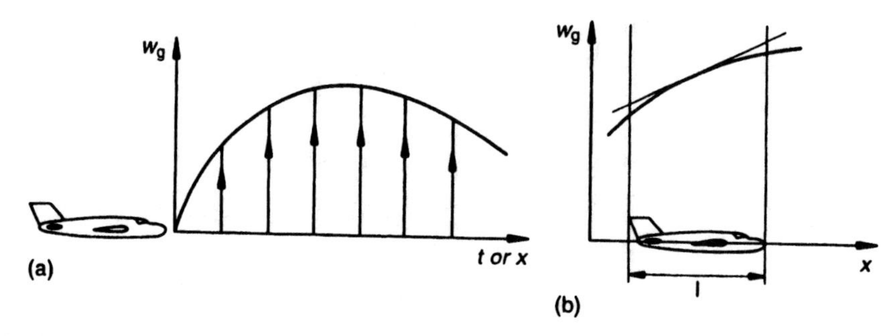

Fig. 10.5 Aircraft entering a graded gust: (a) velocity distribution, (b) definition of terms in derivation of better model

We now need to modify the stability equations to model the effect of a vertical gust. The forces and moments produced by an upgust will all be of aerodynamic origin and will be the same as for a downward motion of the aircraft. Hence we add the upgust velocity w_g to the aircraft vertical velocity w for the aerodynamic terms only; these are readily identified as they are factored by a stability derivative. Making these adjustments to (9.28) to (9.31) and assuming level flight we obtain

$$\left.\begin{aligned}
(\hat{D}+\hat{x}_u)\hat{u} \quad & + x_w(\hat{w}+\hat{w}_g) \quad + x_q\hat{q} + \hat{g}_1\theta = 0 \\
z_u\hat{u} + \left[(1+z_{\dot{w}})\hat{D}+z_w\right]\hat{w} & + (z_{\dot{w}}\hat{D}+z_w)\hat{w}_g + (z_q - 1)\hat{q} = 0 \\
m_u\hat{u} \quad & + (m_{\dot{w}}\hat{D}+m_w)(\hat{w}+\hat{w}_g) + (\hat{D}+m_q)\hat{q} = 0 \\
& \hat{q} - \hat{D}\theta = 0
\end{aligned}\right\} \qquad (10.28)$$

where $\hat{w}_g = w_g/V_e$ is the normalized upgust velocity. The use of the stability derivatives in this way imposes the same restrictions on $\hat{w}_g$ as those on $\hat{w}$, as discussed in Section 8.4.2.

Again the initial motion imposes the greater normal acceleration and so is normally of the most interest and we therefore analyze this using the simplified treatment for the SPPO of Section 9.4.2. Since we are now assuming constant speed, we have $x = U_e t$ and so we substitute into $w_g(x)$ to find $w_g(t)$. From (10.28) we then find

$$(\hat{D} + \hat{z}_w)\hat{w} \qquad - \hat{q} = -z_w\hat{w}_g(\hat{t}) \qquad (10.29)$$

$$(m_{\dot{w}}\hat{D} + m_w)\hat{w} + (\hat{D} + m_q)\hat{q} = -(m_{\dot{w}}\hat{D} + m_w)\hat{w}_g(\hat{t}) \qquad (10.30)$$

where the terms in w_g are now regarded as forcing terms. The normal acceleration is again given by (10.21) and we can solve (10.29) to find

$$\hat{D}\hat{W} - \hat{q} = -z_w(\hat{w} + \hat{w}_g)$$

and hence on substituting we obtain

$$n = \frac{\hat{z}_w}{C_{L_e}}(\hat{w} + \hat{w}_g) \qquad (10.31)$$

We proceed to transform (10.29) and (10.30) giving

$$\left.\begin{aligned}(s + \hat{z}_w)\bar{w} \qquad\qquad\qquad - \bar{q} &= -z_w \bar{w}_g \\ (m_{\dot{w}}s + m_w)\bar{w} + (s + m_q)\bar{q} &= -(m_{\dot{w}}\hat{D} + m_w)\bar{w}_g\end{aligned}\right\}$$

assuming that w and q were zero initially. Then using Cramer's rule we can solve these equations for the transformed vertical velocity in the form

$$\bar{w} = \frac{G_1''(s)}{F'(s)}\bar{w}_g = H_{ww_g}(s)\bar{w}_g(s) \tag{10.32}$$

where H_{ww_g} is the transfer function connecting the vertical velocity change to the upgust velocity. In this equation

$$G_1''(s) = -\left[(\hat{z}_w + m_{\dot{w}})s + (m_w + z_w m_q)\right] \tag{10.33}$$

and $F'(s)$ is again the characteristic equation corresponding to (10.29) and (10.30) and given by (9.80), with s replacing λ. The response in pitching velocity can be found similarly. We also take the Laplace transform of (10.31) to give

$$\bar{n} = \frac{\hat{z}_w}{C_{L_e}}(\bar{w} + \bar{w}_g) \tag{10.34}$$

Then substituting for $\bar{w}$ from (10.32) we find

$$\bar{n} = \frac{\hat{z}_w K}{C_{L_e} F'}\bar{w}_g = H_{nw_g}(s)\bar{w}_g(s) \tag{10.35}$$

where $K(s) = G_1'' + F'$, and H_{nwg} is the transfer function connecting normal acceleration to upgust velocity. Substituting for G_1'' from (10.33) and for F' from (9.80), $K(s)$ is given by

$$K(s) = s(s + \hat{m}_q) \tag{10.36}$$

The response in normal acceleration as a function of time or frequency can be found from (10.35) using the methods already described.

Worked example 10.1

An aircraft is flying at 120 m s^{-1} when it meets a sharp-edge upgust of speed 10 m s^{-1}. Find the equation for the normal acceleration given the following characteristics: $C_{Le} = 0.08$, $\tau = 2.3$ s, $z_w = 2.1$, $m_w = 4.8$, $m_{\dot{w}} = 0.8$, $m_q = 1.3$, $z_{\dot{w}} = 0 = z_q$.

Solution

The Laplace Transform of a unit step is $1/s$ so that $\bar{w}_g = |\hat{w}_g|/s$, where $|\hat{w}_g|$ is the magnitude of the step. We then have from (10.35), (10.36) and (9.80) with s replacing λ

$$\bar{n} = \frac{\hat{z}_w}{C_{L_e}}\left[\frac{(s + m_q)s}{s^2 + (m_q + z_w + m_{\dot{w}})s + m_q z_w + m_w}\right]\frac{|\hat{w}_g|}{s}$$

Cancelling out an s top and bottom and substituting numbers gives

$$\bar{n} = 26.25\frac{s + 1.3}{s^2 + 4.2s + 7.53}|\hat{w}_g| \tag{i}$$

We shall be using the transforms

$$\mathcal{L}\left[e^{-at}\cos bt\right] = \frac{s + a}{(s + a)^2 + b^2} \text{ and } \mathcal{L}\left[e^{-at}\sin bt\right] = \frac{b}{(s + a)^2 + b^2}$$

to invert (i). Completing the square in the denominator and adjusting the numerator to match the transforms gives

$$n = 26.25\frac{(s + 2.1) - 0.453 \times 1.766}{(s + 2.1)^2 + 1.766^2}|\hat{w}_g|$$

and inverting,

$$n = 26.25e^{-2.1\hat{t}}\left[\cos 1.766\hat{t} - 0.453\sin \hat{t}\right]|\hat{w}_g|$$

Finally substituting $\hat{t} = t/\tau$ and $\hat{w}_g = w_g/V_e = 10/120 = 0.0833$ then

$$n = 2.18e^{-0.913\hat{t}}\left[\cos 0.768t - 0.453\sin 0.768t\right]$$

More realistic gust velocity profiles are often used for various purposes. Examples are the ramp, trapezoidal and (1 − cosine) profiles as shown in figure 10.6.

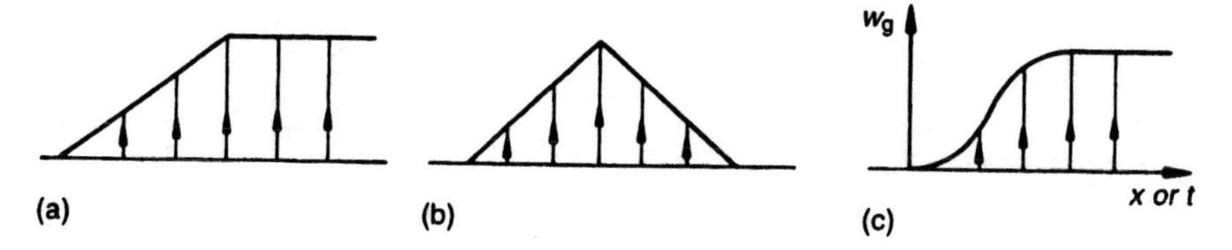

Fig. 10.6 Various types of gust: (a) ramp gust, (b) trapezoidal gust, (c) 1 − cosine gust

At the beginning of this section it was pointed out that the aircraft was being treated as though it were a point; in fact we can improve on this treatment and allow for the rate of change of gust velocity over the length of the aircraft. Considering figure 10.5(b) the increase of upward air velocity over the length l of the aircraft is $l.\mathrm{d}w_g/\mathrm{d}x$, to first order. The increase of relative upward air velocity over the length of the aircraft due to a pitching velocity q is $-ql$. Hence the effective pitching velocity produced by a gust is $q_g = -\mathrm{d}w_g/\mathrm{d}x$. The aircraft must still be fairly short compared with the length of the gust; one-tenth of the gust length has been suggested as the minimum. The modifications to the equations to allow for this can be incorporated in the same way as for the constant upgust. We still have not taken into account the very short-lived initial effects where the aircraft has only partly entered the gust.

10.3.2 Introduction to random variable theory

A random quantity can be considered to consist of the sum of a mean value and a fluctuating part of zero mean value taken over a long time. In the context of an aircraft flying through natural turbulence in the atmosphere, only the fluctuating part of the air velocity is of interest as the mean value simply deflects the path of the aircraft through space.

Various results of random variable theory will be needed in the subsequent development. We start by considering a periodic function of time, $F(t)$, with period T. This can be represented by a Fourier series thus:

$$F(t) = a_0 + \sum_{n=1}^{\infty} a_n \cos n\omega_0 t + \sum_{n=1}^{\infty} b_n \sin n\omega_0 t \tag{10.37}$$

where $\omega_0 = 2\pi/T$ is the fundamental circular frequency, and the coefficients a_0, a_n and b_n are given by

$$a_0 = \frac{1}{T}\int_0^T F(t)\mathrm{d}t \tag{10.38}$$

$$a_n = \frac{2}{T}\int_0^T F(t)\cos n\omega_0 t \cdot \mathrm{d}t \tag{10.39}$$

$$b_n = \frac{2}{T}\int_0^T F(t)\sin n\omega_0 t \cdot \mathrm{d}t \tag{10.40}$$

Using the complex variable (10.37) can be written in the alternative form

$$F(t) = \sum^{\infty} c_n \mathrm{e}^{in\omega_0 t} \tag{10.41}$$

where $c_0 = a_0$, $c_n = (a_n - ib_n)/2$ for $n > 0$ and $c_n = (a_n + ib_n)/2$ for $n < 0$. Corresponding to (10.41) we have

$$c_n = \frac{1}{T}\int_0^T F(t)\mathrm{e}^{-in\omega_0 t}\mathrm{d}t \tag{10.42}$$

where the c_n are complex. The magnitudes of the coefficients c_n can be plotted as a series of vertical lines at equal spacings as in figure 10.7(a) and are known as a spectrum.

Since the function is periodic we can displace the range of integration in (10.42) to other ranges, for example to $-T/2$ to $T/2$, without affecting the results.

A function $F(t)$ with discontinuities in slope will require an infinite number of coefficients c_n for an exact representation. We therefore let $T \to \infty$ in a manner such that $n\omega_0 \to \omega$, $\omega_0 \to \mathrm{d}\omega$ and define

$$C(\omega) = \lim_{\omega_0 \to 0} \frac{c_n}{\omega_0}$$

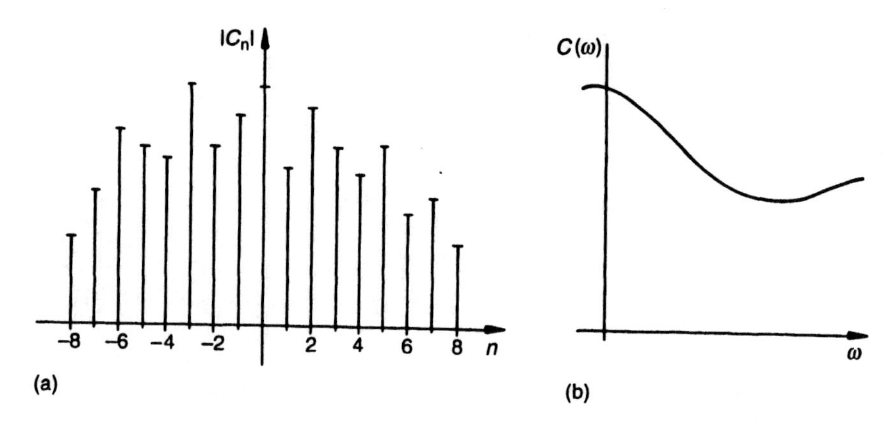

Fig. 10.7 Frequency spectra: (a) discrete, (b) continuous

then (10.41) and (10.42) become

$$F(t) = \lim_{\omega_0 \to 0} \sum_{n=-\infty}^{\infty} \frac{C_n}{\omega_0} e^{in\omega_0 t} \cdot \omega_0 = \int_{-\infty}^{\infty} C(\omega) e^{i\omega t} d\omega \tag{10.43}$$

and using $\omega_0 = 2\pi/T$ in (10.42)

$$C(\omega) = \lim_{\omega_0 \to 0} \frac{C_n}{\omega_0} = \lim_{\omega_0 \to 0} \left[\frac{1}{2\pi} \int_{-T/2}^{T/2} F(t) e^{-in\omega_0 t} dt \right] = \frac{1}{2\pi} \int_{-\infty}^{\infty} F(t) e^{-i\omega t} dt \tag{10.44}$$

Since the fundamental period of the function has become infinitely large the harmonics are only separated in frequency by $d\omega$. The frequency spectrum of the harmonic components is now continuous instead of discrete, as shown in figure 10.7(b). The relations (10.43 and (10.44) can be regarded as a transform and its inverse (compare with the Laplace transform and its inverse) and are known as a Fourier transform pair. They can be written symbolically as

$$C(\omega) = \mathscr{F}\{F(t)\} \text{ and } F(t) = \mathscr{F}^{-1}\{C(\omega)\} \tag{10.45}$$

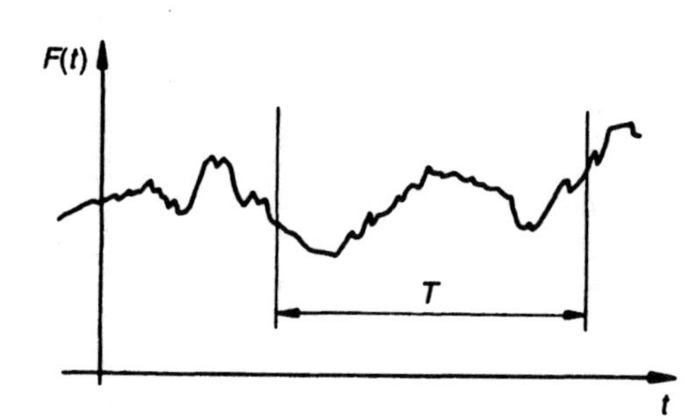

Fig. 10.8 Sketch of time variation of a random variable

The function $F(t)$ can now be aperiodic or totally random as shown in figure 10.8. $C(\omega)$, however, remains continuous and analytic provided that

$$\int_{-\infty}^{\infty} |F(t)| \mathrm{d}t$$

exists and that $F(t)$ is of bounded variation in every finite interval in t; see for instance reference (10.2).

We consider now the Fourier transform of the derivative of a function, written $F'(t)$, that is

$$\begin{aligned}\mathcal{F}\{F'(t)\} &= \int_{-\infty}^{\infty} F'(t)\mathrm{e}^{-\mathrm{i}\omega t}\mathrm{d}t \\ &= \left[F(t)\mathrm{e}^{-\mathrm{i}\omega t}\right]_{-\infty}^{\infty} - \mathrm{i}\omega\int_{-\infty}^{\infty} F(t)\mathrm{e}^{-\mathrm{i}\omega t}\mathrm{d}t\end{aligned}$$

using integration by parts. If $F(t)$ is zero at the limits, the first term vanishes and

$$\mathcal{F}\{F'(t)\} = \mathrm{i}\omega \cdot \mathcal{F}\{F(t)\} \tag{10.46}$$

providing $\mathcal{F}\{F(t)\}$ exists. Repeating this we find

$$\mathcal{F}\{F^{n}(t)\} = (\mathrm{i}\omega)^{n} \cdot \mathcal{F}\{F(t)\} \tag{10.47}$$

We now consider an nth order linear system described by the variable x and subject to a periodic force $F(t)$, the differential equation can be written

$$\left(a_n D^n + a_{n-1}D^{n-1} + \ldots + a_0\right)x = F(t) \tag{10.48}$$

where $D = \mathrm{d}/\mathrm{d}t$. Taking the Fourier transform of both sides we have

$$\left(a_n(\mathrm{i}\omega)^n + a_{n-1}(\mathrm{i}\omega)^{n-1} + \ldots + a_0\right)\mathcal{F}\{x\} = \mathcal{F}\{F(t)\}$$

using (10.47). Then solving

$$\mathcal{F}\{x\} = H(\mathrm{i}\omega)\cdot\mathcal{F}\{F(t)\} \tag{10.49}$$

where

$$H(\mathrm{i}\omega) = \frac{1}{a_n(\mathrm{i}\omega)^n + a_{n-1}(\mathrm{i}\omega)^{n-1} + \ldots + a_0} \tag{10.50}$$

If we had applied the Laplace transform to (10.48), solved to find the transfer function and then substituted $\mathrm{i}\omega$ in place of s we would obtain the result (10.50); evidently $H(\mathrm{i}\omega)$ is the frequency response as defined in Section 10.2.2. If the Fourier transform of $F(t)$ is $C(\mathrm{i}\omega)$ then the inverse of (10.49) is

$$x(t) = \frac{1}{2\pi}\int_{-\infty}^{\infty} C(\mathrm{i}\omega)H(\mathrm{i}\omega)\mathrm{e}^{\mathrm{i}\omega t}\,\mathrm{d}\omega \tag{10.51}$$

We now consider the integral of the square of $F(t)$. Rewriting (10.43) and (10.44) with $2\pi f$ substituted for ω we have

$$F(t) = \int_{-\infty}^{\infty} C(\mathrm{i}f)\mathrm{e}^{\mathrm{i}2\pi ft}\,\mathrm{d}f \tag{10.52}$$

and

$$C(if) = \int_{-\infty}^{\infty} F(t)\mathrm{e}^{-\mathrm{i}2\pi ft}\,\mathrm{d}t \tag{10.53}$$

Denoting the complex conjugate of $C(if)$ by $C^*(if)$ we have

$$C^*(if) = \int_{-\infty}^{\infty} F(t)\mathrm{e}^{\mathrm{i}2\pi ft}\,\mathrm{d}t \tag{10.54}$$

and

$$C(if)\cdot C^*(if) = |C(if)|^2 \tag{10.55}$$

Then the integral of the square of $F(t)$ is

$$\int_{-\infty}^{\infty} F^2(t)\mathrm{d}t = \int_{-\infty}^{\infty} F(t)\cdot F(t)\mathrm{d}t = \int_{-\infty}^{\infty} F(t)\left[\int_{-\infty}^{\infty} C(if)\mathrm{e}^{\mathrm{i}2\pi ft}\mathrm{d}f\right]\mathrm{d}t$$

and on changing the order of integration

$$\int_{-\infty}^{\infty} F^2(t)\mathrm{d}t = \int_{-\infty}^{\infty} C(if)\left[\int_{-\infty}^{\infty} F(t)\mathrm{e}^{\mathrm{i}2\pi ft}\mathrm{d}t\right]\mathrm{d}f$$

Then using (10.56)

$$\int_{-\infty}^{\infty} F^2(t)\mathrm{d}t = \int_{-\infty}^{\infty} C(if)\cdot C^*(if)\mathrm{d}f = \int_{-\infty}^{\infty} |C(if)|^2\,\mathrm{d}f$$

Since $|C(if)|^2$ is an even function of f

$$\int_{-\infty}^{\infty} F^2(t)\mathrm{d}t = 2\int_{0}^{\infty} |C(if)|^2\,\mathrm{d}f \tag{10.56}$$

The mean value of $F^2(t)$ over a time of sampling T, written as $\langle F^2(t)\rangle$, is

$$\langle F^2(t)\rangle = \frac{1}{T}\int_{-T/2}^{T/2} F^2(t)\mathrm{d}t$$

as T tends to infinity $\langle F^2(t)\rangle$ remains finite and

$$\langle F^2(t)\rangle = \lim_{T\to\infty}\frac{1}{T}\int_{-T/2}^{T/2} F^2(t)\mathrm{d}t$$

or from (10.56) and interchanging the operations of integration and taking the limit

$$\langle F^2(t)\rangle = \int_0^\infty \lim_{T\to\infty}\left[\frac{2}{T}|C(if)|^2\right]\mathrm{d}f$$

We write this as

$$\langle F^2(t)\rangle = \int_0^\infty s(f)\mathrm{d}f \tag{10.57}$$

where

$$S(f) = \lim_{T\to\infty}\left[\frac{2}{T}|C(if)|^2\right] \tag{10.58}$$

$S(f)$ is known as the 'spectral density' of the random function $F(t)$; corresponding to the narrow frequency band df, $S(f)\mathrm{d}f$ is the associated amount of $\langle F^2(t)\rangle$. It is also often called the 'power spectral density', reflecting its origin in electronic engineering where a current equal to $F(t)$ dissipates a power $\langle F^2(t)\rangle$ when passing through unit resistance. The abbreviation 'PSD' is often used for power spectral density. In practice experiments to measure real quantities, in a windtunnel say, last a finite time and there may be a small error due to not being able to take the limits in the above expressions to infinity. However, the results will generally be within the limits of engineering accuracy.

We now need to find an expression for the spectral density of the output of a linear system $S_x(f)$ given the spectral density $S(f)$ of an input $F(t)$. From (10.51) we have the output as

$$x(t) = \int_{-\infty}^{\infty} C(if)H(if)\mathrm{e}^{\mathrm{i}2\pi ft}\mathrm{d}f$$

where $2\pi f$ has been substituted for ω. Then following an analysis similar to the previous one we have

$$\langle x^2(t)\rangle = \int_0^\infty \lim_{T\to\infty}\left[\frac{2}{T}|C(if)|^2|H(if)|^2\right]\mathrm{d}f$$

that is, the special density of $x(t)$ is

$$S_x(f) = S(f)|H(if)|^2 \tag{10.59}$$

and

$$\langle x^2(t)\rangle = \int_0^\infty S_x(f)\mathrm{d}f = \int_0^\infty S(f)|H(if)|^2 \mathrm{d}f \tag{10.60}$$

which are the primary results needed.

We introduce one further new concept, that of the 'auto-correlation function' of $F(t)$ defined by

$$R(\tau) = \lim_{T\to\infty} \frac{1}{T}\int_{-T/2}^{T/2} F(t)F(t+\tau)\mathrm{d}t \tag{10.61}$$

Evidently it is the average of the product of the signal and its value at a time later by the amount τ. If we put $\tau = 0$ then

$$R(0) = \lim_{T\to\infty} \frac{1}{T}\int_{-T/2}^{T/2} F^2(t)\mathrm{d}t = \langle F^2(t)\rangle \tag{10.62}$$

It can also be shown that the auto-correlation and the spectral density form a Fourier transform pair. As τ increases we expect $R(\tau)$ to decrease and fall off to zero as shown in figure 10.9 as the velocities at well separated times will be completely uncorrelated.

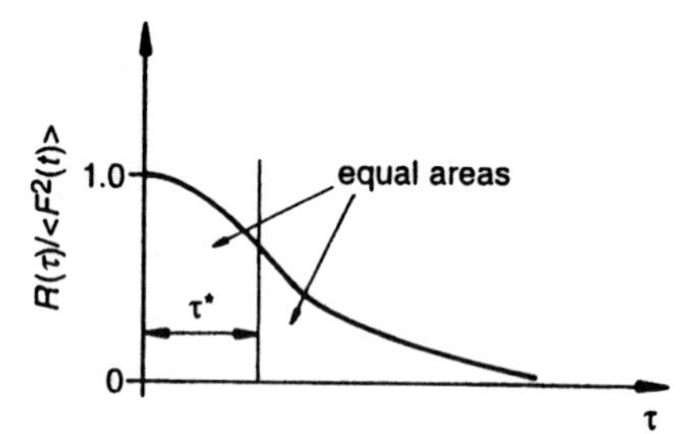

Fig. 10.9 Definition of characteristic time from autocorrelation function

We can define a 'characteristic time' τ^* such that $\tau^*.\langle F^2(t)\rangle$ equals the area under the correlation function as shown in figure 10.9.

10.3.3 Application of random variable theory, the 'PSD method'

The main adaptation required to the theory of the previous section in describing atmospheric turbulence is to adjust it to the fact that it is a variation of atmospheric velocity with distance rather than time that is involved. In the functions and integrals of the previous section we replace time by distance and frequency by a quantity known as 'wave number'. Frequency has the dimensions of $(\text{time})^{-1}$ and wave number Ω, defined as $2\pi/\lambda$ where λ is the wavelength, therefore has the dimensions of $(\text{length})^{-1}$. To emphasize this change of viewpoint we change the symbol for spectral density to Φ. We assume that the statistical properties of the turbulence are constant with time, known as 'stationarity', and over a suitably large region of space, i.e. homogeneous. The statistical properties of turbulence in a particular case may be measured over a long period of time, or on a number of separate occasions over a fixed time and the results then averaged. The latter average is known as an 'ensemble average'. If a case of turbulence is both stationary and homogeneous it is described as 'ergodic'; in such a case the ensemble average will be equal to the time average.

Turbulence is a highly three-dimensional phenomenon so that not only can spectral density and correlation functions be defined for the three velocity components along their respective axes, but also cross spectra and cross correlation functions which involve two axes in their definition.

We assume that the aircraft velocity is much larger than any component of turbulence so that turbulence velocities are effectively constant during the time the aircraft flies through any region of turbulence. This is referred to as 'frozen' turbulence and the concept is known as Taylor's hypothesis. We again concentrate our discussion on the variation of the vertical gust velocity along the flight direction and the normal acceleration produced. The method of course can be applied to any turbulence velocity component and any variable of the aircraft motion or something derived from it, including structural parameters such as stress or bending moment. The steps in the calculation are as follows:

1. Choose a suitable atmospheric turbulence model and hence an expression for the spectral density.
2. Deduce the spectral density of the aircraft response from the equivalent of (10.59).
3. From the latter deduce the probability of the variable of interest having a value outside certain predetermined limits.

A common choice for expressions to describe the atmospheric turbulence spectral densities are the von Kármán expressions which fit the experimental data well. For the vertical velocity we have

$$\Phi(\Omega) = \frac{\sigma^2 L\left[1 + (8/3)(aL\Omega)^2\right]}{2\pi\left[1 + (aL\Omega)^2\right]^{11/6}} \tag{10.63}$$

In this expression L is the 'integral scale length', defined in a similar manner to the characteristic time τ^* through a correlation function, and is a measure of the average wavelength, and frequently taken to be 750 m. Also σ is the root mean square turbulence velocity and ranges from 0.5 m s^{-1} in clear air to over 4 m s^{-1} in storms; a is a purely numerical constant and is taken to be 1.339. Figure 10.10 shows a plot of the resulting spectral density.

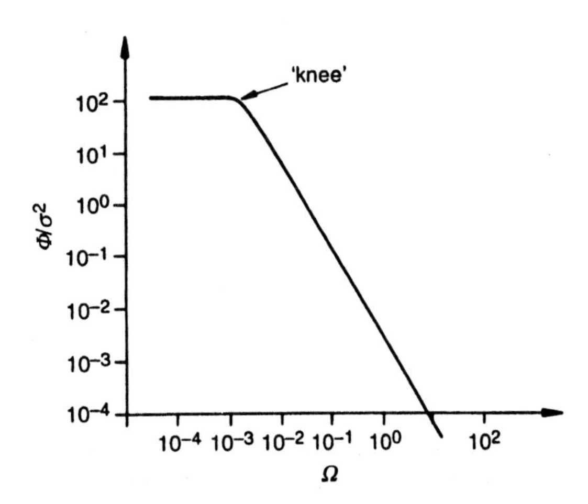

Fig. 10.10 Kármán spectral density variation with wave number

At high frequencies Φ tends to zero like $\Omega^{-5/3}$, a region known as the 'inertial sub-range', whilst at low frequency $\Phi = \sigma^2 L/2\pi$, a constant. The bend in the curve separates the two and is known as the 'knee'. This expression for the spectral density is only valid away from the ground and possibly away from certain meteorological phenomena.

The input turbulence spectral density is converted into the spectral density of normal acceleration using the frequency response function of Section 10.2.2. As the aircraft flies through the turbulence it is subjected to an upgust variation with time, so as in Section 10.3.1 we convert a variation in space to one in time. A wave of length in space of λ is converted to a frequency of $f = U_e/\lambda$, then substituting for λ in terms of the wave number gives $\Omega = 2\pi f/U_e$. As indicated in Section 10.3.1 the maximum normal acceleration is determined primarily by the response in the SPPO mode and its frequency response is determined from (10.35). The natural frequency of the SPPO mode is usually in the inertial sub-range and frequencies close to it are amplified relative to the others, giving a peak in the normal acceleration frequency response at this point. The mean square of the normal acceleration can then be calculated from the output spectral density using (10.60) suitably adapted.

The final step is to calculate the probability of exceeding a certain value of the normal acceleration corresponding to the maximum that the aircraft is designed to withstand. Atmospheric turbulence is approximately a Gaussian process with zero mean and so it is common to assume that the normal acceleration output is also Gaussian. The probability that the normal acceleration lies between the values n and $n + \mathrm{d}n$ is then

$$P(n) = \frac{1}{\sigma_n\sqrt{2\pi}} \mathrm{e}^{-n^2/2\sigma_n^2} \tag{10.64}$$

where σ_n^2 is the mean square of the normal acceleration found in the previous step. A plot of the probability density function is sketched in figure 10.11.

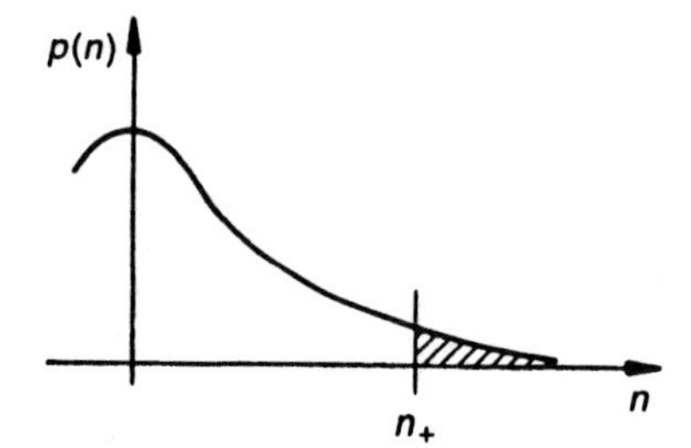

Fig. 10.11 Definition of n_+ from probability density function

The probability that the normal acceleration will exceed a critical value n_+ is equal to the area under the curve and to the right of n_+. This is

$$P(n_+) = \frac{1}{\sigma_n\sqrt{2\pi}} \int_{n_+}^{\infty} \mathrm{e}^{-n^2/2\sigma_n^2} \mathrm{d}n$$

or

$$P(n_+) = \tfrac{1}{2}\left[1 - erf\left(n_+/\sqrt{2}\sigma_n\right)\right] \tag{10.65}$$

where erf(x) is the probability integral. We may interpret this as the fraction of the total flight time for which n exceeds the critical value. There may be a second (negative) critical value, n_-, so that the total probability of damage to the aircraft is $P(n_+) + P(n_-)$. In fact neither turbulence input nor the aircraft response to it is properly Gaussian and also better methods for connecting the spectral density of the output to the probability of damage have been developed.

10.3.4 Statistical discrete gust method

This method has the same objectives as the PSD method but aircraft responses in the time domain are used; actually it is a set of related methods of which we shall describe one. The method starts from the observation that turbulence, including atmospheric turbulence, in fact contains what are known as coherent structures and is not purely random. Examples of coherent structures are such things as vortices in various shapes and localized jets. Calculations are made of the aircraft response in the variable of interest to a 'patterns' of gusts of the (1 − cosine) form: see figure 10.6(c). In the simplest pattern the vertical gust velocity is defined by two parameters, the maximum gust velocity $\hat{w}_g$ and a gradient distance H_1 which varies from zero to the integral scale length L. The gust velocity variation is then

$$\left.\begin{aligned} w_g(x,H_1) &= \frac{\hat{w}_g}{2}\left(1-\cos(\pi s/H_1)\right) \text{ for } 0 \le x \le H_1 \\ w_g(x,H_1) &= \hat{w}_g \text{ for } H_1 < x \le L \end{aligned}\right\} \tag{10.66}$$

where x is the distance from the start of the gust. The gradient length and the maximum gust velocity are related by the expression

$$\hat{w}_g(H) = U_0 H_1^k \tag{10.67}$$

where k is a constant usually taken to be 1/3 and U_0 is a constant with the dimensions of velocity and taken to be 1 m s^{-1}. The variation of $\hat{w}_g$ and the gust velocity for typical members of the pattern is shown in figure 10.12(a).

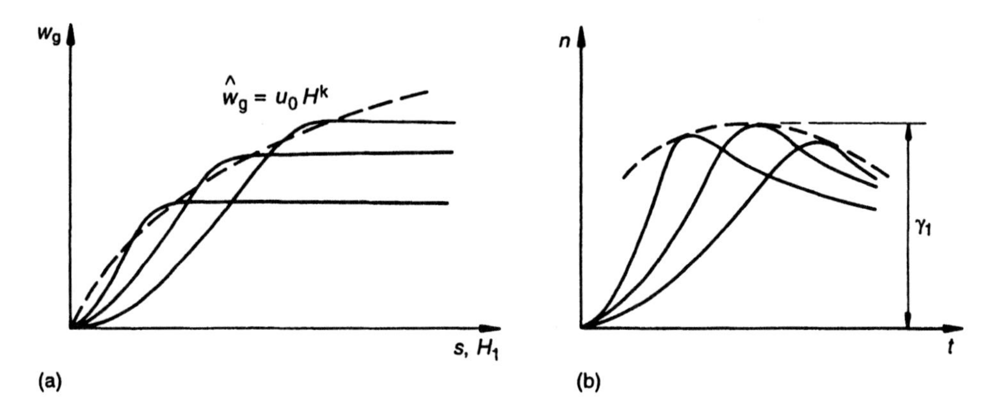

Fig. 10.12 Statistical discrete gust method: (a) pattern of 1 − cosine gusts, (b) envelope of corresponding normal acceleration responses

Figure 10.12(b) sketches the corresponding normal acceleration responses; the peak of the envelope of the peaks of the individual curves is denoted by γ_1. The second pattern consists of a (1 − cosine) ramp followed after a time delay τ_1 by a second ramp in the opposite sense as shown in figure 10.13.

The maximum velocity for the second ramp is related to its gradient distance H_2 by a relation of the form of (10.67) and has the maximum velocity change reduced by a factor p_2. The third pattern is similarly defined with a third ramp in the opposite sense to the second. The aircraft is subjected to (1) all possible single ramps, (2) all possible combinations of two

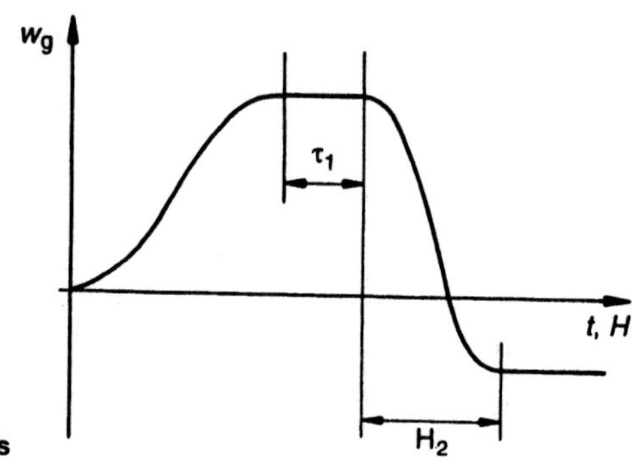

Fig. 10.13 Second pattern of gusts

ramps varying H_1, H_2 and τ_1, (3) all possible combinations of three ramps, and so on. The corresponding γ values are γ_1, γ_2, γ_3 etc. As the number of ramps in a pattern increases the probability of encountering that number decreases and to account for this reduction factors p_1, p_2, p_3 etc. are calculated from

$$p_1 = 1, p_i = 1/0.88\sqrt{i}, \text{ for } i \geq 2 \tag{10.68}$$

Then the maximum peak amplitude of response $\bar{\gamma}$ is taken to be the largest of the products $p_i\gamma_i$ found; this is referred to as the 'tuned case' and the corresponding value of $\bar{H}$ is calculated from

$$\bar{H} = (H_1 H_2 \ldots H_n)^{1/n} \tag{10.69}$$

where n is the corresponding number of component ramps. The average rate N of occurrence of peaks of normal acceleration equal to or greater than n_+ is then given by

$$N = \frac{\alpha}{\lambda \bar{H}} \exp\left(-\frac{n_+}{\beta \gamma}\right) \tag{10.70}$$

In this equation α has been determined from experiment and varies from 0.09 for intermittent turbulence to 0.71 for isolated gusts. The factor β is proportional to gust intensity and taken to be $0.09\sigma_w$ and λ can be taken to be 0.3.

This method is claimed to give comparable results to that of the previous method, but only in the inertial sub-range, which is fortunately the range of most practical interest; this is described as an 'overlap' between the methods. This method involves far more effort to use but is claimed to be more realistic; it deals with encounters with relatively isolated severe gusts and patches of continuous turbulence of lesser intensity within the same framework.

10.3.5 Pilot opinion, handling and flying qualities

The final arbiter as to whether the stability and control characteristics of an aircraft are satisfactory has to be the pilot, regardless of the opinions of the designer and his or her team. Satisfactory in this context means that the pilot can control the aircraft accurately without excessive or over-complicated control movements or large forces and that he or she can fly for a reasonable length of time without becoming mentally or physically fatigued. Satisfactory also means that the airworthiness regulations of the countries that the aircraft may operate in have been complied with, so that Certificates of Airworthiness can be obtained. The first of these matters, referred to as 'handling', is dealt with by specifying various parameters characterizing the stability, control and response of the aircraft. The airworthiness requirements usually refer more directly to pilot assessments of the aircraft, known as 'flying qualities'.

The handling quality requirements have been built up over the years from the opinions of pilots, based on their experiences. These opinions have been compared with the characteristics of the modes as found using the methods described and much effort has been put into finding the parameters which give the best correlation with pilot opinion. Different parameters will give the best correlation depending on the mode considered; the intended function of aircraft also has some effect. This research has been carried out using not only normal aircraft but also aircraft equipped with variable artificial stability systems and ground based simulators. For some specialized types of aircraft specific research using these methods still needs to be undertaken. Passenger comfort is another factor in this picture.

Pilot opinion cannot be reported in any obvious numerical way, but a number of scales of pilot opinion have been devised. The most detailed one is the Cooper–Harper scale in which the pilot evaluates an aircraft in any flight by working through a questionnaire. This is a 10-point scale, ranging from a rating of 1 in which the aircraft can be described as having excellent or highly desirable qualities and the pilot does not need to compensate for any deficiencies to obtain the desired performance. At the other end of the scale is a rating of 10 where the aircraft handling has major deficiencies and control may be lost during some portion of the required operation.

In specifying the characteristics required during the design phase the Cooper–Harper scale is too detailed and too subjective and the shorter scale shown below is used.

Level 1 Cooper–Harper ratings 1, 2 and 3. Flying qualities clearly adequate for the mission flight phase.
Level 2 Cooper–Harper ratings 4 and 5. Flying qualities adequate to accomplish the mission flight phase, but some increase in pilot workload and/or degradation in mission effectiveness exists.
Level 3 Cooper–Harper ratings 6 to between 7 and 8. Flying qualities such that the aircraft can be controlled safely, but pilot workload is excessive or mission effectiveness is inadequate.

To account for the differences between aircraft, four classes of aircraft are specified largely according to their mass as shown in table 10.1.

Table 10.1 Aircraft classes (mass in kg)

I	*II*	*III*	*IV*
<570	570–30 000	>30 000	Highly manoeuvrable

The requirements may also depend on the flight phase of which three are recognized: Category A for rapid manoeuvring expected only of military aircraft, Category B for slow manoeuvres not requiring high precision, and Category C for slow manoeuvres requiring accurate flight path control such as take-off and landing.

For the phugoid mode the handling qualities criteria are simple: for Level 1 a damping ratio of more than 0.04 is required but values higher than 0.15 do not result in any further improvement. A time to double-amplitude greater than 55 s can be coped with by a pilot with some difficulty; if this mode is unstable it presents the pilot with a problem in trimming the aircraft. The airworthiness requirements usually ask that its characteristics be satisfactory, which can be interpreted as meaning that it should be stable and have a frequency well separated from that of the SPPO so as to avoid coupling of the modes.

For the SPPO mode both the natural frequency and the damping ratio are important. Curves of constant pilot opinion are plotted against these parameters. A good combination for light aircraft is a damping ratio of 0.6 with a natural frequency of 3 rad s^{-1}; the plots of constant pilot opinion are closed curves centred roughly about this point. Much lower values of the relative damping give excessive overshoot, whilst much higher ones result in a sluggish aircraft to control. Much lower values of the natural frequency make for difficulties in trimming, whilst much higher ones give too rapid initial response and can lead to pilot induced oscillation.

A parameter describing the controllability of the aircraft is also introduced, an example of which is the 'attitude lead time constant', t_θ. If the elevator is moved by a small amount and then held constant, the rate of pitch will build up rapidly to a steady value as shown in figure 10.14(a), then the pitch angle will be increasingly linearly. As a result of the inclination of the aircraft it will begin to climb and the climb angle will reach a steady rate of increase a little after the pitch rate becomes constant. The lag in time between the pitch angle reaching a certain value and the climb angle reaching the same value in the linear part of the curve is the attitude lead time constant, as shown in figure 10.14(b). It needs to be fairly short for accurate control.

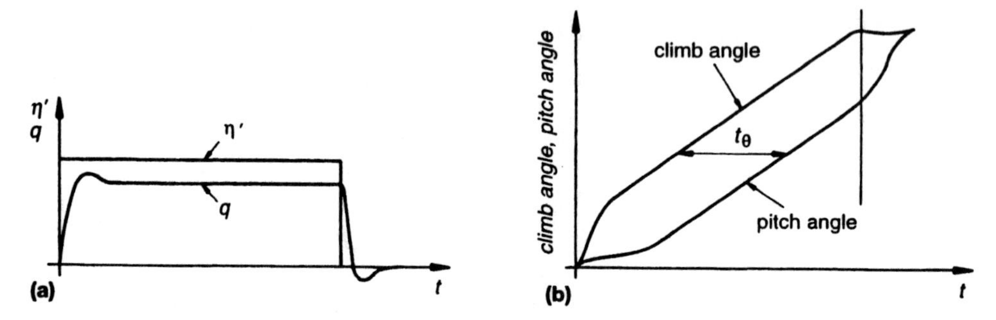

Fig. 10.14 Aircraft response to rectangular pulse input to elevator: (a) pitch rate response, (b) pitch angle and incidence responses

Airworthiness requirements usually state that any short-period oscillation must be heavily damped with the controls both fixed and free. These requirements are in addition to those for static stability discussed in Section 5.7.

Discussion of the handling and flying qualities for the lateral motion is left to Section 11.7.

Student problems

10.1 A stable aircraft in steady level flight has its elevator moved from the initial value η through an arbitrary series of values to a final steady value $\eta + \eta'$, where η' is assumed small. Assuming that the aircraft remains within the range of linear aerodynamics, show that values of the motion variables $\hat{u}_\infty$, $\hat{w}_\infty$ and θ_∞ after the motion has again become steady are

$$\hat{u}_\infty = (\hat{m}_\eta z_w - z_\eta m_w)\eta' / \Delta$$
$$\hat{w}_\infty = (z_\eta m_u - m_\eta z_u)\eta' / \Delta$$
$$\theta_\infty = \{m_\eta(x_w z_u - x_u z_w) - z_\eta(x_w m_u - x_u m_w)\}\eta' / \hat{g}_1\Delta$$

where

$$\Delta = z_u m_w - z_w m_u$$

10.2 Rework the theory of Section 10.3.1 to accommodate the improved representation of an aircraft traversing a gust developed at the end of the section.

10.3 Find the equation for the pitching velocity for the aircraft in Worked example 10.1.

10.4 Find an expression for the normal acceleration produced by a step change in elevator angle of 0.05 rad on an aircraft with the following characteristics: $\hat{z}_w = 5.5$, $m_w = 49.7$, $m_{\dot{w}} = 1.4$, $m_q = 5$, $m_\eta = 49.3$, $z_\eta = 0$, $\tau = 9$ s. (A)

10.5 Find the frequency response function for the pitch due to elevator angle change for the aircraft of the previous problem in the form of (10.16) given the following additional derivatives: $\hat{x}_u = 0.061$, $x_w = -0.355$, $z_u = 0.778$, $z_q = z_{\dot{w}} = m_u = 0$ and $C_{L_e} = 0.53$. Assume level flight.

10.6 Develop a program written in Basic or Fortran to calculate the response of an aircraft to a (1 − cosine) gust adapting the method of Section 10.2.3.

Note

1. The same result can be obtained by adding the centripetal acceleration to the time derivative of vertical velocity.

11
Lateral dynamic stability and response

11.1 Introduction

The main aim of this chapter is to solve the lateral stability equations for the case of free motion, i.e. the homogeneous case, and then using these results to discuss the modes, their dependence on aircraft geometry and the conditions for stability. We start by describing the estimation of the stability derivatives and end with a short section on control and lateral response.

11.2 Lateral stability derivatives

The general remarks on derivatives in Sections 9.2 and 9.2.6 apply equally in this case. There is a difference between these derivatives and the longitudinal ones in that they all have to be estimated by adding together the separate contributions from the various components of the aircraft. A similarity is that many are functions of lift coefficient or incidence. Contributions from the wing usually have to be estimated using a lifting line or lifting surface theory or from empirical data. Some of the aerodynamic mechanisms by which these derivatives are produced have already been discussed in Chapters 6 and 7, but it is worth briefly discussing each of them here in turn.

11.2.1 Derivatives due to sideslip velocity

The derivatives that we are concerned with in this section are

- Y_v, the change in sideforce due to sideslip velocity
- L_v, the change in rolling moment due to sideslip velocity
- N_v, the change in yawing moment due to sideslip velocity.

Section 6.2.2 discussed most of the effects of sideslip on the components of the aircraft and the contributions to the derivatives produced. The important derivative L_v, sometimes known as the dihedral derivative, tends to have both positive and negative contributions, making it one of the most difficult derivatives to estimate accurately. The designer is able to use dihedral angle to adjust the value of L_v to that required. Other contributions discussed in Section 6.2.2 were those from sweepback, which is proportional to lift coefficient, and a contribution from wing position on the fuselage. There is also a small contribution from the fin produced by the sideforce generated.

The derivatives Y_v and N_v are contributed to mainly by the fin and the fuselage. In Section 6.2.2 we saw how sideslip produces an incidence on the fin giving rise to a sideforce and hence a yawing moment. An effect, not mentioned in that section for the sake of simplicity, arises from the trailing vortices from the wing tips. In a sideslip they assume a skewed position as shown in figure 6.2(a). The result is that the horizontal components of the velocities induced by the vortices at the fin do not fully cancel out and a small nett sideways velocity is left. This usually reduces the mean incidence of the fin and hence reduces its lift and therefore the contributions to these derivatives. It was mentioned in Section 6.2.2 that the fuselage produces a sideforce and a destabilizing yawing moment. The changes in the distribution of lift across the wing span on a sweptback wing in a sideslip cause corresponding changes in the trailing vortex drag and hence yawing moments, giving rise to a contribution to the derivative N_v. This is proportional to the square of the lift and can become important at high incidence.

Propellers, if present, produce a sideforce in a sideslip in the same sense as a fin and hence also a yawing moment. Normally they are placed ahead of the aircraft cg and so produce a negative contribution to N_v and therefore are destabilizing. It should be remembered that they also produce a yawing moment about a vertical line through the plane of the disc.

11.2.2 Derivatives due to rate of roll

The derivatives that we are concerned with in this section are

- Y_p, the change in sideforce due to rate of roll
- L_p, the change in rolling moment due to rate of roll
- N_p, the change in yawing moment due to rate of roll.

The most important derivative of this trio is L_p which provides the damping in roll and was discussed in Section 7.2.1. Incidence changes on the wing due to the rolling motion give lift changes producing an opposing moment; since the wing is the largest lifting surface the derivative is relatively large. Similar, but much smaller, contributions are made by the tailplane and fin. The lift changes on the wing due to rolling which produce the damping also cause changes in the trailing vortex drag distribution across the span, giving rise to a yawing moment. This effect is the main source of the derivative N_p. Rolling of the aircraft gives a lateral velocity to the fin which gives rise to a sideforce, producing the derivative Y_p and a small contribution to N_p. The derivative Y_p is usually small and often ignored.

11.2.3 Derivatives due to rate of yaw

The derivatives that we are concerned with in this section are

- Y_r, the change in sideforce due to rate of yaw
- L_r, the change in rolling moment due to rate of yaw
- N_r, the change in yawing moment due to rate of yaw.

Some of the effects of rate of yaw on the components of the aircraft and the derivatives produced were discussed in Section 6.2.3. The most important derivative is N_r which provides the damping in yaw and is contributed to by the wing, fuselage and fin. The main contribution comes from the fin sideforce caused by the fin incidence change discussed in Section 6.2.3; this also causes contributions to the derivatives Y_r and L_r. Rate of yaw causes the wing tips to have different forward speeds giving rise to a rolling moment also discussed in Section 6.2.3. This produces changes in the distribution of drag across the wing span and hence produces a yawing moment which adds to $(-N_r)$. The fuselage also contributes to N_r due to its resistance to sideways motion. Because propellers have a fin-like effect they contribute to Y_r and N_r. The derivative Y_r is usually small and often ignored.

11.2.4 Estimation of the lateral derivatives

In the absence of flight test, computer fluid dynamics or good windtunnel data as in the early stages of the design of an aircraft, derivatives have to be estimated from semi-empirical data such as ESDU data items. Table 11.1 lists the items relevant to each derivative for estimation in subsonic flow. A few of the items are also relevant to supersonic flow and in other cases there are items for supersonic flow.

Table 11.1 Estimation of lateral derivatives using ESDU data items

Force/moment	*Due to sideslip*		*Due to rolling*		*Due to yawing*	
Sideforce	[−]		[−]		[+]	
	79006	W/F	81014	WP	83006	B
	82010	F	85006	WD	82017	F
	93007	F	83006	F	89047	P
	92029	F				
	89047	P				
Rolling moment	[[−]]		[−]		[−]	
	80033	WP	Aero A 06. 01.01	W&T	72021	WP
	Aero A 06.01. 03 & 09	WD	85006	WD	81017	F
	73006	W/F	83006	F		
	81032	T				
	82010	F				
	92007	F				
	93029	F				
Yawing moment	[[+]]		[+]		[−]	
	79006	W/B	81014	WP	71017	WP
	82010	F	85006	WD	83026	P
	93007	F	83006	F	82017	F
	92029	F			89047	P
	89047	P				

Key

[−] or [+] indicates the usual sign of the derivative.
[[−]] or [[+]] indicates that the derivative must be of this sign for static stability.

Letters indicate that the contribution is due to the:

W wing
WP wing planform
WD wing dihedral
B fuselage
W/F wing and fuselage together or wing–fuselage interaction
T tailplane
F fin
P propeller

Data items 81032, 82011, 84002 and 85010 give general information on the estimation of these derivatives.

Table 11.2 Conversions of derivatives to concise forms

Y-force	*Rolling moment*	*Yawing moment*
$y_v = -Y_v$	$l_v = -\mu_2 L_v/i_x$	$n_v = -\mu_2 N_v/i_z$
$y_p = -Y_p/\mu_2$	$l_p = -L_p/i_x$	$n_p = -N_p/i_z$
$y_r = -Y_r/\mu_2$	$l_r = -L_r/i_x$	$n_r = -N_r/i_z$
$y_\xi = -Y_\xi$	$l_\xi = -\mu_2 L_\xi/i_x$	$n_\xi = -\mu_2 N_\xi/i_z$
$y_\zeta = -Y_\zeta$	$l_\zeta = -\mu_2 L_\zeta/i_x$	$n_\zeta = -\mu_2 N_\zeta/i_z$

11.2.5 Control derivatives

The derivatives, with relevant ESDU data items indicated, are

- the sideforce, rolling moment and yawing moment due to aileron, Y_ξ, L_ξ (88013) and N_ξ (88029) and
- the sideforce, rolling moment and yawing moment due to rudder, Y_ζ, L_ζ and N_ζ (87008).

11.2.6 Conversion of derivatives to concise forms

Table 11.2 summarizes the required conversions and is based on table 8.4.

11.2.7 Conversions to derivatives in American notation

The conversions between these derivatives and those expressed in the American notation, defined in Section 8.6.3, have been shown by Babister, reference (11.1), to be as follows.

The stability derivatives $C_{y\beta}$, $C_{l\beta}$ and $C_{n\beta}$ in American notation are equal to their counterparts Y_v, L_v and N_v in the current notation.

The stability derivatives C_{y_p}, C_{l_p}, C_{n_p}, C_{y_r}, C_{l_r} and C_{n_r} in American notation are equal to twice their counterparts Y_p, L_p, N_p, Y_r, L_r and N_r in the current notation.

The control derivatives $C_{y\xi}$, $C_{l\xi}$, $C_{n\xi}$, $C_{y\zeta}$, $C_{l\zeta}$ and $C_{n\zeta}$ are equal to their counterparts Y_ξ, L_ξ, N_ξ, Y_ζ, L_ζ and N_ζ, in the current notation.

11.3 Solution of lateral equations

The concise dynamic-normalized lateral equations are (8.57) to (8.61). For most of this chapter we shall be using wind axes and hence we put $W_e = \hat{W}_e = 0$ and $U_e = V_e$ so that $\hat{U}_e = 1$. The equations are then

$$(\hat{D} + \hat{y}v)\hat{v} + y_p\hat{p} + (1 + y_r)\hat{r} - \hat{g}_1\phi - \hat{g}_2\psi = -\hat{y}(\hat{t}) \quad (11.1)$$

$$l_v\hat{v} + (\hat{D} + l_p)\hat{p} + (e_x\hat{D} + l_r)\hat{r} = -\hat{l}(\hat{t}) \quad (11.2)$$

$$n_v\hat{v} + (e_z\hat{D} + n_p)\hat{p} + (\hat{D} + n_r)\hat{r} = -\hat{n}(\hat{t}) \quad (11.3)$$

$$\hat{p} - \hat{D}\phi = 0 \quad (11.4)$$

$$\hat{r} - \hat{D}\psi = 0 \quad (11.5)$$

In these equations the control terms are, from (8.62) to (8.64),

$$\left.\begin{aligned}\hat{y}(\hat{t}) &= \hat{y}_\xi\xi'(\hat{t}) + \hat{y}_\zeta\zeta'(\hat{t})\\ \hat{l}(\hat{t}) &= \hat{l}_\xi\xi'(\hat{t}) + \hat{l}_\zeta\zeta'(\hat{t})\\ \hat{n}(\hat{t}) &= \hat{n}_\xi\xi'(\hat{t}) + \hat{n}_\zeta\zeta'(\hat{t})\end{aligned}\right\} \quad (11.6)$$

Frequently interest is confined to level flight; then $\hat{g}_2 = 0$ and we can substitute for $\hat{r}$ in (11.1)–(11.3) from (11.5) reducing the number of variables to four.

The product of inertia terms e_x $(= I_{xz}/I_x)$ and e_z $(= I_{xz}/I_z)$ are usually small and can be eliminated between (11.2) and (11.3) without altering the effective values of the derivatives unduly. Thus if we subtract e_z times (11.2) from (11.3) and e_x times (11.3) from (11.2) and divide through by $A_2 = 1 - e_xe_z$ we find

$$\hat{l}_v^*\hat{v} + (\hat{D} + l_p^*)\hat{p} + l_r^*\hat{r} = \hat{l}^*(\hat{t}) \quad (11.7)$$

$$n_v^*\hat{v} \qquad + n_p^*\hat{p} + (\hat{D} + n_r^*)\hat{r} = \hat{n}^*(\hat{t}) \tag{11.8}$$

where

$$\left.\begin{aligned}
l_v^* &= (l_v - e_x n_v)/A_2\\
l_p^* &= (l_p - e_x n_p)/A_2\\
l_r^* &= (l_r - e_x n_r)/A_2\\
l^*(\hat{t}) &= (l(\hat{t}) - e_x n(\hat{t}))/A_2\\
n_v^* &= (n_v - e_z l_v)/A_2\\
n_p^* &= (n_p - e_z l_p)/A_2\\
n_{rr}^* &= (n_r - e_z l_r)/A_2\\
n_v^*(\hat{t}) &= (n(\hat{t}) - e_z l(\hat{t}))/A_2
\end{aligned}\right\} \tag{11.9}$$

Our equations are now (11.1), (11.4), (11.5), (11.7) and (11.8); this form gives a slightly clearer view of matters. Each can readily be solved for the highest derivative thus putting them in state-space form as described in Section 7.5. A schematic representation of these equations is shown in figure 11.1 which illustrates the various interactions between the three freedoms.

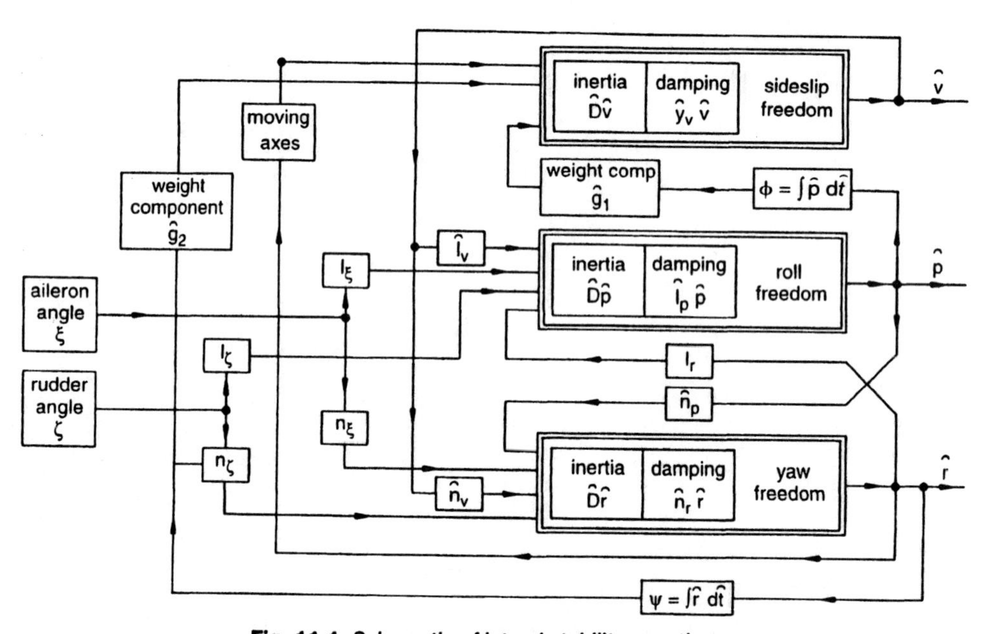

Fig. 11.1 Schematic of lateral stability equations

In this figure we have omitted the effects of the small derivatives $\hat{y}_p$ and $\hat{y}_r$. The same figure would represent the original set of equations if we had simply neglected e_x and e_z as well. It should be noted that none of the freedoms has a term proportional to the particular displacement, that is a term having the function of the spring term in the archetypal mass–damper–spring system. Roll and yaw angles are involved but only as coupling terms with the sideslip equation. To a first approximation there is no significant coupling term from roll into the sideslip freedom. The uses to which we can put these equations are, of course, the same as outlined for the longitudinal equations in Section 9.2.

11.3.1 Solution of the equations of free motion

To find the solution of the equations for the case of free motion with the controls fixed we put the right-hand sides to zero and assume a solution of the form

$$\left.\begin{aligned} \hat{v} &= k_1 e^{\lambda \hat{t}} \\ \phi &= k_2 e^{\lambda \hat{t}} \\ \psi &= k_3 e^{\lambda \hat{t}} \end{aligned}\right\} \qquad (11.10)$$

where k_1, k_2 and k_3 and λ are real or complex. We now substitute for $\hat{p}$ and $\hat{r}$ in (11.1), (11.7) and (11.8) from (11.4) and (11.5), and substitute the assumed solutions (11.10). Each term then contains a factor $\exp(\lambda \hat{t})$ which can be cancelled out, leaving the equations

$$\left.\begin{aligned} (\lambda + \hat{y}_v)k_1 &+ (y_p\lambda - \hat{g}_1)k_2 + [(1 + y_r)\lambda - \hat{g}_2]k_3 = 0 \\ l_v^* k_1 &+ (\lambda^2 + l_p^*\lambda)k_2 + l_r^*\lambda k_3 = 0 \\ n_v^* k_1 &+ n_p^*\lambda k_2 + (\lambda^2 + n_r^*\lambda)k_3 = 0 \end{aligned}\right\} \qquad (11.11)$$

We again employ the argument that, apart from the trivial solution $k_1 = k_2 = k_3 = 0$, these equations are incompatible unless their determinant vanishes. Forming the determinant and equating it to zero gives

$$F(\lambda) = \begin{vmatrix} \lambda + \hat{y}_v & y_p\lambda - \hat{g}_1 & (1 + y_r)\lambda - \hat{g}_2 \\ l_v^* & \lambda^2 + l_p^*\lambda & l_r^*\lambda \\ n_v^* & n_p^*\lambda & \lambda^2 + n_r^*\lambda \end{vmatrix} = 0 \qquad (11.12)$$

Multiplying out the determinant gives a quintic in λ which we write as

$$F(\lambda) = \lambda^5 + B_2\lambda^4 + C_2\lambda^3 + D_2\lambda^2 + E_2\lambda = 0 \qquad (11.13)$$

The coefficients B_2, C_2, D_2 and E_2 are functions of the derivatives and $\hat{g}_1$ and $\hat{g}_2$ as follows:

$$B_2 = y_v + l_p^* + n_r^* \qquad (11.14)$$

$$C_2 = y_v(l_p^* + n_r^*) + (l_p^*n_r^* - l_r^*n_p^*) - l_v^*y_p - n_v^*(1 + y_r) \qquad (11.15)$$

$$\begin{aligned} D_2 = {} & y_v(l_p^*n_r^* - l_r^*n_p^*) - (1 + y_r)(l_p^*n_v^* - l_v^*n_p^*) \\ & + y_p(l_r^*n_v^* - l_v^*n_r^*) + l_v^*\hat{g}_1 + n_v^*\hat{g}_2 \end{aligned} \qquad (11.16)$$

$$E_2 = \hat{g}_1(l_v^*n_r^* - l_r^*n_v^*) + \hat{g}_2(l_p^*n_v^* - l_v^*n_p^*) \qquad (11.17)$$

It may be noted that double product terms like $(l_p^*n_r^* - l_r^*n_p^*)$ can be simplified on substituting from (11.9), in this case to $(l_pn_r - l_rn_p)/A_2$. The result is that only the first term on the right of (11.14) is not inversely proportional to A_2. The quantity A_2 is in fact the coefficient of λ^5 in the characteristic equation of (11.1) to (11.5) as they stand.

11.3.2 Iterative solution of the characteristic quintic

We can immediately extract the root $\lambda = 0$, then like the longitudinal quartic the remaining quartic has a fairly fixed pattern of roots and hence a normal pattern for the coefficients. The lateral quartic normally has two real roots, one large compared to unity, and one small; the remaining roots form a complex pair. The result is normally that the coefficients B_2, C_2 and D_2 are rather larger than unity and increase in size in that order; by contrast the coefficient E_2 is usually much the smallest of all. These facts lead to approximate expressions for the roots which can then be improved by iteration.

11.3.2.1 The large real root

To find this root we note that if λ is rather larger than unity then the inequalities $\lambda^4 \gg \lambda^3 \gg \lambda^2 \gg \lambda \gg 1$ hold, so keeping only the two highest order terms of the quartic we have $\lambda^4 + B_2\lambda^3 \approx 0$, or

$$\lambda \approx -B_2 \tag{11.18}$$

as a first approximation. This root is associated with a mode known as the 'roll subsidence', which describes it fairly accurately in that it is a highly damped motion mainly in the roll freedom. It is often possible to improve this value in a simple manner. We first write the quartic in the form

$$\lambda + B_2 + C_2/\lambda + D_2/\lambda^2 + E_2/\lambda^3 = 0$$

and then approximate the last three terms on the left, which should be the smaller ones, using the value already obtained which we write as λ'. Then solving for the first term, written as λ'', we find

$$\lambda'' = -\left(B_2 + \frac{C_2}{\lambda'} + \frac{D_2}{\lambda'^2} + \frac{E_2}{\lambda'^3}\right) \tag{11.19}$$

which can be used to repeat the iteration process. In some cases this process does not converge, the second or third terms in the brackets becoming comparable to the first, and it is necessary to use some other method such as Newton's to improve the root. The time to halve the amplitude of an initial disturbance can then be found as in Section 9.2.2.

11.3.2.2 The small real root

In this case we note that if λ is less than unity the inequalities that hold now are $1 \gg \lambda \gg \lambda^2 \gg \lambda^3 \gg \lambda^4$; then keeping only the two lowest-order terms of the quartic we have $D_2\lambda' + E_2 \approx 0$, or

$$\lambda' = -E_2/D_2 \tag{11.20}$$

as a first approximation. This root is associated with a mode known as the 'spiral' mode, again a fair description in that typically the motion is a slow, almost correctly banked turning motion. The mode may be stable or slightly unstable; the spiral motion is of increasing radius if stable and vice versa. This root can be improved by using the first approximation to estimate the high-order terms. Denoting the first and second approximations as λ' and λ'' and rewriting the quartic as

$$\lambda'^4 + B_2\lambda'^3 + C_2\lambda'^2 + D_2\lambda'' + E_2 \approx 0$$

and solving for λ'' we obtain

$$\lambda'' = -\frac{1}{D_2}\left(E_2 + C_2\lambda'^2 + B_2\lambda'^3 + \lambda'^4\right) \tag{11.21}$$

which can be used to continue the iteration. Section 9.2.2 can then be used to find the time to halve or double the initial amplitude of a disturbance.

11.3.2.3 The complex pair

Having obtained two roots of the four we can divide out the corresponding factors from the quartic, either one at a time or by multiplying the factors together and dividing by the resulting quadratic. The resulting quadratic is solved by the usual method. An alternative is to use the fact that the coefficient B_2 is equal to minus the sum of the roots and E_2 is equal to their product, but this information is often better left to provide a check on the results. This pair of roots are associated with the 'dutch roll' mode or lateral oscillation. For most aircraft all the freedoms – sideslip, roll and yaw – are excited, and the magnitudes and phase differences between them are strongly dependent on the aircraft configuration. This gives a rather complicated and variable motion for which there are no simple generalizations. Having found the root, Section 9.2.2 enables the frequency and time to halve the initial amplitude of a disturbance to be found. A spreadsheet setup for solving lateral quartics is given at the end of this chapter.

Worked example 11.1

Solve the lateral stability quartic $\lambda^4 + 9.43\lambda^3 + 31\lambda^2 + 192\lambda + 3.18 = 0$ and state whether the aircraft is stable or not.

Solution

Using (11.18) we have the first approximation to the root corresponding to the roll mode as $\lambda' = -B_2 = -9.43$. Then using the improved result (11.19) we find

$$\lambda'' = -\left(B_2 + \frac{C_2}{\lambda'} + \frac{D_2}{\lambda'2} + \frac{E_2}{\lambda'3}\right) = -\left(9.43 + \frac{31}{-9.43} + \frac{192}{(-9.43)2} + \frac{3.18}{(-9.43)3}\right) = -8$$

Repetition of this process gives finally $\lambda = -8.447$.

We next find the root corresponding to the spiral mode; a first approximation is from (11.20) $\lambda' = -E_2/D_2 = -3.18/192 = -0.0166$. Then using the improved result (11.21)

$$\begin{aligned}\lambda'' &= -\frac{1}{D_2}\left(E_2 + C_2\lambda'^2 + B_2\lambda'^3 + \lambda'^4\right) \\ &= -\frac{1}{192}\left(3.18 + 31 \times (-0.0166)^2 + 31 \times (-0.0166)^3 + (-0.0166)^4\right) = -0.0166\end{aligned}$$

so that to this accuracy no improvement is possible.

We then multiply together the linear factors corresponding to these roots, obtaining $(\lambda + 8.447)(\lambda + 0.0166) = (\lambda^2 + 8.464\lambda + 0.1402)$ and divide this factor into the quartic using long division. The other quadratic factor is thus found to be $(\lambda^2 + 0.966\lambda + 22.68)$. Finally solving this gives the complex pair $\lambda = -0.485 \pm i4.738$. Since the roots are all negative or have negative real parts the aircraft is stable.

11.4 Discussion of the lateral modes

Taking each mode in turn we discuss their characteristics, which are the result of the interaction of several influences. These are the aerodynamics, the components of the weight along the axes, the effects of inertia, and kinematic effects such as the fact that rotation of the aircraft in yaw combined with the forward speed has the effect of generating a negative sideslip velocity. If an aircraft is given a small rotation ψ around the z-axis then there is a component of the forward speed $V_e\psi$ along $-Oy$. That is $v = -V_e\psi$ or $\hat{v} = \beta = -\psi$. As it is easier to visualize an angle, in this case the sideslip angle β defined in Section 6.2.2, than the normalized sideslip velocity, $\hat{v}$, we use the former in the following discussions, although they are approximately equal.

The moments and products of inertia about the three axes change if we rotate the axes relative to the aircraft. In this way it is possible to find an orientation for the axes in which the products vanish; these axes are then the 'principal axes of inertia' and the moments of inertia are then at a maximum or minimum. An aircraft of span rather less than its length will have a principal axis of inertia roughly pointing in the forward direction and about this axis the inertia will be a minimum.

11.4.1 The zero root

A zero value for an eigenvalue implies a mode with neutral stability, so that any displacements left after the other modes have decayed to zero never die away. Using (11.10) we find

$$\hat{v} = k_1,\ \phi = k_2,\ \psi = k_3 \tag{11.22}$$

so that all the values are constant. Then substituting into (11.1) to (11.3) gives

$$\left.\begin{aligned} \hat{g}_1\phi + \hat{g}_2\psi &= 0 \\ \hat{v} &= 0 \end{aligned}\right\} \tag{11.23}$$

so that there is no sideslip, and the second equation of (8.21) shows that there is no sideways component of the weight. Then substituting $\hat{g}_1 = C_{L_e} = \hat{g}_2 = C_{L_e} \tan \Theta_e$ gives

$$\phi \cos \Theta_e + \psi \sin \Theta_e = 0 \tag{11.24}$$

From this we deduce that if the initial climb angle is non-zero then the aircraft, if stable, reaches a combination of roll and yaw angles such that the y-axis is horizontal. If the climb angle is zero then the final roll angle must be zero and the yaw angle, from (11.24), is then indeterminate. The physical reason for this behaviour is that in the final state there is no moment tending to turn the aircraft back to the original heading. By contrast, in longitudinal stability the components of the weight along the axes provide forces which ultimately restore the aircraft, if stable, to the original trimmed climb angle.

11.4.2 The spiral mode

Referring to the approximate value for the corresponding root (11.20) we see that this mode is stable if D_2 and E_2 are both positive, as would be expected from the work of Section 9.3.3 on test functions. In fact D_2 is always positive and the effective condition is

$$E_2 > 0 \tag{11.25}$$

then from (11.17) we have, assuming level flight,

$$E_2 = \hat{g}_1(\hat{l}_v^* n_r^* - n_v^* l_r^*) = C_{Le}(l_v n_r - n_v l_r)/A_2$$

On substituting for the concise derivatives using table 11.2 the condition becomes

$$(-L_v)(-N_r) - (-N_v)(-L_r) > 0$$

Now the derivatives L_r and N_v are always positive and L_v and N_r are always negative. We can then rewrite the inequality as

$$(-L_v)(-N_r) - N_v L_r > 0$$

or

$$\frac{-L_v}{N_v} > \frac{L_r}{-N_r} \tag{11.26}$$

This result shows that the stability of the mode is solely determined by the balance of the aerodynamic rolling moment derivatives L_v and L_r and the yawing moment derivatives N_v and N_r, that is by purely aerodynamic effects. It would not be expected that inertia effects would be much involved as the motion is a very slow one. The stability of the mode can be improved by increasing $(-L_v)$, perhaps by increasing the dihedral, or by decreasing N_v which means decreasing the fin area. The latter method has two disadvantages: $(-N_r)$ is also decreased and fin area may be fixed by other considerations such as control requirements. The mechanics of the mode can be explained by reference to figure 11.2.[1]

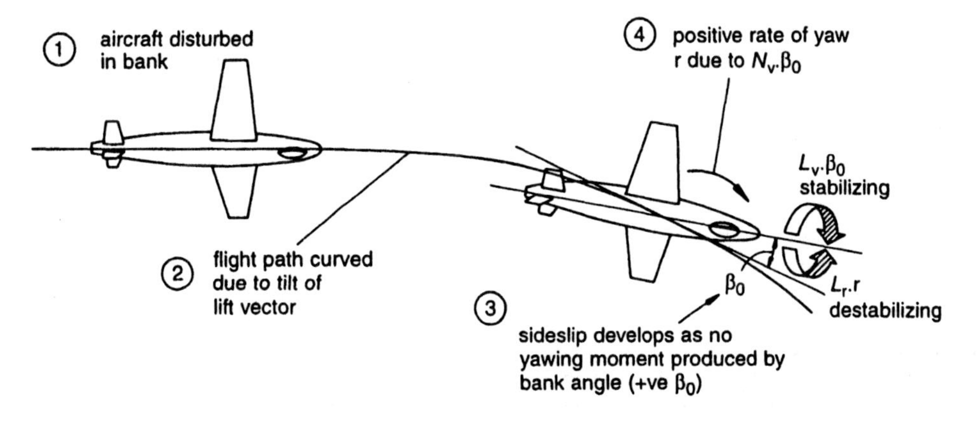

Fig. 11.2 Motion of aircraft in spiral mode

At position (1) in the figure the aircraft is shown disturbed in bank angle; the tilt of the lift vector gives a horizontal component of force causing the flight path at (2) to be curved, in effect a turn. There is no yawing moment on the aircraft to keep its longitudinal axis tangential to the flight path and it develops a sideslip angle β_0 as shown at position (3). This sideslip generates a yawing moment due to the derivative N_v and hence a yaw rate appears as shown at (4). This yaw rate and the sideslip also generate rolling moments through the derivatives L_v and L_r and the motion will be stable if the nett rolling moment is in the opposite sense to the original disturbance. Since the motion is very slow we can neglect inertia effects so that the yawing moments are approximately in balance; then $N_v\beta_0 + N_r\hat{r} = 0$, or

$$\hat{r} = -\beta_0 \frac{N_v}{N_r} \tag{11.27}$$

The mode will be stable if

$$-L_v\beta_0 > L_r\hat{r}$$

and substituting from (3) for $\hat{r}$ we find

$$-L_v\beta_0 > -L_r\beta_0 \frac{N_v}{N_r}$$

which leads immediately to the result (11.26).

A typical shape for the mode is shown in figure 11.3. As the root is real, the eigenvectors are also real and they can only be in phase or in antiphase. It will be noted that the sideslip is very small and so the motion is almost correctly banked.

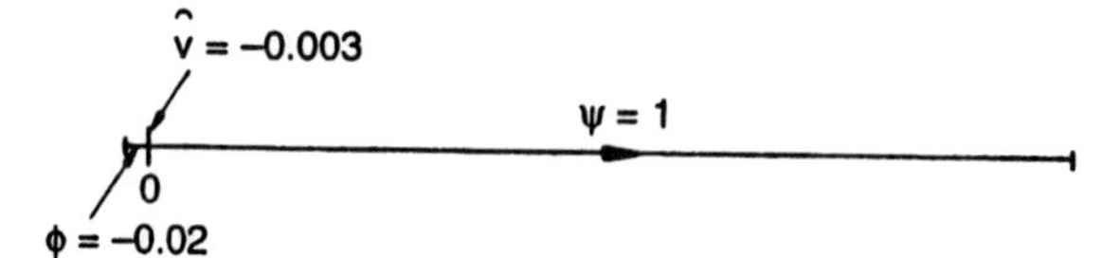

Fig. 11.3 Typical shape of spiral mode

11.4.3 The roll subsidence mode

Referring to the approximate value for the corresponding root (11.18) we see that this mode is stable if B_2 is positive, which is consistent with the work of Section 9.3.3 on test functions. From (11.14) we have

$$B_2 = \hat{y}_v + l_p^* + n_r^* = y_v + (l_p + n_r - e_x n_p - e_z l_r)/A2$$

using (11.9). This expression is nearly always dominated by $\hat{l}_p = -L_p/i_x$, which combines what are normally the largest lateral derivative and the smallest inertia. The derivative L_p, discussed in Section 7.2.1, is always negative below the stall, so $\hat{l}_p$ is positive and hence this mode is always stable. A typical shape of the mode is shown in figure 11.4.

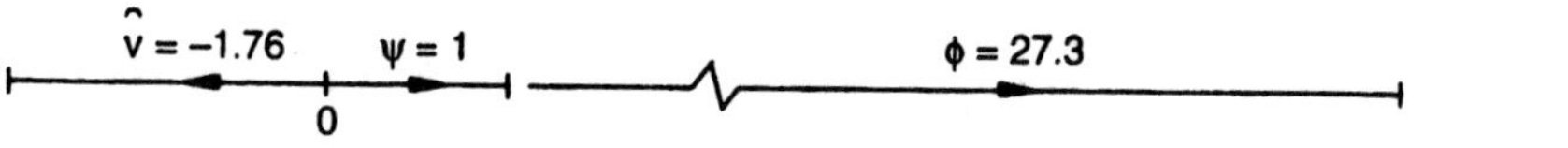

$\hat{v} = -1.76$ $\psi = 1$ 0 $\phi = 27.3$

Fig. 11.4 Typical shape of roll subsidence mode

To discuss this mode further and also the dutch roll mode we need to consider the kinematics of rolling; in general an aircraft will tend to follow the path of least resistance. There are two basic possibilities: these are rolling about the wind axes, that is the flight direction, and rolling about the forward minimum inertia axis. With conventional configurations having modest inertias and relatively strong aerodynamic stiffnesses, that is the weathercock and longitudinal static stabilities discussed in Sections 6.3 and 5.3, rolling will tend to be about the flight path. Rolling exactly around the flight path is illustrated in figure 11.5(a) and it can be seen that the aircraft remains in trim, the incidence remaining at the trimmed value and the sideslip angle at zero.

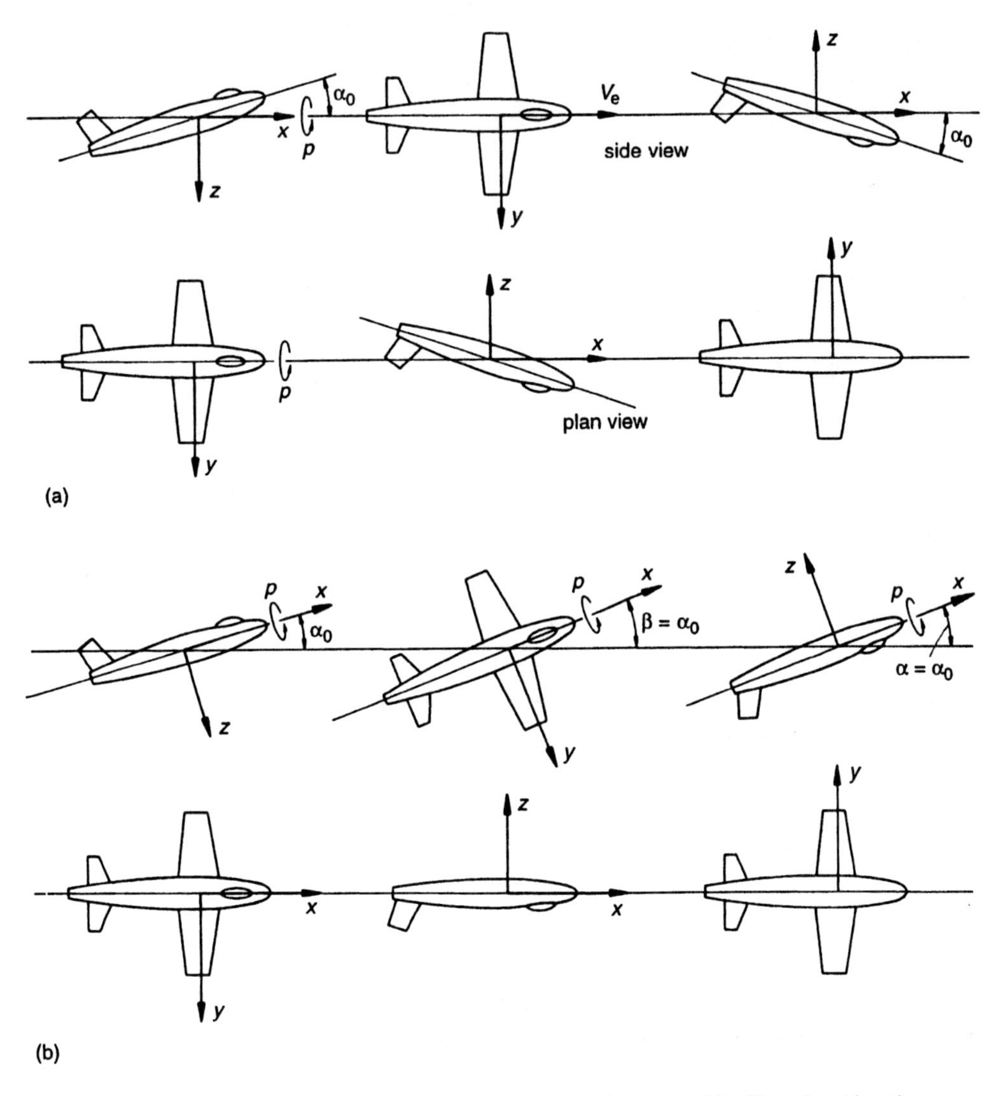

Fig. 11.5 Extreme types of roll motion: (a) rolling about wind axes, (b) rolling about inertia axes

By contrast highly loaded, more slender aircraft have relatively weaker static stabilities and tend to roll around the minimum inertia axis; rolling exactly around this axis is shown in figure

11.5(b). The most obvious example is Concorde, as shown in figure 11.6. In this case, as the aircraft rolls the incidence and sideslip angles interchange at every 90° of rotation; the motion relative to wind axes is clearly oscillatory. Aircraft for which the roll inertia is much less than the pitch and yaw inertias are said to be 'inertially slender' and are discussed further in Section 11.4.4.6. Analysis of the motion of such aircraft will have more physical significance if the equations of motion are referred to axes aligned with the principal axes of inertia.

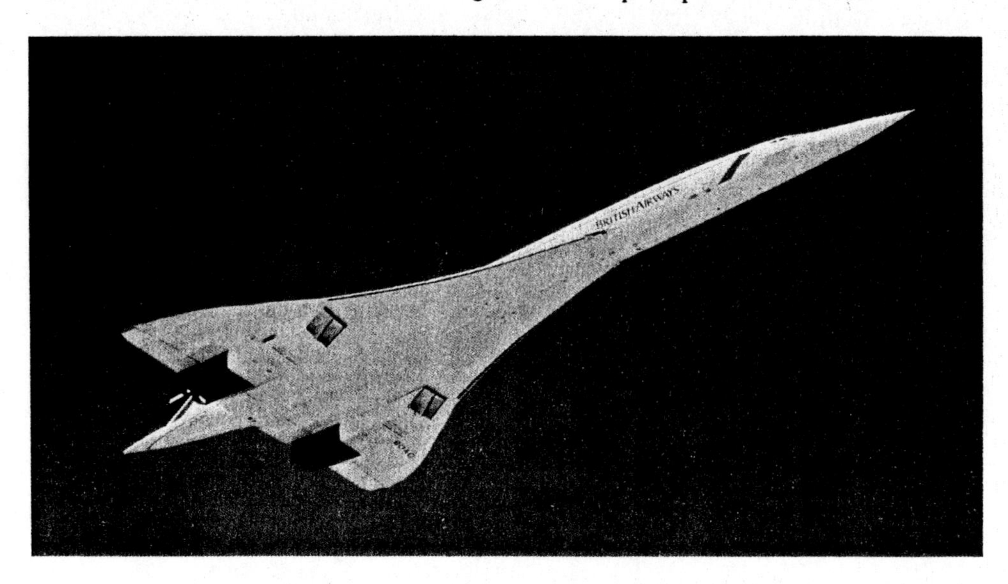

Fig. 11.6 Concorde in flight. (Courtesy British Aerospace)

Returning to the motion in the roll subsidence mode this is clearly non-oscillatory and therefore is essentially rolling about the flight direction. In fact the motion must also contain some yawing as yawing moments are generated by the derivative N_p. The motion is illustrated in figure 11.7 which has been drawn assuming N_p positive and neglecting the effect of the weight.

At position (1) in the figure, the aircraft is shown rolling with a yawing moment being generated by the derivative N_p. At position (2) the aircraft has rotated in yaw as well as roll and the yaw angle generates negative sideslip in conjunction with the forward velocity. Then due to the derivative L_v (a negative quantity) a further rolling moment is generated in the same sense as the direction of rolling. If the yawing velocity is small we can neglect the inertia effect and then the balance of yawing moments gives approximately

$$N_v\beta + N_p\hat{p} \approx 0 \text{ or } \beta \approx -\frac{N_p}{N_v}\hat{p}$$

The rolling moment generated is then proportional to

$$L_v\beta \approx -L_v \cdot \frac{N_p}{N_v}\hat{p}$$

so the effective damping is proportional to

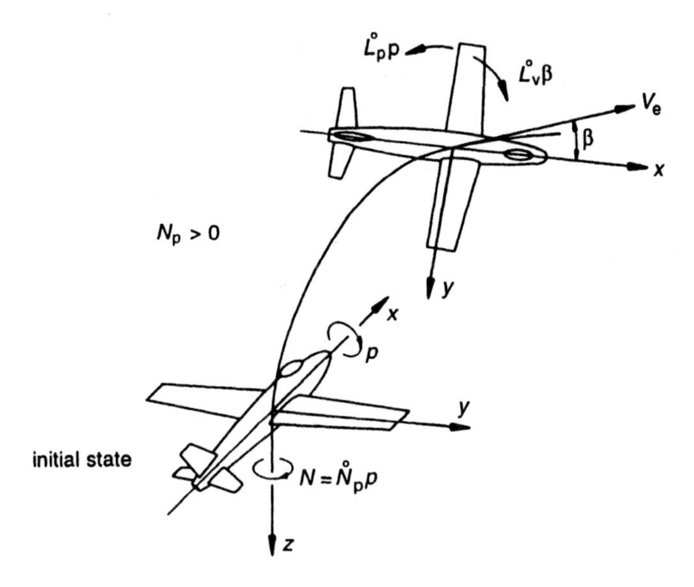

Fig. 11.7 Motion in roll subsidence mode

$$L_p - L_v \cdot \frac{N_p}{N_v}$$

and this can be confirmed by suitable analysis. The sideslip generates a sideforce due to the derivative Y_v and this provides a centripetal force making the motion a barrel roll.

11.4.4 The dutch roll

The criterion for the stability of this mode is that Routh's discriminant, R, is positive since we found in Section 9.2.3 that this quantity determined the stability of oscillatory modes. It has already been stated that the motion in the dutch roll is a rather complicated one but a first impression can be obtained from a typical shape of the mode. Figure 11.8(a) shows the shape of the mode for a conventional aircraft.

We note that sideslip and yaw angles are very nearly equal in magnitude and in antiphase, which is a common feature of this mode and can readily be interpreted physically. If the aircraft cg were constrained to move in a straight line, we would have $\beta = -\psi$ as discussed in Section 11.3. If we now superimpose a sideways velocity $\dot{y}$ on the cg, where y is the displacement from the mean flight path, we now have

$$\beta = \chi - \psi \text{ where } \chi = \dot{y}/V_e \tag{11.28}$$

The observation that the sideslip and yaw angles are nearly equal and in antiphase means that χ is small; we can find χ by solving (11.28) to give $\chi = \beta + \psi$ and treating this as a vector equation as in figure 11.8(b). Three, rather than two, parameters are often used to summarize the characteristics of this mode. These are likely to be the periodic time or the natural frequency, a measure of the damping such as the time to halve the initial amplitude, and the ratio of amplitude of the roll angle to that of the sideslip angle. The latter is loosely referred to as the 'roll/sideslip ratio'.

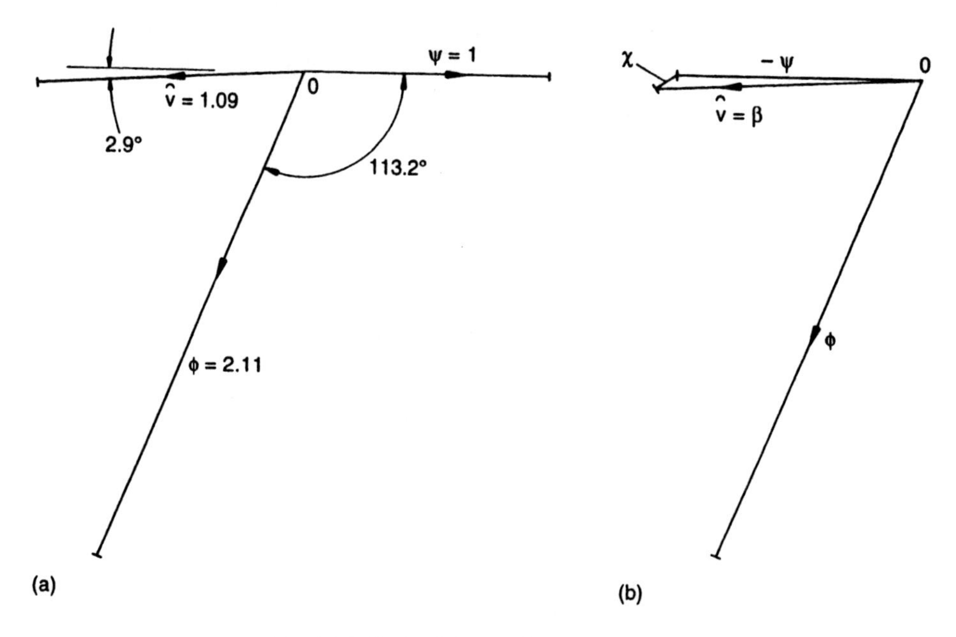

Fig. 11.8 Typical shape of dutch roll mode

For extreme configurations the motion becomes simpler and to achieve some understanding of this mode we will examine some of these cases in the next two sections. We can then proceed to slightly more complex cases when it is possible to consider one of the freedoms as the main one and to treat the other freedoms as subordinate. As already noted in Section 11.2 the equations have no spring term and a key question is therefore where this effect comes from to produce the oscillation found.

11.4.4.1 The directional oscillation

If we do not allow the aircraft freedom in sideslip then $\hat{v} = \beta = -\psi$, and if we further restrict the aircraft to yawing only, the yawing equation (11.3) can be written

$$(\hat{D}^2 + \hat{n}_r\hat{D} - n_v)\psi = 0 \text{ or } \left(I_x\frac{d^2}{dt^2} - \mathring{N}_r\frac{d}{dt} + \mathring{N}_v\right)\psi = 0$$

This represents an oscillation with the damping provided by the damping in yaw derivative N_r, a negative quantity, and the spring term provided by the weathercock stability through the derivative N_v. The motion is simply an oscillation in yaw about the cg, which travels in a straight line. We can build up a picture of the type of aircraft which will tend to behave in this manner if we consider the dimensional rolling moment equation (8.29), omitting the time dependent input and rewritten in the form

$$\mathring{L}_v\beta + \mathring{L}_r r = I_x\dot{p} - \mathring{L}_p p \qquad (11.29)$$

where the cross product of inertia I_{xz} has also been neglected. Terms on the left-hand side of this equation are inputs to the roll freedom and those on the right tend to restrain the roll motion. Then for a small response in roll we require

- L_v to be small, implying no sweepback and little dihedral;
- L_r to be small, implying a fairly low value of C_L which means low altitude or high speed;
- L_p to be large, implying a large aspect ratio;
- I_x to be relatively large, implying a large aspect ratio and engines mounted on the wings.

The picture of the aircraft built up is that of a large transport aircraft with an unswept wing of high aspect ratio flying at low altitude or fairly high speed. Having solved the yaw equation the roll equation could be used to find the roll response.

11.4.4.2 The directional oscillation with lateral freedom

The assumptions of the last section were somewhat unrealistic and we can include sideways motion for this aircraft type with little difficulty, whilst still assuming that rolling is negligible. Considering the sideslip motion first, we have from (8.27)

$$m(\dot{v} + rV_e) = \mathring{Y}_v v \tag{11.30}$$

where we have assumed level flight, used wind axes and neglected the small derivative Y_r. Now differentiating (11.28) with respect to time gives

$$\dot{\chi} = \dot{\beta} + r = \frac{\dot{v} + rV_e}{V_e}$$

then substituting into (11.30) results in

$$\dot{\chi} = \frac{\mathring{Y}_v}{m}\beta \tag{11.31}$$

Now since $\mathring{Y}_v$ is almost always negative, $\dot{\chi}$ is in antiphase to the sideslip angle β and therefore in phase with the yaw angle ψ. Now, we know that for the constant amplitude oscillation $y = a \sin \omega t$, y, $\dot{y}$ and $\ddot{y}$ are successively shifted in relative phase by 90°. For a damped oscillation the same is true provided that the damping is small. Applying this to the case in hand we see that $\dot{\chi} \propto \ddot{y} \propto -y$ and hence y is in phase with the sideslip β. To determine the phase of r relative to χ, we note that $\dot{\chi}$ is in phase with ψ. Hence $\ddot{\chi} \propto -\chi \propto \dot{\psi} = r$, so that r is in antiphase with χ. Now normally N_r is negative and N_v is positive, so that χ produces a yawing moment proportional to $N_v\chi$ which is in the same sense as that produced by the damping in yaw, N_r. Hence the effect of the lateral freedom is to increase the damping of the mode.

The motion that has been derived is shown in figure 11.9, in which the lateral displacements have been exaggerated for clarity. At position (1) the aircraft is shown at the maximum displacement from the mean flight path, hence the lateral acceleration is at a minimum. The yaw angle ψ is at a minimum and the sideslip angle β at a maximum giving a positive yawing moment and a negative sideforce. The yaw rate is zero. In position (2), a quarter of a cycle later, the yaw angle is zero as is the sideslip it produces. The lateral velocity is at minimum and the yaw rate is at its maximum so both produce yawing moments opposing the yaw rate. Position (3) reverses the state of all the quantities in position (1). We have found that the lateral displacement and the yaw angle are almost in phase; this implies that the oscillation can be reduced to a simple oscillation about a point fixed in the aircraft a small distance ahead of the cg. The exaggeration of the displacements in figure 11.9 erroneously suggests that it is ahead of the aircraft. The existence of this centre has implications in the design of autocontrol systems and the perception of the motion by the pilot.

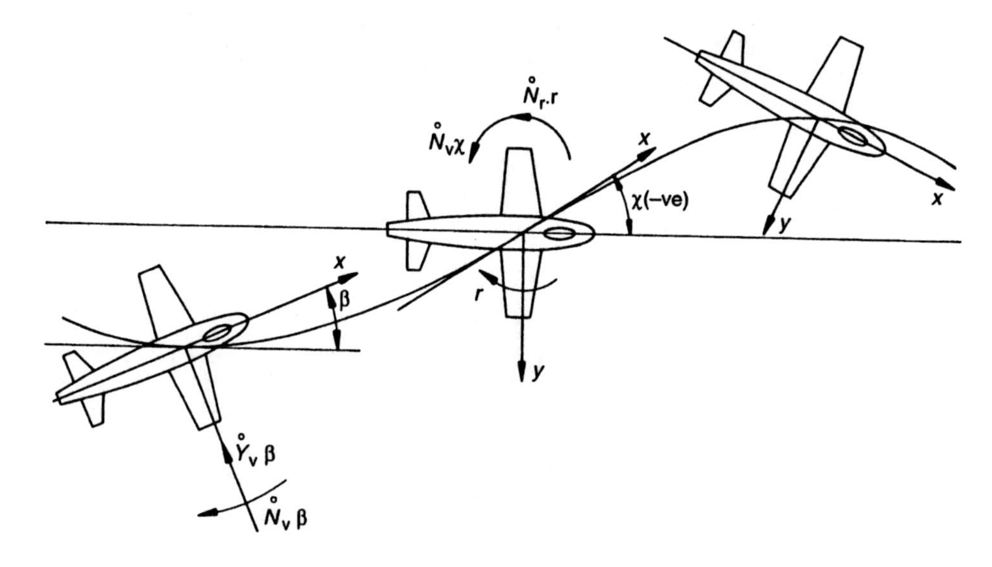

Fig. 11.9 Motion in directional oscillation with lateral freedom

We now briefly deduce the characteristics of the mode for an aircraft conforming to the approximations made, so that (11.1) to (11.3) become

$$(\hat{D} + \hat{y}_v)\hat{v} \qquad\qquad + \hat{r} = 0 \tag{11.32}$$

$$n_v\hat{v} + (\hat{D} + n_r)\hat{r} = 0 \tag{11.33}$$

The characteristic equation corresponding to these equations is

$$\lambda^2 + \lambda(\hat{n}_r + y_v) + n_r y_v - n_v = 0 \tag{11.34}$$

which leads to a time to halve the initial amplitude of

$$t_H = \frac{2\tau \ln 2}{\hat{n}_r + y_v} \tag{11.35}$$

and a periodic time of

$$t_p = \frac{2\pi\tau}{\sqrt{-\hat{n}_v + y_v n_r}} \tag{11.36}$$

where we have assumed the damping to have negligible effect on the period.

For use later in Chapter 12 we also note the normalized natural (circular) frequency

$$\hat{\omega}_\psi = \omega_\psi \tau = \sqrt{-\hat{n}_v + y_v n_r} \tag{11.37}$$

11.4.4.3 Conventional dutch roll with large damping in roll

We now allow motion in all three freedoms, regarding the basic oscillation as that in yaw and the roll and sideslip freedoms to be driven by the coupling with yaw. This is a type of motion that was associated in the past with this mode and so is referred to here as the conventional dutch roll. For the roll response to be fairly small there must still be some resistance to the motion; this can take one of two extreme forms:

- inertia in roll negligible compared to the damping in roll;
- damping in roll negligible compared to the inertia in roll.

The first of these cases will be discussed in this section and the second in the next section. High aspect ratio aircraft and aircraft at low altitudes with no heavy masses on the wings will tend to conform with the motion described in this section.

The equation for the rolling moments is (11.29) and with the inertia and the yaw rate terms omitted we obtain

$$\mathring{L}_v \beta = -\mathring{L}_p p \tag{11.38}$$

Now both of these derivatives are normally negative so that the rate of roll p is antiphase to the sideslip β produced as a result of the rotation in yaw. The latter is equal to $-\psi$ and hence p is in phase with ψ. The sideforce equation is now dominated by the lateral component of the weight and is approximately

$$mg\phi = m\ddot{y} \tag{11.39}$$

This gives a lateral motion so generating an extra sideslip β_L, say. Differentiating (11.39) we have $p \propto d^3y/dt^3 \propto -\dot{y}$, and the extra sideslip is in antiphase with p and hence in phase with β, and the restoring moment on the aircraft is proportional to $N_v(-\psi + \beta_L)$, changing the frequency. No moments are produced in phase with the rate of yaw and the damping is unchanged from the previous case.

The motion that has been derived is shown in figure 11.10, in which the lateral displacements have again been exaggerated for clarity. At position (1) the aircraft is shown at the position of maximum roll angle and lateral acceleration while the lateral displacement is at a minimum

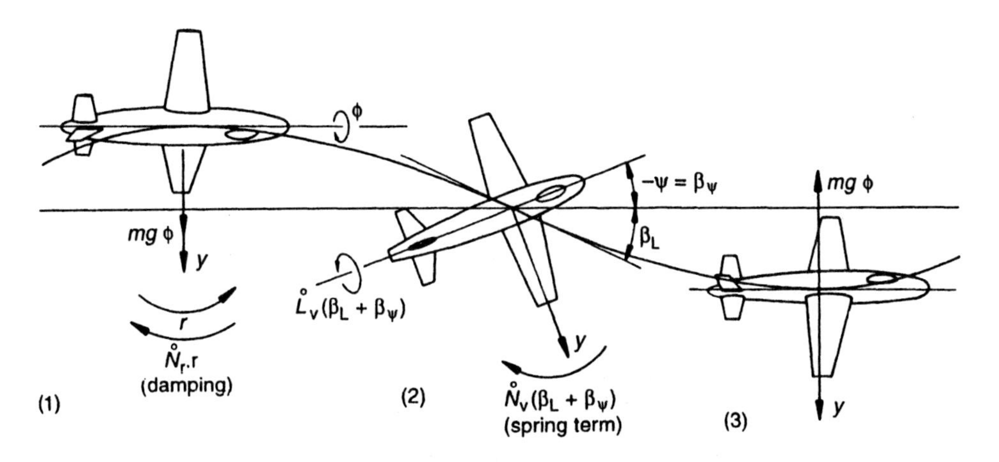

Fig. 11.10 Motion in conventional dutch roll, large damping in roll

consistent with (11.38). At position (2), a quarter of a cycle later, the lateral velocity reaches its maximum in phase with the sideslip generated by the yaw rotation. This produces a yawing moment due to the derivative N_v and a negative rolling moment due to the derivative L_v tending to decrease the roll angle. The result is that the aircraft reaches the position shown at (3).

We can readily deduce the effect of the coupling derivative N_p which has an important influence on this mode. Since the rate of roll is in antiphase with the sideslip β the yawing moment due to $N_p p$ is in phase with $N_v \beta$ if N_p is negative, which further increases the frequency.

11.4.4.4 Conventional dutch roll with large inertia in roll

This is the second of the two extreme cases listed at the start of the last section. Highly loaded aircraft with low aspect ratio wings flying at high altitudes will tend to conform with the motion described in this section.

The equation for the rolling moments is (11.29) with the damping in roll and the yaw rate terms omitted giving

$$\overset{\circ}{L}_v \beta = I_x \dot{p} \tag{11.40}$$

In this case, as L_v is negative the acceleration in roll is in antiphase to the sideslip β produced by rotation in yaw, and the roll angle is in phase with this sideslip. The lateral motion caused by the component of the weight along the y-axis now lags by 90° compared with the last case and so the effects produced also lag by this amount.

The motion that has been derived is shown in figure 11.11, and again the lateral displacements have been exaggerated. Comparison with the previous figure shows that the maximum roll angle is still shown in position (1) but that the maxima of the roll rate and yaw angle occur a quarter cycle earlier, consistent with the deduction from (11.40). The yawing moment due to β_L now opposes the damping term $N_r r$, instead of assisting the term $-N_v \psi$. This effect is proportional to the roll angle which is proportional to L_v/I_x from (1); from (11.39) it is also proportional to the weight, that is to C_L. The derivative L_v is either roughly constant, in the case of straight wings, or roughly proportional to C_L in the case of swept wings. The result is that the effect of β_L increases like C_L or faster and at high incidence the mode becomes unstable. Roll rate also lags by 90° by comparison with the last case and so the effect of N_p is also detrimental to the damping if it is negative. The magnitude of N_p increases with C_L, reinforcing the tendency for the mode to become unstable at high incidence.

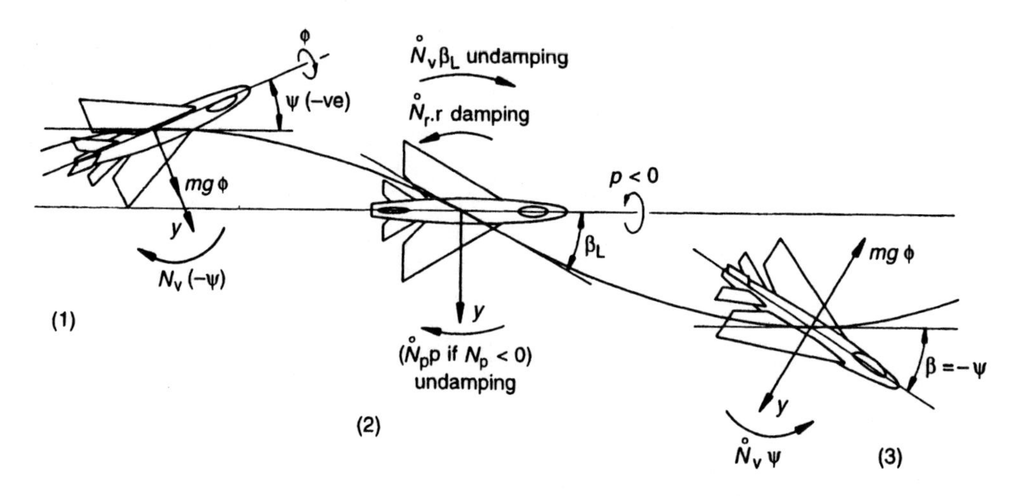

Fig. 11.11 Motion in conventional dutch roll, poor damping in roll, modest I_x/I_z

We can put forward an approximate analysis to cover all the cases discussed so far by assuming level flight and by neglecting small and cross-coupling derivatives except those due to sideslip. Using (11.1) to (11.4) we obtain

$$\left.\begin{aligned} (\hat{D} + \hat{y})\hat{v} \qquad & \hat{g}_1\phi \qquad && + \hat{r} = 0 \\ l_v\hat{v} + (\hat{D}^2 + l_p\hat{D})\phi \qquad & && = 0 \\ n_v\hat{v} \qquad & && + (\hat{D} + n_r)\hat{r} = 0 \end{aligned}\right\} \tag{11.41}$$

These equations lead in the usual way to the characteristic equation

$$\begin{aligned} \lambda^4 + \lambda^3\left[\hat{n}_r + y_v + l_p\right] + \lambda^2\left[y_v n_r - n_v + l_p(y_v + n_r)\right] \\ + \lambda\left[l_p(y_v n_r - n_v) - \hat{g}_1 l_v\right] - \hat{g}_1 l_v n_r = 0 \end{aligned} \tag{11.42}$$

This can be factorized approximately into quadratic factors as follows:

$$\begin{aligned} \left\{\lambda^2 + \lambda\left(\hat{n}_r + y_v + \frac{\hat{g}_1 l_v}{l_p^2 - n_v}\right) + \left(l_p^2 - n_v + n_r y_v\right)\right\} \\ \times \left\{\lambda^2 + \lambda\left(l_p - \frac{\hat{g}_1 l_v}{l_p^2 - n_v}\right) - \left(l_p^2 + \frac{\hat{g}_1 l_v n_r}{l_p^2 - n_v}\right)\right\} \approx 0 \end{aligned} \tag{11.43}$$

where the first factor corresponds to the dutch roll mode. Within this factor many of the terms found in (11.34) can be seen.

11.4.4.5 The rolling oscillation

We come now to the case of aircraft for which both the inertia and the damping in roll are small; usually these will be of the type we called inertially slender in Section 11.4.3. Such aircraft tend to roll around the axis of minimum inertia, which will be at an angle α_e to the flight direction. As noted in Section 11.2 there is no apparent spring term in the rolling moment equation (11.2); however, it was noted in Section 11.4.4 that using principal axes was the natural choice in this case. We now examine rolling about these axes, as shown in figure 11.12. In the figure the components of the forward velocity U_e along the axis of inertia and W_e normal to it will remain in the same vertical plane as the aircraft rolls about the axis. The y-axis rotates with the aircraft and hence there is a component of W_e along the y-axis of amount $W_e \sin\phi = V_e \sin\phi \sin\alpha_e$. The result is that there is a sideslip angle of amount

$$\beta_\phi = \sin\phi \sin\alpha_e \approx \phi \sin\alpha_e \tag{11.44}$$

and the roll sideslip ratio is

$$\phi/\beta = 1/\sin\alpha_e \tag{11.45}$$

The roll equation (11.2) on neglecting yawing and using (11.44) becomes

$$(\hat{D} + \hat{l}_p\hat{D} + l_v \sin\alpha_e)\phi = 0 \tag{11.46}$$

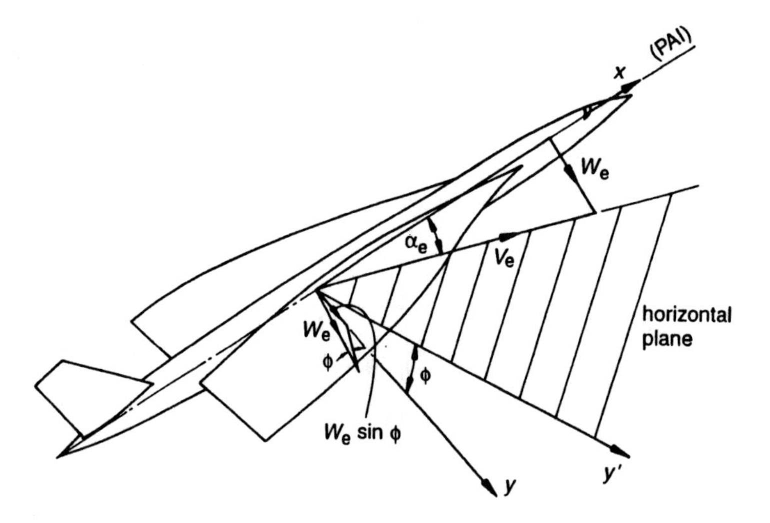

Fig. 11.12 Generation of sideslip due to roll of aircraft about inertia axis

and we now have a spring term provided by the kinematics of the motion. We can find the characteristics of the motion in the usual manner, giving

$$t_H = \frac{2\tau \ln 2}{\hat{l}_p} \tag{11.47}$$

and

$$t_p = \frac{4\pi\tau}{\sqrt{4\hat{l}_v \sin\alpha_e - l_p^2}} \tag{11.48}$$

Since for this type of aircraft the damping in roll is small, $\hat{l}_v \gg l_p$ and the second term under the square root can be neglected. It can be seen that the period decreases with incidence.

For there to be a very small response in yaw we need the derivatives N_v and N_p to be small, as well as needing the yaw inertia to be much larger than the roll inertia already mentioned. It can be shown (86041) that the value of N_v expressed in inertia axes N_v^I is given by

$$N_v^I = N_v \cos\alpha_e + L_v \sin\alpha_e$$

where N_v and L_v are here the values in wind axes. Now N_v will be positive and L_v will be negative so that N_v^I will decrease with incidence; as this type of aircraft is also likely to have a highly sweptback wing for which $-L_v$ increases with incidence the effect is accentuated. The span of this type of aircraft will be small to produce small values of the inertia and damping in roll; these will also tend to produce a small value of N_p.

11.4.4.6 The rolling oscillation with lateral freedom

The previous section has illustrated the principle by which an oscillation might be produced for this type of aircraft. It was somewhat impractical as rolling the aircraft gives a component of the weight along the y-axis, which generates a further contribution to the sideslip. To investigate this further we first write down the stability equations referred to principal axes.

Choosing principal axes makes the cross product of inertia terms e_x and e_z vanish, then as $W_e = V_e \sin\alpha_e$ and $U_e = V_e \cos\alpha_e$ (8.57) to (8.61) become

$$(\hat{D} + \hat{y})\hat{v} + (y_p - \sin\alpha_e)\hat{p} + (y_r + \cos\alpha_e)\hat{r} - \hat{g}_1\phi - \hat{g}_2\psi = \hat{y}(\hat{t}) \quad (11.49)$$

$$l_v\hat{v} + (\hat{D} + l_p)\hat{p} + l_r\hat{r} = \hat{l}(\hat{t}) \quad (11.50)$$

$$n_v\hat{v} + n_p\hat{p} + (\hat{D} + n_r)\hat{r} = \hat{n}(\hat{t}) \quad (11.51)$$

$$\hat{p} - \hat{D}\phi = 0 \quad (11.52)$$

$$\hat{r} - \hat{D}\psi = 0 \quad (11.53)$$

It should be noted that although the same symbols have been used, ϕ, ψ, $\hat{p}$ and $\hat{r}$ have different values when expressed relative to inertia axes compared with their values relative to wind axes. Most stability derivatives also have different values as noted in Section 9.2.6. Any given problem can be solved using these equations or using equations referred to wind axes, and the solutions will be identical in essence. We have simply introduced these equations as they are more closely related to the physics of this problem.

We now neglect the yaw freedom, substitute for $\hat{v} = \beta$ from (11.44), assume level flight and that the weight component is much larger than the sideforce due to sideslip. The equations then reduce to

$$\hat{D}\hat{v} - (D\sin\alpha_e + \hat{g}_1)\phi = 0 \quad (11.54)$$

$$\hat{l}_v\hat{v} + (\hat{D}_2 + l_p)D\phi = 0 \quad (11.55)$$

The corresponding characteristic equation is

$$\lambda^3 + \lambda^2\hat{l}_p + \lambda l_v \sin\alpha_e + l_v\hat{g}_1 = 0 \quad (11.56)$$

which factorizes approximately into

$$\left\{\lambda^2 + \left(\hat{l}_p - C_L \operatorname{cosec}\alpha_e\right)\lambda + l_v \sin\alpha_e\right\}\cdot\left(\lambda + C_L \operatorname{cosec}\alpha_e\right) \approx 0 \quad (11.57)$$

where the first factor represents the dutch roll mode and we have replaced $\hat{g}_1$ by C_L. This leads to almost the same periodic time as in the last section, but the time to halve the initial amplitude of a disturbance is

$$t_H = \frac{2\tau \ln 2}{\hat{l}_p - C_L \operatorname{cosec}\alpha_e} \quad (11.58)$$

hence the aircraft becomes unstable if

$$C_L \operatorname{cosec}\alpha_e > \hat{l}_p = -L_p / i_x \quad (11.59)$$

The damping in roll L_p is almost constant with incidence whilst C_L cosec α_e increases with incidence; hence the aircraft becomes unstable at high incidences.

The additional sideslip β_L generated by the lateral motion induced by the component of the weight will be 90° behind the roll angle, and therefore in antiphase to the rate of roll. The rolling moment produced by this sideslip is proportional to $L_v\beta_L$, and hence opposes the damping in roll $L_p p$. This explains the loss in stability found from (11.58). The motion is shown in figure 11.13.

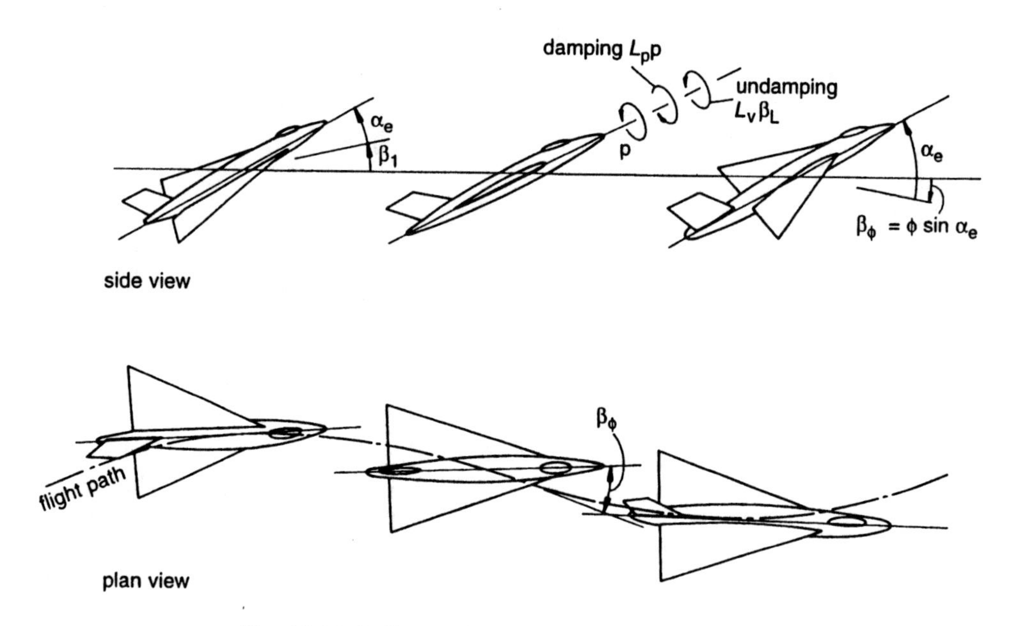

Fig. 11.13 Rolling oscillation of inertially slender aircraft

11.4.4.7 Discussion of dutch roll characteristics

We have seen the various types of extreme dutch roll mode vary from oscillation in yawing and sideslip to oscillation in rolling and sideslip. In between came the conventional dutch roll where the oscillation was in yawing which drives the sideslip and rolling freedom. The roll angle in this case will lag the yaw angle by up to 90° if the damping in roll dominates over the inertia. In the opposite case of the inertia in roll dominating, the lag can approach 180°. No aircraft can exactly match these extreme cases, only approach them. A single aircraft can in one flight condition be close to one extreme, but approach another extreme in a different flight condition. At low altitude and high speed the damping in roll may dominate over the inertia, but the situation may reverse at low speed and high altitude. The use of the wings to store fuel in and its use to mount stores can add to the effects of changes in speed and height.

An aircraft can also change from having a conventional dutch roll oscillation to behaving as an inertially slender aircraft. We have seen that the basic spring terms in the two kinds of dutch roll are $-\hat{n}_v$ and $\hat{l}^{\text{I}}_v \sin \alpha_e$ and a critical incidence α_{crit} at which these are equal can be defined as

$$\alpha_{crit} = \sin^{-1}\left(\frac{-\hat{n}_v}{\hat{l}^{\text{I}}_v}\right) = \sin^{-1}\left(\frac{N_v I_x}{-L^{\text{I}}_v I_z}\right)$$

An empirical investigation made by Pinsker, reference (11.2), suggested that if $\alpha_e < 2\alpha_{crit}/3$ an aircraft has a conventional type of dutch roll and if $\alpha_e > 4\alpha_{crit}/3$ it behaves as an inertially slender configuration. Between these limits the motion is an intermediate one. If the yawing spring is relatively large for a given aircraft then α_e is large and actual incidences are likely to lessen and a conventional dutch roll oscillation will result; conditions are reversed if the rolling spring is the larger. This can be generalized to stating that the aircraft tends to oscillate in the freedom with the larger spring, that is the one with the higher frequency of oscillation. Inertially slender aircraft are likely to have wings of low aspect ratio and therefore low lift curve slopes, have a large change of incidence over the flight envelope, and will change from a conventional dutch roll to a nearly purely rolling one. Aircraft with wings of high aspect ratio will have a small change of incidence over the flight envelope and are likely to remain with a conventional dutch roll.

11.5 Effects of speed

The effect of variation of speed on the lateral stability has been calculated by solving the quartic equations for the same aircraft and altitude as were used in Section 9.4.5. Again the actual aircraft has a automatic control system but none has been assumed. The results are shown in figure 11.14.

Figure 11.14(a) shows the effect of speed on the time to half-amplitude of the roll subsidence mode. It can be seen that t_H decreases steadily with speed; this is in agreement with the approximate theory of Section 7.2.1. From (7.5) we see that t_H is inversely proportional to the damping in roll derivative which is itself proportional to speed.

The spiral mode for this aircraft is unstable at low speeds and becomes stable as the speed is increased, a common situation, and it is then more convenient to plot the reciprocal of the time to half or double the initial amplitude as shown in figure 11.14(b). Figure 11.14(c) shows the variation of the characteristics of the dutch roll mode with speed; it can be seen that the damping is poor at low speeds. The poor stability of this mode and the spiral mode at low speeds, i.e. high values of C_L, is discussed further in the next section.

11.6 Stability diagrams and some design implications

We have discussed many of the effects of the most important derivatives but have not dealt adequately with the two static stability derivatives L_v and N_v. These are also important because the designer has some control over their values through the dihedral angle and fin area. A convenient way of discussing the effect of these derivatives is by means of a stability diagram. In this the boundaries between unstable and stable conditions are plotted on a graph having axes for the two derivatives. We need not consider the roll subsidence which is always stable below the stall. We are left with the spiral mode, for which the condition for stability (Section 11.4.2) is $E_2 > 0$, and the dutch roll mode for which the condition (Section 11.4.4) is that Routh's discriminant is positive. Calculations have been made for the aircraft of Section 9.4.5 assuming the role of the designer, varying the dihedral angle to vary derivative L_v and varying the fin area to vary N_v. Other derivatives are also affected but to a much lesser degree.

The boundary for the dutch roll mode is also a function of the relative density parameter μ_2, which it should be remembered increases with altitude. Both boundaries are also functions of C_L, that is, of speed. Figure 11.15(a) shows the stability diagram for a C_L value of 0.15. The spiral mode is stable below the curve marked $E_2 = 0$, and the dutch roll mode is stable to the left of the boundaries marked $R_2 = 0$. It is evident that the stability of the dutch roll mode deteriorates with increase of altitude; this is due to the reduction of the aerodynamic damping terms with decrease of density.

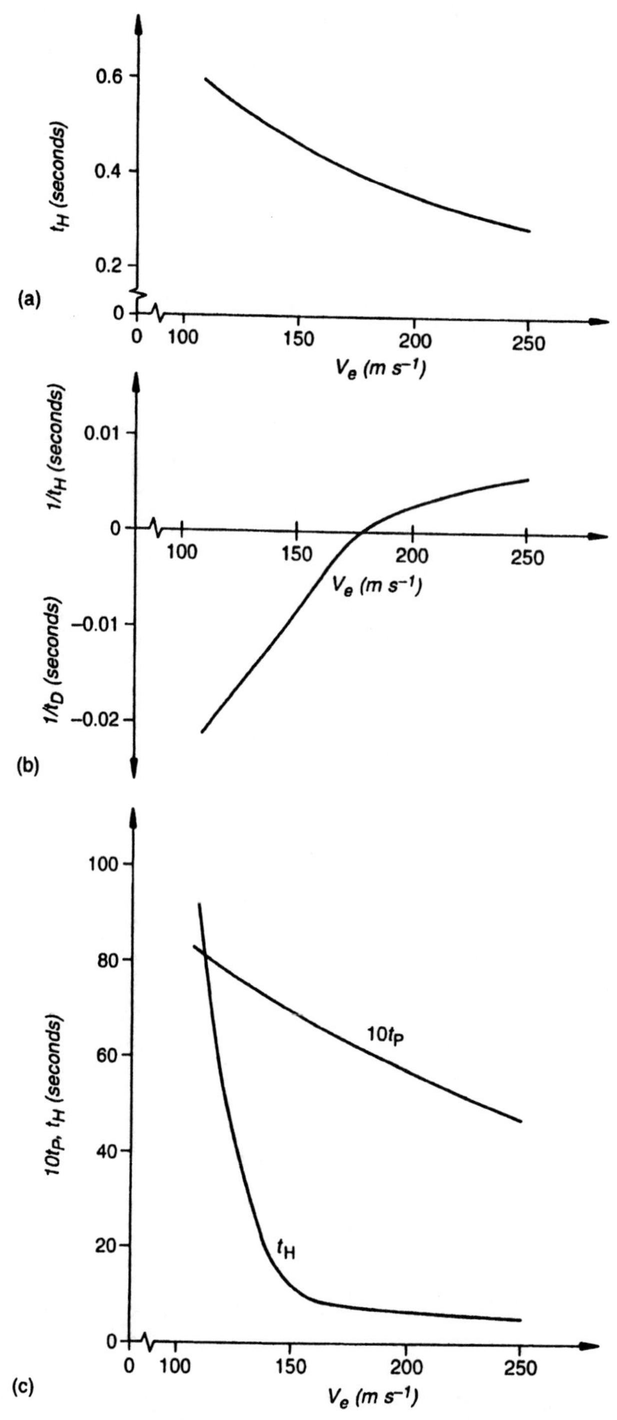

Fig. 11.14 Variation of lateral stability characteristics with speed: (a) variation of time to half amplitude of roll subsidence mode, (b) variation of time to half/double amplitude of spiral mode, (c) variation of periodic time and time to half amplitude of dutch roll mode

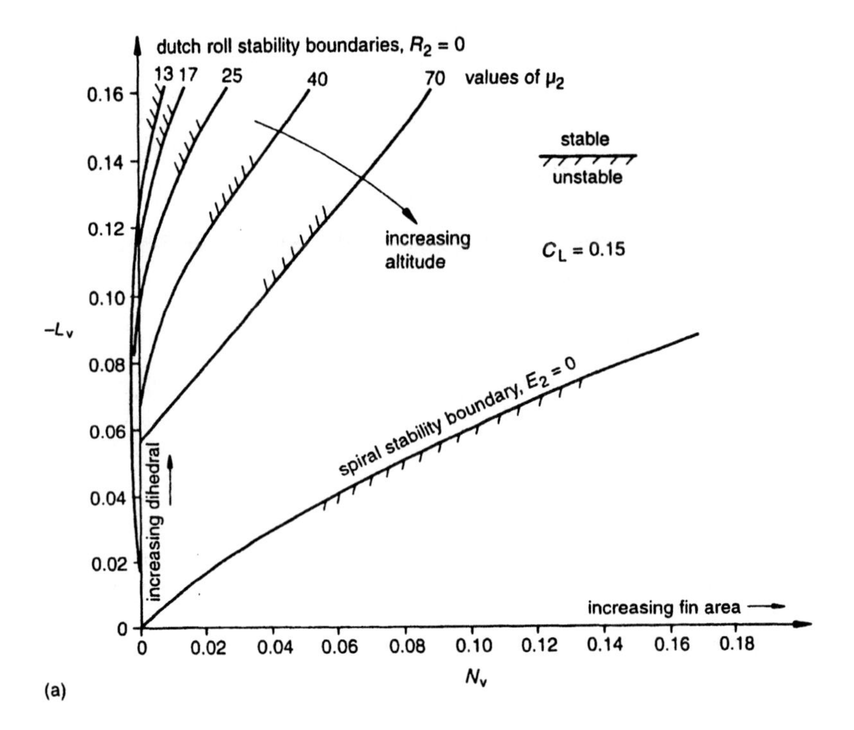

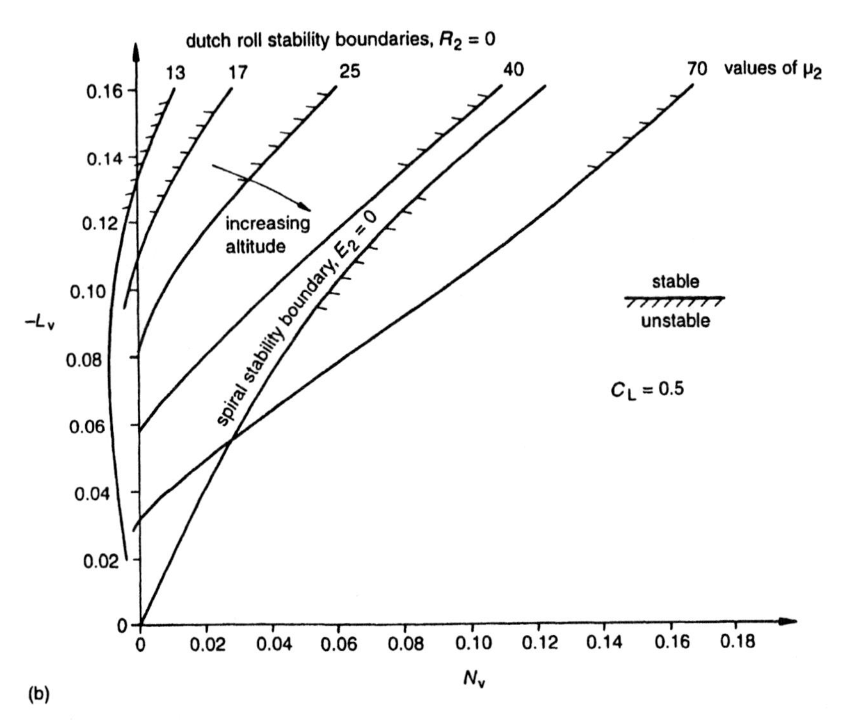

Fig. 11.15 Lateral stability diagram for static stability derivatives: (a) $C_L = 0.15$, i.e. high speed, (b) $C_L = 0.5$, i.e. lower speed

Figure 11.15(b) shows the stability diagram for a C_L value of 0.5 where it is clearly seen that the spiral mode boundary has moved upwards and the dutch roll boundaries have moved to the right. Whereas, in the previous case, there was a large region in which both modes are stable, in this case there is hardly any such region for μ_2 having a value greater than about 50. This means that the designer may have to make the choice of which mode to make unstable. In practice an aircraft is still controllable with an unstable spiral mode but is seriously unsatisfactory with an unstable dutch roll. The designer will therefore choose the fin area and dihedral angle combination to place the aircraft below the spiral mode boundary in the figure. Fin area is also chosen to satisfy the requirements of Section 6.3.3.

11.7 Control and response

In the lateral control of aircraft probably the most important manoeuvre is rolling since before the aircraft can turn at a high rate it must acquire an appreciable bank angle and hence the most important control is the aileron. The rudder has important functions in producing sideslip in such manoeuvres as take-off and landing and in counteracting the adverse yawing moment due to aileron and the yawing moment in the case of engine failure on a multi-engined aircraft. The initial response due to aileron angle deflection is in the roll subsidence mode for aircraft having a conventional dutch roll mode. Since roll manoeuvres are usually fairly rapid it is the first few seconds of the response that are important and this can take several extreme forms. If the damping in roll is much larger than the inertia in roll the aircraft very rapidly reaches a steady rate of roll and the aileron is effectively a 'rate control'. This type of aircraft is therefore that described in Section 11.4.4.3. On the other hand, if the inertia is dominant then the time to halve the amplitude of an initial disturbance is longer than that of the typical manoeuvre and for practical purposes the aileron controls the acceleration in roll. This type of aircraft is therefore that described in Section 11.4.4.4; most conventional aircraft will fall somewhere between these extremes for the dutch roll mode, so they will also fall between the extremes for roll response.

In the case of inertially slender aircraft above the critical incidence, the time to half-amplitude in the roll subsidence mode is likely to be longer than that for the rolling oscillation, and the response to aileron is then determined more by the characteristics of the latter mode. Since roll angle generates a sideslip β_ϕ and this in turn generates a rolling moment due to the derivative L_v, the aircraft can assume a steady roll attitude in which the rolling moment due to aileron is balanced by that due to the sideslip. This is to be expected as this effect provided the spring term for the rolling oscillation. The response to rapid aileron movement in this case tends to be a rapid acceleration to this steady rate of roll plus a lesser response in the roll subsidence mode, and the aileron behaves more nearly as a position control. As the spring effect stiffens with increasing incidence, see (11.44), so the effectiveness of the aileron decreases, and at some incidence the rudder may become the more effective of the two. The rudder acts by producing some sideslip through the yaw angle mechanism discussed in Section 11.4 countering that produced by the roll.

There is evidence that a pilot expects the aileron to produce roll about the flight direction and the rudder to produce sideslip. The inertially slender aircraft produces some difficulty for the pilot by tending to roll about the axis of minimum inertia; the loss of aileron effectiveness at high incidence adds to the problems. A further undesirable effect is the reduction in incidence as the aircraft rolls as shown in figure 11.5(b). One solution to the problems is to couple the rudder to the aileron circuit in order to produce more normal control characteristics. This coupling should ideally be dependent on incidence: this can be done by a physical connection or in the AFCS.

11.7.1 Response to control action

In this section we will treat the response to changes of aileron and rudder angles in the same manner as the treatment of response to changes of elevator angle in Section 10.2.1. We will assume level flight and that we can neglect the sideforce control derivatives $\hat{y}_\xi$ and $\hat{y}_\zeta$. The form of the equations will be the same whether aileron or rudder deflections are considered and we can write one set of equations. The equations we require are (11.1), (11.7) and (11.8), in these we substitute for $\hat{p}$ from (11.4) and obtain

$$(\hat{D} + \hat{y}_v)\hat{v} + (y_p\hat{D} - \hat{g}_1)\phi + (1 + y_r)\hat{r} = 0 \tag{11.60}$$

$$l_v^*\hat{v} + (\hat{D} + l_p^*)\hat{D}\phi + l_r^*\hat{r} = -l_\delta^*\delta(\hat{t}) \tag{11.61}$$

$$n_v^*\hat{v} + n_p^*\hat{D}\phi + (\hat{D} + n_r^*)\hat{r} = -n_\delta^*\delta(\hat{t}) \tag{11.62}$$

where δ stands for ξ' or ζ'. In these equations we have from (11.6) and (11.9)

$$\hat{l}_\delta^* = (\hat{l}_\xi - e_x n_\xi)/A_2 \text{ and } \hat{n}_\delta^* = (n_\xi - e_z l_\xi)/A_2 \tag{11.63}$$

in the case of aileron angle, or

$$\hat{l}_\delta^* = (\hat{l}_\zeta - e_x n_\zeta)/A_2 \text{ and } \hat{n}_\delta^* = (n_\zeta - e_z l_\zeta)/A_2 \tag{11.64}$$

in the case of rudder movement. In the case of the controls being moved together the results of these equations can be added together. We now take the Laplace transform of equations (11.60) to (11.62) with the result

$$(s + \hat{y}_v)\bar{v} + (y_p s + \hat{g}_1)\bar{\phi} + (1 + y_r)\bar{r} = 0 \tag{11.65}$$

$$l_v^*\bar{v} + (s2 + l_p^* s)\bar{\phi} + l_r^*\bar{r} = -l_\delta^*\bar{\delta}(s) \tag{11.66}$$

$$n_v^*\bar{v} + n_p^* s\bar{\phi} + (s + n_r^*)\bar{r} = -n_\delta^*\bar{\delta}(s) \tag{11.67}$$

where we have written $\bar{v}$, $\bar{\phi}$, $\bar{r}$ and $\bar{\delta}$ for the Laplace transform of v, ϕ, r and δ and assumed their initial values to be zero. The solution is

$$\frac{\bar{v}}{G_1(s)} = \frac{\bar{\phi}}{G_2(s)} = \frac{\bar{r}}{G_3(s)} = \frac{\bar{\delta}}{F(s)} \tag{11.68}$$

In these equations $F(s)$ is the characteristic equation and is given by (11.13). The polynomials G_1, G_2 and G_3 are obtained as in Section 10.2.1 and are found to be

$$
\begin{aligned}
G_1 &= s^2\left[\hat{y}_p l_\delta^* + (1 + y_r)n_\delta^*\right] \\
&\quad + s\left[\left\{y_p n_r^* - \hat{g}_1 - (1 + y_r)n_p^*\right\}l_\delta^* + \left\{(1 + y_r)l_p^* - y_p l_r^*\right\}n_\delta^*\right] \\
&\quad + \hat{g}_1\left[-n_r^* l_\delta^* + l_r^* n_\delta^*\right]
\end{aligned} \tag{11.69}
$$

$$
\begin{aligned}
G_2 &= -s^2 l_\delta^* + s\left[-\left\{y_v + n_r^*\right\}l_\delta^* + l_r^* n_\delta^*\right] \\
&\quad + \left[\left\{(1 + y_r)n_v^* - y_v n_r^*\right\}l_\delta^* + \left\{y_v l_r^* - (1 + y_r)l_v^*\right\}n_\delta^*\right]
\end{aligned} \tag{11.70}
$$

$$
\begin{aligned}
G_3 &= -s^3 n_\delta^* + s^2\left[n_p^* l_\delta^* - \left\{y_v + l_p^*\right\}n_\delta^*\right] \\
&\quad + s\left[\left\{y_v n_p^* - y_p n_v^*\right\}l_\delta^* + \left\{y_p l_v^* - y_v l_p^*\right\}n_\delta^*\right] \\
&\quad + \hat{g}_1\left[n_v^* l_\delta^* - l_v^* n_\delta^*\right]
\end{aligned} \tag{11.71}
$$

The rest of the solution then follows the treatment in Section 10.2.1.

11.7.2 Typical results

Calculations have been made of the response to step changes in aileron and rudder angles for the typical aircraft using the method of Section 10.2.3. All modes are stable at the speed chosen of 194 m s^{-1}.

Figure 11.16 shows the responses in sideslip velocity $\hat{v}$, roll rate $\hat{p}$, yaw rate $\hat{r}$, yaw angle ψ, and roll angle ϕ for a step change of aileron angle of 0.04 rad. Figure 11.16(b) shows the initial response most clearly with the roll rate building up in about a second to a fairly steady final value, this is the response in the roll subsidence mode. The oscillations seen are the response in the dutch roll mode whilst the spiral mode is seen most clearly in the build-up of yaw rate and angle in figures 11.16(c) and (d).

Figure 11.17 shows the responses in sideslip velocity $\hat{v}$, roll rate $\hat{p}$, yaw rate $\hat{r}$, yaw angle ψ, and roll angle ϕ for a step change of rudder angle of 0.08 rad. Again the oscillations are the response in the dutch roll mode and the slow increase in the yaw rate, roll and yaw angles in figures 11.17(c), (d) and (e) are the response in the spiral mode. It will be seen that the initial response in roll rate in figure 11.17(b) is positive; this is due to the rolling moment produced by the rudder, the response being more rapid in roll due to the roll inertia being rather smaller than that in yaw. The sign of the roll rate rapidly reverses; this is due to the build-up of sideslip which gives a powerful rolling moment in the opposite sense through the derivative L_v and leads to the build-up of roll angle seen in figure 11.17(e).

11.7.3 Response to gusts

Gust effects can also be important for the lateral characteristics of an aircraft. Apart from vertical gusts which can have a variation of velocity across the span which would produce the effect of rolling the aircraft, we must also consider horizontal gusts coming from the side of the aircraft. If constant in magnitude along the length of the aircraft the effect is of an added sideslip and if there is a gradient lengthwise the effect is similar to yawing. These effects can be dealt with in a manner similar to that in Section 10.4. In the case of a steady side gust there will be inputs into the roll and yaw freedoms proportional to $L_v v_g$ and $N_v v_g$, which will produce effects similar to aileron and rudder deflections. For aircraft with conventional dutch roll modes the response will be initially in the roll subsidence mode followed by the dutch roll and

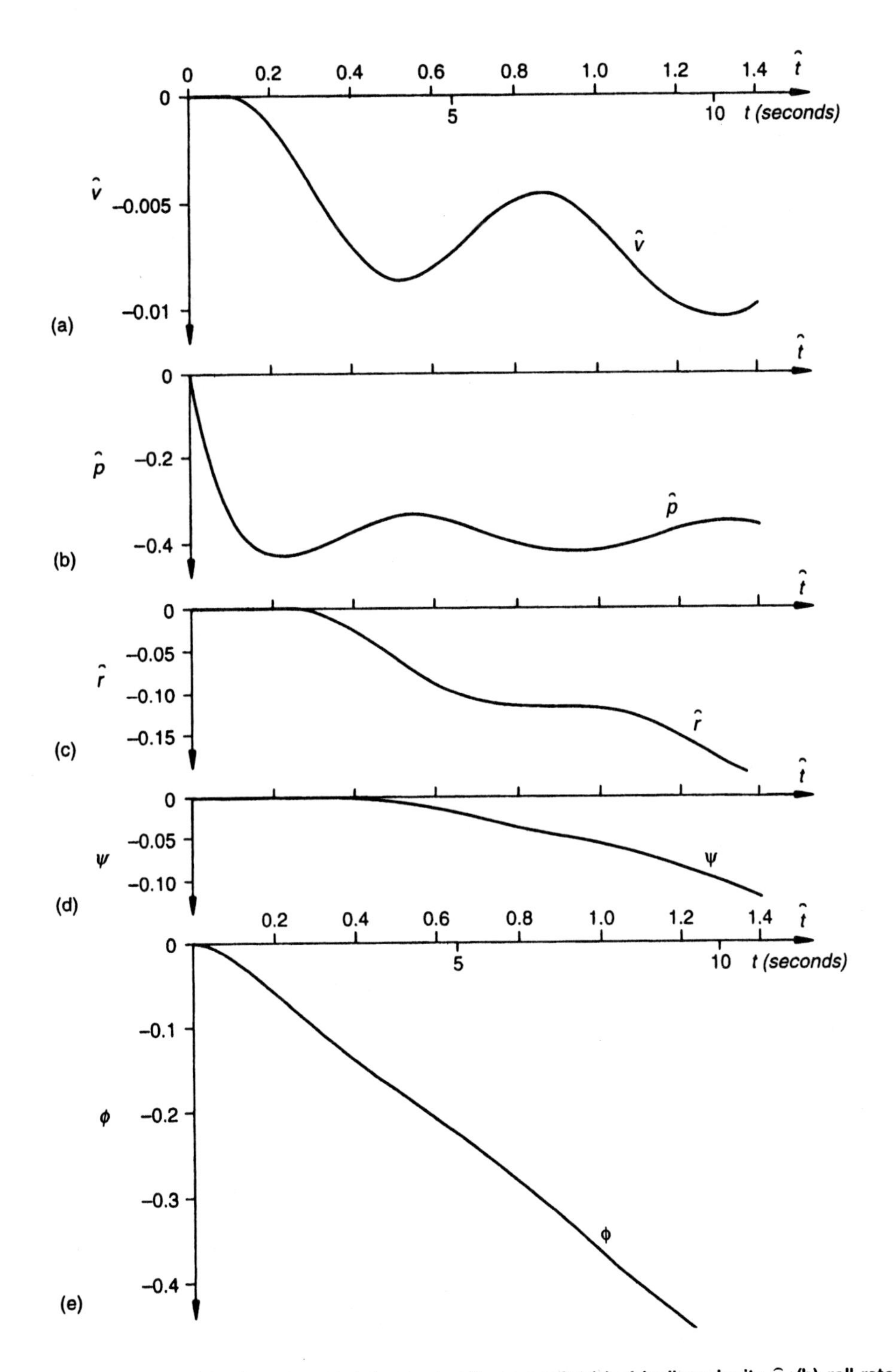

Fig. 11.16 Lateral response to step change in aileron angle: (a) sideslip velocity $\hat{v}$, (b) roll rate $\hat{p}$, (c) yaw rate $\hat{r}$, (d) yaw angle ψ, (e) roll angle ϕ

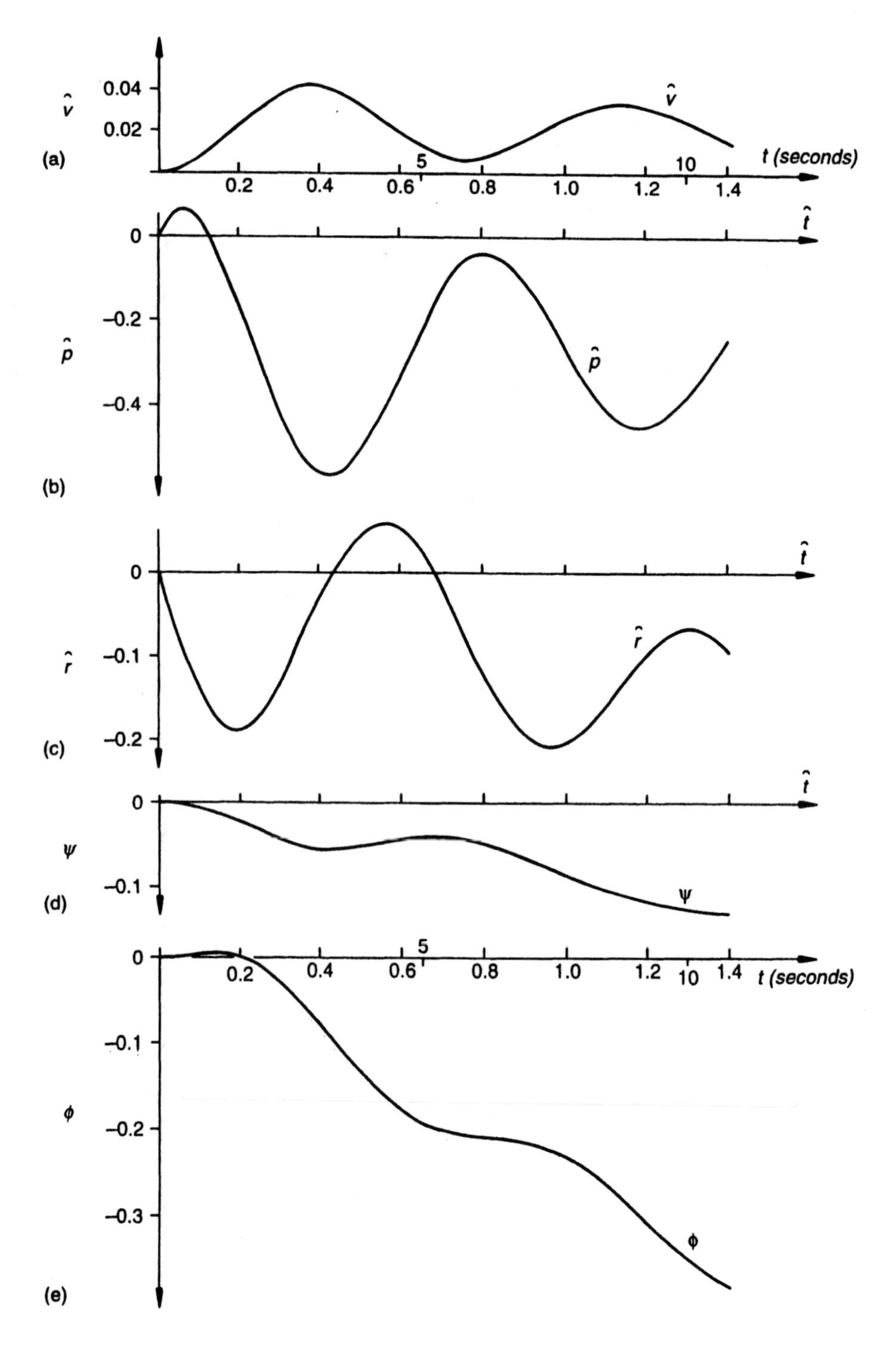

Fig. 11.17 Lateral response to step change in rudder angle: (a) sideslip velocity $\hat{v}$, (b) roll rate $\hat{p}$, (c) yaw rate $\hat{r}$, (d) yaw angle ψ, (e) roll angle ϕ

spiral modes. For inertially slender aircraft the response may be in the rolling oscillation first, followed by the other modes.

11.8 Lateral handling and flying requirements

This subject was discussed in general in Section 10.5.3 in relation to longitudinal motion and we add here only some remarks specific to the lateral modes. In the case of the roll mode satisfactory handling characteristics are obtained if the time constant is not less than 1.5 s and the roll rate achieved with full aileron is not less than 60° s^{-1}. The flying quality requirement for the spiral mode is that the time to double the initial amplitude of a disturbance shall not exceed 12 s. As may be expected, the flying quality requirements for the dutch roll mode are more complicated than for other modes, and only a typical requirement is quoted (92006). For rapid manoeuvring (flight phase A), level 1 flying qualities and for class II or III aircraft the relative damping ζ shall be greater than 0.19, the natural frequency ω_n greater than 0.5 rad s^{-1} and the product $\zeta\omega_n$ greater than 0.35.

Appendix: Solution of lateral quintic using a spreadsheet

This is a simple routine for finding the roots of the lateral quintic with the zero root divided out and is based mostly on the work of Section 11.3.2. It was developed using the Quatro®-Pro spreadsheet and has been tested on other popular spreadsheets. It should run on any spreadsheet that has the facility to turn off automatic calculation and allows the use of IF statements. We assume that the lower-order coefficients have been divided by the coefficient of λ^4.

The first step is to turn off automatic calculation; this is necessary as there are circular cell references again. We then enter some headings as follows:

into D1: Solution of the lateral quartic
into H2: Roll root
into N2: Spiral root
into R2: Dutch roll roots
into D3: Coefficients
into both H3 and N3: Approximations

We leave some columns on the left for any preliminary calculations that may be necessary and proceed to label the individual columns. Assuming that the character '^' is used to centre text in a cell, we label cells D4 to M4 as follows:

^B, ^C, ^D, ^E, FIRST, ITER, FINAL, F(LAMBDA), ^F', ^F/F'

and cells N4 to V4 as

FIRST, FINAL, F(LAMBDA), ^a2, ^b2, DISCRIM, TYPE, ROOT1, ROOT2

Then in cells D5 to G5 we insert the following test values: 5, 10, 20, 0.1.

We now start putting formulae into the cells, starting with the first approximation to the roll root into H5, namely −D5. The next cell is used to count the number of iterations used, so we enter +I5+1 into I5. We are going to use Newton's method to improve the first approximation; unfortunately, however, simply inserting the relevant formula can result in a division by zero, and we have to resort to a trick. Into J5 we insert the correct formula and add a term which becomes very small after a few iterations so that the results will converge. The first two terms evaluate to zero at the first iteration because of zeros in other cells and the third to the first approximation. This third term is equal to the first approximation divided by 10 raised to the power of the square of the iteration number minus one and rapidly decreases with repeated iterations. The formula we insert into J5 is:

J5−M5+H5/10^((I5−1)*(I5−1))

where H5 is the first approximation. Into the next cell, K5, we insert a formula to calculate the value of the quartic with the current value of the root:

(((J5+D5)*J5+E5)*J5+F5)*J5+G5

which evaluates to 150.1 with the test data. Then in the next, L5, we calculate the value of the derivative using

((4*J5+3*D5)*J5+E5

which should evaluate to −205. This sequence ends by calculating the correction according to Newton's formula, by putting +K5/L5 into M5. This cell value was used by the formula in J5.

We continue by finding the spiral root following the same procedure as for the roll root except that we use (11.21) instead of Newton's formula. Into N5 we insert −G5/F5, which is the first approximation. Into O5 we insert

−(O5*O5*((D5+O5)*O5+E5)+G5)/F5

This corresponds to (11.21). Into P5 we insert

(((O5+D5)*O5+E5)*O5+F5)*O5+G5

which calculates the value of the quartic for the current value of the root. We need to do this as a check since we will calculate the coefficients of the remaining quadratic using the fact that the coefficient B_2 is equal to minus the sum of the roots and the coefficient E_2 is equal to their product. If the process for finding the spiral root does not converge the results for the dutch roll root will be incorrect.

The rest of the program finds the coefficients a2 and b2 of the remaining quadratic and solves it using the same process as for the phugoid and SPPO in the appendix to Chapter 9.

Into Q5 we insert +D5+J5+O5, finding a2
Into R5 we insert +G5/J5/O5, finding b2
Into S5 we insert +Q5*Q5−4*R5, finding the discriminant
Into T5 we insert @IF(S5>0,"REAL","COMPLEX"), indicating the type of the roots
Into U5 we insert @IF(S5>0,(−Q5+@SQRT(S5))/2,−Q5/2), finding the first real root or the real part of a complex pair
Into V5 we insert @IF(S5>0,−(Q5+@SQRT(S5))/2,@SQRT(−S5)/2), finding the second real root or the imaginary part of a complex pair

The function key to cause a calculation to take place (usually F9) should be pressed a few times; the values of F(LAMBDA) should become very small and the final roots rapidly converge to the values −3.7535, −0.005 01, −0.620 74, 2.220 296.

It is important to note that if a blunder is made when inserting the formulae, or detected after the whole procedure has been entered, the error should be corrected and the whole line copied to the line below. This is to ensure that the expected initial values exist in the cells. Later the corrected line can be moved into its proper place if desired. Some spreadsheets indicate when the calculated contents of a cell become smaller than the smallest number which can be stored (an 'underflow'); further iteration can cause this to be flagged as an error. This particularly affects cells containing F(LAMBDA).

To use the setup insert the values of the coefficients into the appropriate columns and copy the formulas in cells H5 to V5 to the block below. Press the F9 key a few times and check that the results have converged.

Student problems

11.1 Consider a wing rolling about the wind direction as in figure 7.1. Assuming that the changes in local wing incidence are small, write down an expression for the increment in lift on a chordwise strip of the wing. Hence derive an equation for the damping in roll derivative L_p. Assuming that the local lift curve slope can be approximated by the overall value, find a value of L_p for the wing in problem (9.1).

11.2 Using the approximations to the dutch roll mode given by (11.34) and (11.43) find expressions for the relative damping and the natural frequency. Find values for these quantities using the data of problem (11.3).

11.3 An aircraft has the following concise lateral stability derivatives: $\hat{y}_v = 0.34$, $l_v = 248$, $l_p = 8.3$, $l_r = 16.2$, $n_v = -35$, $n_p = 0.39$, $n_r = 1.66$, $\hat{g}_1 = 0.16$ and all other terms in the stability equations are zero. The values lead to the quartic

$$\lambda^4 + 10.14\lambda^3 + 45.79\lambda^2 + 431.7\lambda + 156.6 = 0$$

Find the roots and comment on the values. If $\tau = 5.7$ s find the times to half-amplitude and the periodic time. Find the eigenvectors corresponding to the dutch roll and roll subsidence modes. (A)

11.4 In Section 11.4.3 (last paragraph) does the centre of the barrel lie in the $+y$ direction or not?

11.5 Redraw the figure 11.6 for a negative value of N_p.

11.6 Derive equations (6.12), (6.13), (6.18) and (6.19) directly from the lateral stability equations. Show that an aircraft which has neutral spiral stability requires the same aileron and rudder angles to trim in a correctly banked turn as in straight and level flight at the same speed.

Note

1. Figures 11.2, 11.5, 11.6, 11.8, 11.9 and 11.10 have been redrawn from reference (11.2).

12
Effects of inertial cross-coupling

12.1 Introduction

The treatment of dynamic stability that has been presented so far has been physically fairly straightforward, if somewhat involved mathematically. It served the aeronautical world well until the later 1940s and early 1950s when inexplicable incidents began to occur. These took the form of sudden divergent motions in yaw or unexpectedly high rates of roll. The advent of the jet engine had caused a rapid increase of fighter aircraft speeds resulting in sweptback wings of smaller span than previously. The availability of rockets enabled guided weapons and test vehicles to reach supersonic speeds at low altitude. All these were capable of much higher rates of roll due to the much lower roll inertias than had been common in the past. We now know that the cause of the incidents was the effect of gyroscopic terms, i.e. products of the roll rate with the other angular velocities, that had been eliminated in the process of the linearization of the Euler equations. The theory of the phenomena had been put forward in 1948, but no notice was taken by designers or airworthiness authorities; with the benefit of hindsight various incidents can now be explained including possibly one with which the author was loosely connected.

12.2 Roll–yaw and roll–pitch inertia coupling

The appearance of these two types of coupling effect in their purest forms depends on the aircraft having particular static stability characteristics as well as the capability for high rates of roll. In particular the aircraft requires one or other of the following:

- small longitudinal and large directional static stabilities; an example would be a slender supersonic aircraft flying at subsonic speed;
- small directional and large longitudinal static stabilities; the example in this case is a supersonic aircraft at supersonic speed.

The compressibility effects causing these changes are

- the rather lower fin lift curve slope supersonically compared with its subsonic value, thus reducing the value of N_v;
- the cg margin being usually larger at supersonic speed increasing the value of $-M_w$, unless measures are taken to move the cg rearward. Various parameters are involved in this increase of cg margin, the most important of which is likely to be the more rearward wing aerodynamic centre position supersonically compared with subsonically.

We will first discuss the physical cause of these phenomena and later give an analysis which leads to stability criteria. Principal axes of inertia are the most appropriate ones to use for these problems. Taking the case of large longitudinal and small directional static stabilities first, we consider the slender aircraft shown in figure 12.1(a) which is rolling at a high rate, when it receives a disturbance in angular velocity in pitch. The same situation is shown in plan

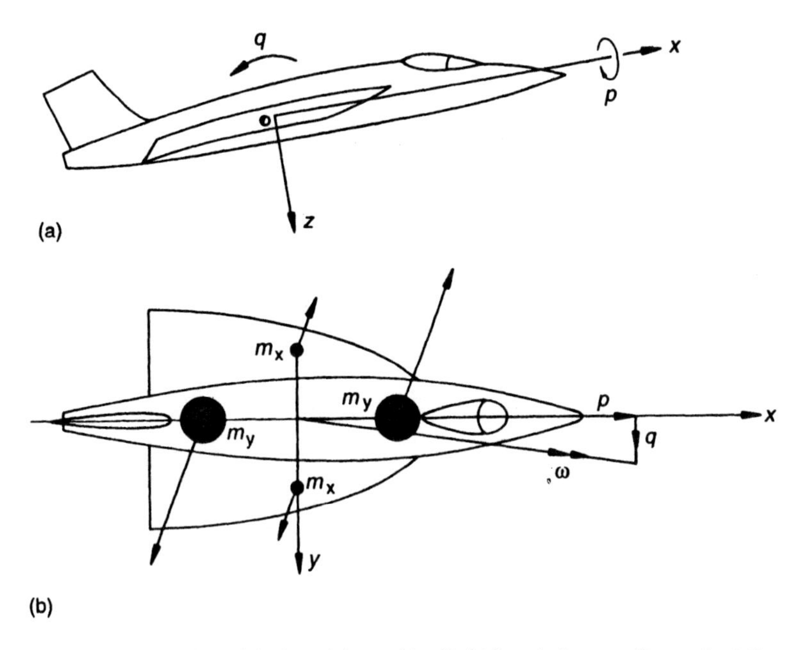

Fig. 12.1 Inertia coupling: (a) aircraft in rapid roll, (b) inertia forces after a short time

in figure 12.1(b), where we have represented the inertias by suitably sized concentrated masses: m_x for the roll inertia and m_y for the pitch inertia. The yaw inertia is contributed to by all the masses. We add together vectorially the rolling and pitching angular velocities, giving a resultant angular velocity inclined downwards as shown. We can now see that the centrifugal inertia forces[1] acting on the masses m_y give a large yawing moment away from the flight direction whilst the masses m_x give a small moment towards it. This is resisted by the small directional stability and the result is that the aircraft diverges in yaw if the inertial moment is large enough. Any tendency for divergence in pitch is resisted by the much larger longitudinal static stability. This form of inertial coupling is known as 'roll–yaw inertial coupling'.

The aircraft can achieve the combination of roll and pitching angular velocities in various ways, one in particular being fairly likely. The aircraft is an inertially slender one and will tend to roll about the axis of minimum inertia as shown in figure 11.5(b). Assume that it is rolling and has an angle of sideslip as in the second position; this would be converted into a nearly equal incidence angle after rolling through 90°. However, due to the large longitudinal static stability a large pitching moment appears which produces an angular velocity in pitch.

'Roll–pitch inertial coupling' is dynamically the same process but involving an aircraft with large directional and small longitudinal static stabilities and starting with angular velocities in yaw and roll. As the name implies, the divergence is in pitch in this case. Vehicles with a cruciform wing and stabilizing surfaces will have equal static stabilities in the two planes and so will have no tendency to suffer from these problems.

12.2.1 Equations of motion

To attack this problem we need to derive the equations again, allowing for the situation that products of the angular velocity in roll with the other disturbance quantities can no longer be assumed small. We will use inertial axes again so avoiding the appearance of the products of inertia. Combining (7.49) with (8.1) and (7.50) with (8.2) we find

$$\left.\begin{aligned} m(\dot{U} - rV + qW) &= X_a + X_g \\ m(\dot{V} - pW + rU) &= Y_a + Y_g \\ m(\dot{W} - qU + pV) &= Z_a + Z_g \\ I_x\dot{p} - (I_y - I_z)qr &= L_a \\ I_y\dot{q} - (I_z - I_x)rp &= M_a \\ I_z\dot{r} - (I_x - I_y)pq &= N_a \end{aligned}\right\} \tag{12.1}$$

where use has been made of (7.42) with $I_{xy} = I_{yz} = I_{zx} = 0$. These equations cannot be solved analytically without simplifying assumptions. In this case we assume that the motion of interest takes place in short time so that the forward speed and rate of roll do not change significantly. We assume that the aircraft, before it is disturbed, is flying at a speed V_e, the x-axis is at an angle α_e to the flight direction and rate of roll is p_e; the latter becomes an important parameter in the equations. We also neglect the components of the weight, assuming their effect cancels out due to the rapid rate of roll, and neglect squares and products of disturbance quantities not involving the rate of roll. From (8.15) and (8.16) we have

$$\left.\begin{aligned} U &= U_e + u \\ V &= v \\ W &= W_e + w \end{aligned}\right\} \tag{12.2}$$

where

$$U_e = V_e \cos \alpha_e \simeq V_e \tag{12.3}$$

and

$$W_e = V_e \sin \alpha_e \simeq V_e\alpha_e \tag{12.4}$$

Then omitting the forward force and roll equations, (12.1) become

$$m(\dot{v} - p_eV_e\alpha_e - p_ew + rV_e) = \mathring{Y}_v v + \mathring{Y}_p p_e \tag{12.5}$$

$$m(\dot{w} - qV_e + p_ev) = \mathring{Z}_w w \tag{12.6}$$

$$I_y\dot{q} - (I_z - I_x)p_er = \mathring{M}_w w + \mathring{M}_{\dot{w}} \dot{w} + \mathring{M}_q q \tag{12.7}$$

$$I_z\dot{r} - (I_x - I_y)p_eq = \mathring{N}_v v + \mathring{N}_r r + \mathring{N}_p p_e + \mathring{N}_\xi \xi' \tag{12.8}$$

The small derivatives Y_r, Y_ξ and Z_η have been neglected. We assume that the rudder and elevator remain at their trimmed values and that the roll is the result of aileron deflection. The aileron angle can be determined from the balance of rolling moments

$$\mathring{L}_v v + \mathring{L}_p p_e + \mathring{L}_r r + \mathring{L}_\xi \xi' = 0 \tag{12.9}$$

and hence on solving

$$\xi' = -\overset{\circ}{L}_{\mathrm{v}} v / \overset{\circ}{L}_{\xi} - \overset{\circ}{L}_{\mathrm{p}} p_{\mathrm{e}} / \overset{\circ}{L}_{\xi} - \overset{\circ}{L}_{\mathrm{r}} r / \overset{\circ}{L}_{\xi}$$

The last equation of (12.5) then becomes on substituting for ξ'

$$I_z \dot{r} - (I_x - I_y) p_e q = (\overset{\circ}{N}_{\mathrm{v}} - f \overset{\circ}{L}_{\mathrm{v}}) v + (\overset{\circ}{N}_{\mathrm{r}} - f \overset{\circ}{L}_{\mathrm{r}}) r + (\overset{\circ}{N}_{\mathrm{p}} - f \overset{\circ}{L}_{\mathrm{p}}) p \quad (12.10)$$

where

$$f = \overset{\circ}{N}_{\xi} / \overset{\circ}{L}_{\xi}$$

which should ideally be a small quantity. We now convert these to concise aero-normalized equations by multiplying the force equations (12.5) and (12.6) by the factor $1/\frac{1}{2}\rho V_e^2 S$, the pitching moment equation (12.7) by $\mu_1 / \frac{1}{2}\rho V_e^2 S \bar{\bar{c}} i_y$, and the yawing moment equation (12.10) by $\mu_2 / \frac{1}{2}\rho V_e^2 S b i_z$ and substitute for concise quantities.

We also write

$$\hat{p}_e = \tau p_e \quad (12.11)$$

where, as before,

$$\tau = m / \tfrac{1}{2}\rho V_e S$$

and also

$$\left.\begin{aligned} i_1 &= (I_z - I_x)/I_y \\ i_2 &= (I_y - I_x)/I_z \end{aligned}\right\} \quad (12.12)$$

If the aircraft were a plane lamina we would have $I_z = I_y + I_y$ and then $i_1 = 1$ and $i_2 < 1$ but in practice $i_1 \simeq 1$ and $i_2 < 1$. The results of these operations are

$$\left.\begin{aligned} (\hat{D} + y_v)\hat{v} - \hat{p}_e \hat{w} + \hat{r} &= (\alpha_e - y_p)\hat{p}_e \\ \hat{p}_e \hat{v} + (\hat{D} + z_w)\hat{w} - \hat{q} &= 0 \\ (m_{\dot{w}} \hat{D} + m_w)\hat{w} + (\hat{D} + m_q)\hat{q} - i_1 \hat{p}_e \hat{r} &= 0 \\ n'_v \hat{v} + i_2 \hat{p}_e \hat{q} + (\hat{D} + n'_r)\hat{r} &= -n'_p \hat{p}_e \end{aligned}\right\} \quad (12.13)$$

where

$$\left.\begin{aligned} n'_v &= n_v - f l_v \\ n'_p &= n_p - f l_p \\ n'_r &= n_r - f l_r \end{aligned}\right\} \quad (12.14)$$

The corresponding state-space equations are then

$$\bar{D}\begin{bmatrix}\hat{v}\\ \hat{w}\\ \hat{q}\\ \hat{r}\end{bmatrix}=\begin{bmatrix}-\hat{y}_v & \hat{p}_e & 0 & -1\\ -\hat{p}_e & -z_w & 1 & 0\\ \hat{p}_e m_{\dot{w}} & m_{\dot{w}}z_w - m_w & -(m_q + m_{\dot{w}}) & i_1\hat{p}_e\\ -n'_v & 0 & -i_2\hat{p}_e & -n'_r\end{bmatrix}\begin{bmatrix}\hat{v}\\ \hat{w}\\ \hat{q}\\ \hat{r}\end{bmatrix}+\begin{bmatrix}\alpha_e - y_p\\ 0\\ 0\\ -n'_p\end{bmatrix}\hat{p}_e \quad (12.15)$$

These equations are a fourth-order set and consequently their characteristic equation is a quartic, the coefficients of which have $\hat{p}_e$ as a parameter. The conditions for stability are given, as before, in Section 9.3.3. Like the longitudinal set there are usually modes corresponding to a pair of roots of large modulus and a pair of small modulus. The pair of large modulus correspond to the SPPO and the other pair are, depending on the value of the roll rate, either

- a complex pair corresponding to a damped long period oscillation or
- a real pair, one negative and one positive, that is a divergence.

If the quartic is written

$$F(\lambda) = \lambda^4 + B\lambda^3 + C\lambda^2 + D\lambda + E$$

the approximate factorization from (9.64) is

$$F(\lambda) \simeq (\lambda^2 + B\lambda + C)\cdot\left(\lambda^2 + \frac{CD - BE}{C^2}\lambda + \frac{E}{C}\right) \quad (12.16)$$

The first factor corresponds to the SPPO and the second to the mode(s) of interest in this case, the condition for the stability for which is

$$E > 0 \quad (12.17)$$

The coefficient E is found in the usual way, and is

$$\begin{aligned}E &= i_1 i_2\hat{p}_e^4 + (\hat{m}_q n'_r - i_1 n'_r - i_2 m_w + i_1 i_2 y_v z_w)\hat{p}_e^2\\ &\quad - (n'_v - y_v n'_r)(m_w + z_w m_q)\end{aligned} \quad (12.18)$$

from which the stability can be determined.

12.2.2 Stability diagram and 'tuning'

The results obtained so far are useful in the design office but, as so often occurs, give little feel for how this relates to the real world. Fortunately we can put a little more meaning into this. Firstly we note that the expression for E can be approximately factorized into

$$E \simeq (i_1\hat{p}_e^2 - \hat{m}_w - z_w m_q)(i_2\hat{p}_e^2 + n'_v - y_v n'_r \quad (12.19)$$

where on multiplying out it can be seen that the missing terms involve the damping derivatives y_v, z_w, m_q and n'_r. We now can introduce the frequencies of the pitch and directional oscillations into (12.19). From (9.82) we have the normalized natural frequency in pitch

$$\hat{\omega}_\theta = \omega_\theta \tau = \sqrt{\hat{z}_w m_q + m_w} \tag{12.20}$$

and from (11.37) the normalized natural frequency of the directional oscillation

$$\hat{\omega}_\psi = \omega_\psi \tau = \sqrt{-\hat{n}_v' + y_v n_r'} \tag{12.21}$$

To be exact we are now using $\hat{\omega}_\theta$ and $\hat{\omega}_\psi$ to have slightly different meanings than before since the derivatives used were relative to wind axes and we are now using inertial axes; also we have used the modified derivatives $\hat{n}_v'$ and $\hat{n}_r'$. With this slight change of meanings we can now write E in the form

$$E = (i_1 p_e^2 - \omega_\theta^2)(i_2 p_e^2 - \omega_\psi^2)\tau^4$$

using (12.11). We can now express the conditions for stability in a more direct form; we require either

$$\left(\frac{p_e}{\omega_\theta}\right)^2 < \frac{1}{i_1} \text{ and } \left(\frac{p_e}{\omega_\psi}\right)^2 < \frac{1}{i_2}$$

or

$$\left(\frac{p_e}{\omega_\theta}\right)^2 > \frac{1}{i_1} \text{ and } \left(\frac{p_e}{\omega_\psi}\right)^2 > \frac{1}{i_2}$$

We can now construct a stability diagram as a plot of $(\omega_\theta/p_e)^2$ against $(\omega_\psi/p_e)^2$ with stability boundaries as the lines $(\omega_\theta/p_e)^2 = i_1 = (I_z - I_x)/I_y$ and $(\omega_\psi/p_e)^2 = i_2 = (I_y - I_x)/I_z$. The result is shown in figure 12.2(a).

A given aircraft is represented on the plot as a point which moves along a straight line through the origin as the roll rate varies, as shown. As roll rate increases the point moves towards the origin. At low roll rates the aircraft is stable, then as roll rate increases the point representing the aircraft generally passes through one or other of the unstable regions. If there is sufficient rolling power available from the ailerons then at still higher roll rates the aircraft becomes stable again. This stable region nearest the origin corresponds to a spin-stabilized vehicle. Aircraft A, likely to be a subsonic aircraft with small static stability in pitch and large directional stability, suffers roll–yaw divergence; this then is the more common form of instability found. The reverse case, aircraft B, suffers roll–pitch divergence.

If the effect of the terms omitted in the approximate factorization is now included the change in the boundaries is small except near the intersection. A sketch of the form of the stability boundaries with the two damping terms included is shown in figure 12.2(b), which shows that in fact the two stable regions in figure 12.2(a) are connected by a narrow 'corridor'. This opens up the possibility of designing an aircraft such that as the roll rate increases the point on the diagram representing the aircraft moves down a line which remains continuously in the stable region. This is known as 'tuning' and involves adjusting the frequencies of the pitching and directional oscillations appropriately. The gains actually available are small, because as the connecting corridor is approached the stability deteriorates markedly and small

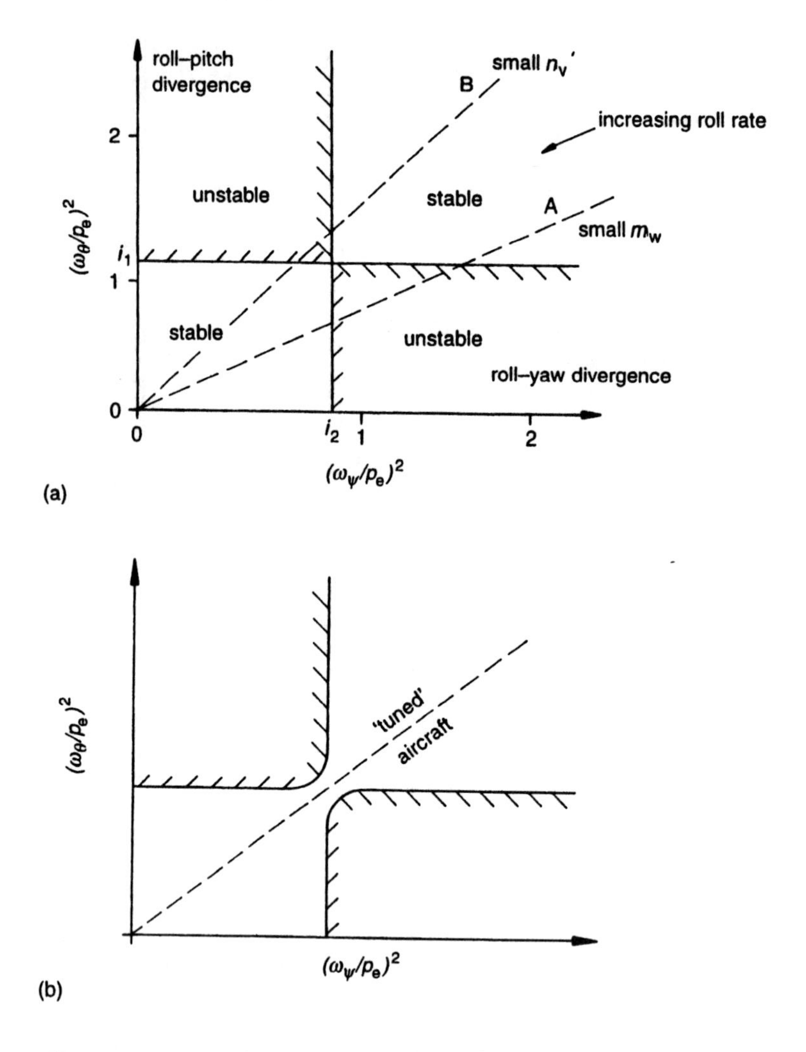

Fig. 12.2 Stability diagram: (a) general case, (b) tuned aircraft

disturbances can cause large displacements in pitch or yaw leading to unacceptably large loads on the aircraft.

To raise the roll rate at which these inertial coupling effects become apparent we need to increase ω_θ and ω_ψ and decrease i_1 and i_2. The first of these implies increasing N_v and $-M_w$, the aerodynamic stiffnesses in pitch and yaw, and decreasing the inertias in pitch and yaw. The inertia ratios can be reduced by decreasing the pitch and yaw inertias and increasing the roll inertia.

12.3 Other inertial coupling problems

Two other problems traceable to inertial coupling effects are known. One, autorotational inertial coupling, has occurred in practice; the other, rudder induced pitching, has apparently only been detected as a result of calculations of the response to rudder movement of inertially slender aircraft.

12.3.1 Autorotation inertial coupling

Autorotation is the maintenance of rolling without the use of ailerons or other means of producing a rolling moment. It is possible to demonstrate a form of autorotation using a wing mounted on a spindle aligned with the airflow in a windtunnel. Setting the wing at or just above its stalling incidence will cause it to autorotate when the windtunnel is run, the effect being caused by the downgoing half-wing stalling whilst the upgoing half-wing is not stalled. See Section 7.2.1 for a discussion on the effect of rolling on local wing incidence changes and Section 1.3.2 for the typical shape of a lift curve. The effect in that case is due to separation of the airflow and is related to the spin; in the case under discussion the effects are caused by the interaction of aerodynamic and inertial coupling terms in the equations of motion.

Consider the aircraft shown at position (a) in figure 12.3 which is rolling rapidly around its minimum inertia axis which is inclined downwards as shown.

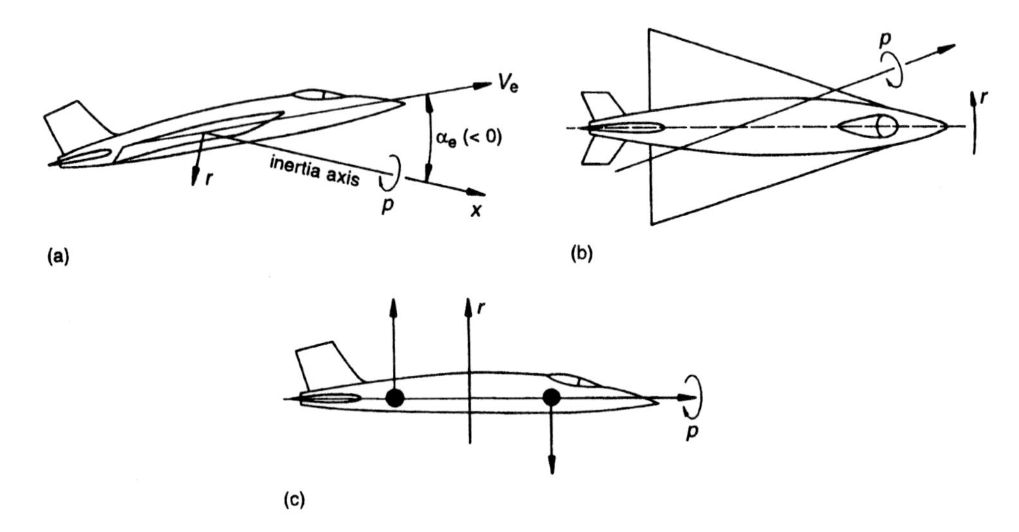

Fig. 12.3 Autorotational inertial coupling

The aircraft is inertially slender and so rolls around the axis and after 90° of rotation would tend to have converted this incidence into a negative angle of sideslip. However, due to the derivative N_v, the aircraft achieves a negative rate of yaw as shown at position (b), but some negative sideslip remains. A side view of the aircraft at this stage is shown in (c); the resultant of the rates of roll and yaw is inclined upwards as shown. The pitch inertia is simulated by masses as before and it can be seen that a negative inertial pitching moment is produced, resulting in a negative rate of pitch. Rate of pitch and rate of roll combine to produce an inertial yawing moment which increases the (negative) sideslip and so the derivative L_v produces a rolling moment in the same sense as the rolling. If this is large enough to overcome the damping in roll the rolling can be maintained without use of the ailerons. This process is independent of any flow separations; however, if the downgoing wing tip should happen to stall this would reduce the damping in roll and increase the roll rate still further. The motion is a barrel roll as the incidence is increased above the trim value, increasing the lift.

Inertial autorotation can be investigated theoretically by simply assuming that a steady rolling state exists and using (12.13), and (12.9) in its aero-normalized form. The acceleration terms are omitted and the aileron angle taken as zero; also elevator terms should be added to the vertical force and pitching moment equations. The five resulting equations are ordinary simultaneous equations linear in $\hat{v}$, $\hat{w}$, $\hat{q}$ and $\hat{r}$; any four can be solved for these variables. Substituting into the fifth gives a quadratic equation in the square of the roll rate. Two real solutions are found for negative and small positive incidences of the minimum inertia axis, corresponding to two rates of autorotational rolling in either direction. Only the lower of the two roll rates is of practical significance. The vanishing of the discriminant gives a critical incidence above which the phenomenon does not exist and a single roll rate in either sense. The coefficients of the quadratic are complicated functions of the derivatives and the inertia numbers i_1 and i_2 are difficult to interpret physically. A full solution of the equations leads to a stability diagram as sketched in figure 12.4, which shows roll rate plotted against incidence of the minimum inertia axis.

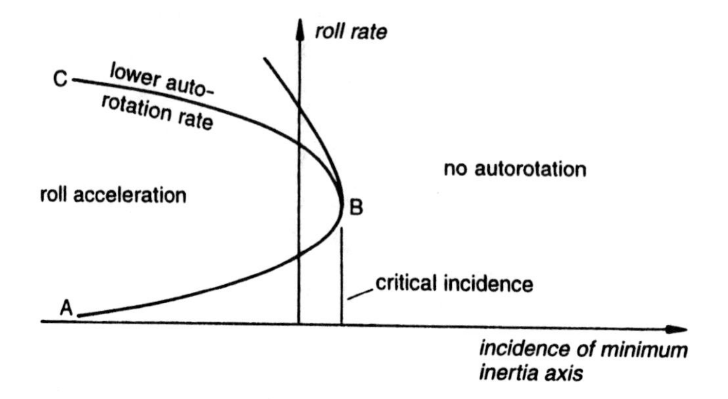

Fig. 12.4 Stability diagram for autorotational inertial coupling

For incidences above the critical incidence and for combinations of incidence and roll rate below the lower branch AB of the curve, autorotation cannot occur. The branch BC shows the lower steady autorotation rate as a function of the incidence. Between these branches an aircraft will accelerate in roll until it reaches the lower autorotation rate; the lower the incidence the lower the roll rate from which it will accelerate. Application of aileron to counter the roll may well recover the situation, the real problem caused by this inertia coupling is the uncertain relationship between roll rate and aileron angle, making precise manoeuvring difficult.

12.3.2 Rudder-induced pitching

This inertial coupling effect is rather like roll–pitch inertial coupling except that the aircraft is not initially rolling. Consider the inertially slender aircraft shown in figure 12.5.

In position (a) the rudder is shown deflected and the aircraft has started to yaw thereby gaining a sideslip angle. At position (b) the sideslip acting through the derivative L_v produces a rolling moment causing a rapid acceleration in roll due to the small inertia and damping. As the aircraft is inertially slender it rolls about the minimum axis of inertia converting some of the yaw angle into incidence. A side view of the situation is shown at (c) and it can be that there is an inertially produced pitching moment generated also tending to increase the incidence. The incidence will shortly afterwards be reduced towards the trim value by the rapid response in the SPPO mode. The result is a large, short-lived peak in the normal acceleration leading to possible stressing problems.

Because of the low usage of the rudder this effect is less important than the others; perhaps its importance is to show that there may yet be other types of inertial coupling to be discovered after the introduction of some new configuration or increase in manoeuvrability.

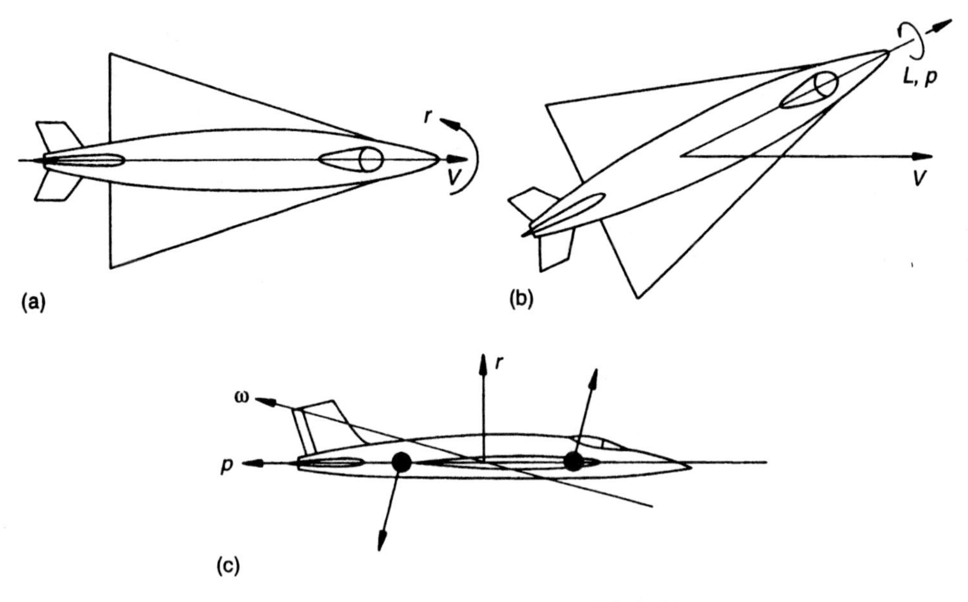

Fig. 12.5 Rudder induced inertial pitching

12.4 Design, development and airworthiness implications

Ideally considerations of inertial coupling should enter the design process at a very early stage because it can be much harder to correct a design later than to get it right, or nearly so, at the start. Most of the rules that might be laid down for the design of an aircraft run directly counter to the whole point of considering an inertially slender aircraft. Almost certainly the aircraft is intended to be capable of supersonic speeds, in which case for the lowest wave drag the designer is forced to use a low aspect ratio wing of the smallest thickness/chord ratio structurally possible and a fuselage of the smallest length/diameter ratio. These requirements make it difficult for the designer to follow the requirements on inertias for increasing the roll rate at which roll–yaw and roll–pitch inertia coupling effects appear given at the end of Section 12.2.2. Fortunately low wing aspect ratios imply long chords close to the fuselage which provide some storage space in spite of the low thickness/chord ratios. Equipment rather than fuel should be placed there to avoid problems at low fuel weights. Most of the disposable masses probably have to be arranged in a long line along the fuselage, but should be as close to the cg as possible to keep the pitch and yaw inertias as low as possible. There is also the problem of the shift in the relative sizes of $-M_w$ and N_v between subsonic and supersonic speeds referred to at the start of Section 12.2. One possibility is to move the cg in flight thereby controlling the cg margin and with it $-M_w$; this might be done by pumping the fuel along the fuselage from one tank to another. It is easier perhaps to avoid a downward inclination of the axis of minimum inertia, thus helping to avoid autorotational problems. Care has to be taken to specify only sufficient roll performance at each point of the flight envelope to enable the aircraft to fulfil its intended function.

Before the first flight extensive calculations of the response to control application will have to be made with particular emphasis on rolling. Rolling combined with other motions will have to be investigated, in particular the rolling pull-out, and all parts of the flight envelope have to be covered. After the first few flights, measurements of the stability derivatives will have to be made and compared with estimates and the calculated responses repeated in the case of disagreements. Actual responses will have to be compared with calculations and modifications made to the aircraft when problems appear. One of the complications of this work is that this is a nonlinear phenomenon: the response to 50 per cent of full aileron angle is not simply twice that due to 25 per cent as it is in linear theory, so that a series of increasing aileron angles and say normal accelerations have to be used. Another difficulty is the sensitivity of inertial coupling effects to small changes in derivatives, so that values including aeroelastic effects have to be used. Only gradually will the safety of rolling manoeuvres be checked and improved. The complete solution of the problems by the use of the AFCS is not currently possible as the amount of control angle that can be applied is limited, usually to about 10 per cent of the maximum, from safety considerations.

Note

1. The reader is reminded that an inertia force is a fictitious force which enables a dynamic situation to be envisaged as a static one. Accelerated masses are given a force equal to the product of the mass and the acceleration, and in the opposite direction to that of the acceleration. These forces are sometimes referred to as 'd'Alembert forces'.

13
Introduction to automatic control and stabilization

13.1 Introduction

All aircraft excepting those carrying only a very few passengers will have some form of automatic control or stabilization to assist the pilot in flying the aircraft. The history of aircraft autocontrol systems is actually older than that of successful powered flight; Hiram Maxim, who attempted to fly a steam driven aircraft in 1891, equipped it with a flight control system also driven by steam but in all other ways resembling modern designs. The Wright brothers, recognizing that their aircraft was tiring to fly due to its inherent instability, went on to design and develop an automatic stabilization system which was able to stabilize their aircraft. However, the design and development of the first really successful system is credited to the Sperrys, father and son, between 1910 and 1912. This was a two-axis system and by 1914 they had produced a three-axis system. The development of these systems has continued unabated ever since and is now a subject large enough to have had whole books written about it; unfortunately space can only be afforded in this book for a brief introduction.

13.2 'Open loop' and 'closed loop' systems, the feedback principle

We start by introducing the practice of drawing a box to represent a component or a subsystem in a system as shown in figure 13.1(a). The component is assumed to have a single input and a single output.

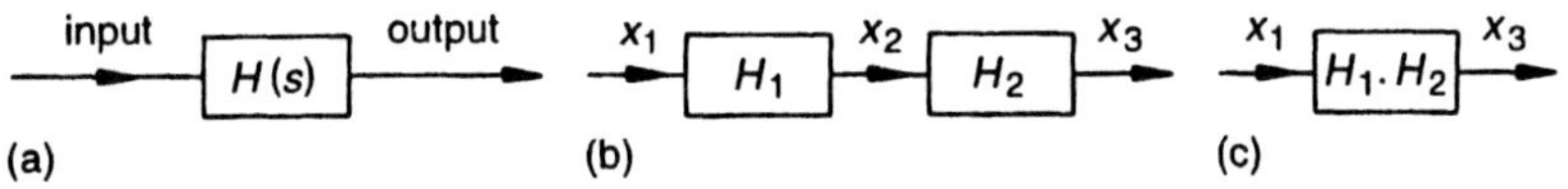

Fig. 13.1 Box representation of components: (a) single component, (b) two components in series, (c) equivalent box

Assuming that we are dealing with linear components the relation between the input and output is given by the transfer function $H(s)$ as introduced in Section 10.2.1. Now consider two such components connected together as shown in figure 13.1(b). Let x_1 be the input to the first and x_2 be the output; the latter is also the input to the second and x_3 is its output. Suppose that the transfer functions are $H_1(s)$ and $H_2(s)$ then we have

$$\bar{x}_2 = H_1(s)\bar{x}_1 \text{ and } \bar{x}_3 = H_2(s)\bar{x}_2$$

From these relations we immediately obtain

$$\bar{x}_3 = H_1(s)H_2(s)\bar{x}_1 \tag{13.1}$$

which implies that we can replace a series of components or subsystems in a linear system by a single box having a transfer function equal to the product of the separate transfer functions, as shown in figure 13.1(c). Equally for linear components we can reverse the order of the boxes. What actually 'flows' along the connecting lines in a real system varies from system to system and from point to point within systems. The boxes often convert one type of quantity to another, for example from an electrical signal to a physical movement. We can, however, think of a physical quantity flowing along a line, even if actually it is an electrical signal which is proportional to it. A system such as that shown in figure 13.1(b) is known as an 'open loop' system; an example might be an engine and the gearbox it is connected to. The position of the throttle might be considered as the input and the rotational speed of the final drive shaft as the output.

With many systems it is possible to measure the output, that is to convert it to a different physical quantity with a magnitude in proportion, subtract it from a desired value of the output and use this to adjust a control. In the previous example it is possible to measure the speed of the output shaft, subtract it from a desired value and use the difference to control the throttle position; this could provide the basis of a simple constant speed control. The essential point is that the output is measured and subtracted from the input to produce an error signal; this error is then used to adjust a control in such a manner as to tend to reduce the error as shown in figure 13.2.

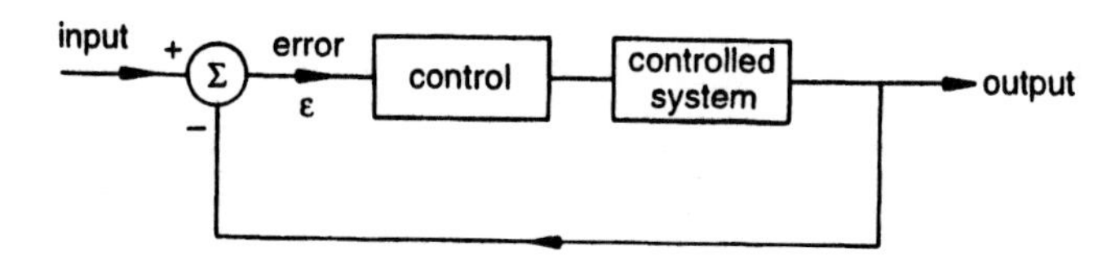

Fig. 13.2 Schematic of simple feedback system

This is a 'closed loop' system. The signal fed back may be processed in some way in order to improve the characteristics of the combined system. The error signal may also be modified for the same purpose and may then be referred to as the 'actuating' signal. Two fairly similar systems using the feedback principle can be distinguished.

The input in an 'autocontrol' system is intended to cause the aircraft to hold some quantity constant or to perform some manoeuvre. It may try to keep quantities such as the height or Mach number of the aircraft constant or to force the aircraft to turn, follow a radio beam or land automatically. When the autopilot is engaged the pilot controls the aircraft by changing the demanded value of any controlled motion variable.

The purpose of an 'autostabilizer' is to augment a stability derivative. For instance a yaw damper augments the damping in yaw derivative, N_r, so the quantity measured and fed back is the yaw rate and the input is zero as we require zero yaw rate. The system applies the rudder in such a sense as to oppose the yaw rate. The augmentation to the yaw derivative appears in body rather than wind axes and if the rudder produces a rolling moment then a rolling derivative is also produced. Pitch and roll dampers can also be produced in a similar manner using the elevator or ailerons. In sophisticated aircraft a number of both sorts of system will be installed and the whole forms the 'automatic flight control system' or AFCS. Even the lightest aircraft are likely to incorporate the feedback principle upon which these systems depend somewhere in their main or auxiliary power systems or instruments. In a sense aircraft have always had a sophisticated AFCS in the form of the pilot, without which an aircraft is an open loop system.

13.2.1 Example of a simple feedback system

As a simple and important example of a feedback system we describe the hydraulic servomotor. Servomotors were introduced initially because, in spite of measures taken to balance control surfaces aerodynamically, the forces required from pilots to manoeuvre became too large. The purpose of a servomotor is to use power from an auxiliary source to move a control surface under the control of the pilot. A schematic diagram is shown in figure 13.3(a).

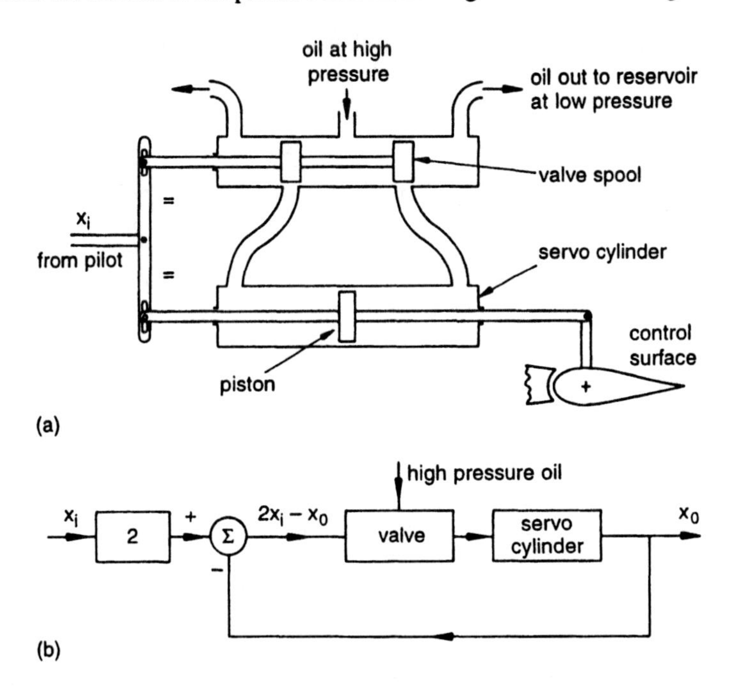

Fig. 13.3 Hydraulic servomotor: (a) physical layout, (b) equivalent circuit

It consists of a cylinder having a piston which is connected via a rod to the control surface. Located nearby is the valve body; inside it has a cylindrical bore housing the valve spool and connections are provided to the hydraulic system. The valve spool consists of two small pistons which can just cover two ports leading to the bore of the main cylinder and joined by a rod which passes out of the end of the valve body. The rods in the two cylinders are connected by a lever; in this example the input from the pilot is connected to the centre of this lever. The servomotor is shown in its stationary position. If the pilot moves the control rod to the right, say, the piston in the main cylinder is initially unable to move and the motion is transmitted to the valve spool. This causes high pressure oil to be admitted through the valve body to the left side of the main piston. At the same time oil can flow from the right side of the piston through the bore of the valve body to the reservoir. The piston of the servo then moves rapidly to the right moving the control surface, the speed being roughly proportional to the distance the valve spool is from its neutral position. This motion is fed back to the valve spool via the lever stopping the motion when the ports are again covered; the output movement in this case is twice the input. Virtually no force is required from the pilot.

Figure 13.3(b) shows a functional diagram of the hydraulic servomotor in which most of the components are represented by boxes linked by lines. The lever is represented by a summation sign in a circle as it effectively performs this arithmetical operation. A minus sign appears next to the signal fed back from the main cylinder, to indicate that its input is subtracted, and a plus sign next to that from the pilot. The nett operation performed by the lever is the subtraction of the signal fed back from twice the pilot input. Servomotors powered electrically are also manufactured; see reference (13.1), which also describes much of the 'hardware' used in AFCS's. We have seen that the servomotor is itself a feedback system and therefore forms an 'inner loop' in the main system; this is very common and most systems in fact feature a number of inner loops; autostabilizers often form an inner loop when part of a more complex system.

13.2.2 Advantages of AFCS's

The provision of servomotors to move the control surfaces immediately allows more rapid movements against larger hinge moments because of the higher power available. Servomotors can be used simply to assist the pilot, in which case he or she still has stick forces which provide a natural feel to the control stick. Alternatively the servo can take over the whole task of moving the control surface and 'artificial feel' given to the pilot by a spring, possibly arranged to vary in rate with the dynamic pressure. The removal of the need for a solid connection between the pilot and control surface enables the use of electrical signals to transmit the pilot's input, known as 'fly by wire', or the use of light transmitted by optical fibres. These save weight but three or four independent channels must be used to ensure reliability.

The addition of an autocontrol system gives higher accuracy of control because of its error sensing characteristic and relieves the pilot of the need to control the aircraft continuously, still further reducing pilot fatigue and enhancing safety. As we have seen in Section 11.4.1 aircraft possess neutral stability in heading and a prime target for the application of autocontrol has to be to correct this. When the task is simple an autocontrol system can achieve better accuracy than a pilot. A pilot tends to use more control surface deflection in manoeuvring or correcting the effects of atmospheric disturbances and so causes more drag on average. One result of installing an AFCS can be a reduction in fuel consumption. Autocontrol systems have been developed that can land an aircraft thus permitting all-weather operations, with further consequent economic benefits.

The addition of an AFCS enables the use of designs of conventional layout with negative cg margins giving improved response or makes possible the use of a smaller tailplane or a wider cg range. Tailless aircraft become a more reasonable option as the poor damping in pitch can be augmented. However, there is a limit to the amount of instability allowed because the authority of the AFCS is limited for the sake of safety. Typically the maximum control surface deflection allowed to be produced by the AFCS is limited to 10–50 per cent of the maximum travel.

13.3 General theory of simple systems

We come now to consider how the feedback system modifies the stability and response characteristics of the aircraft. The discussion is limited to simple systems with a single input and single output.

13.3.1 Effect of feedback

Figure 13.4(a) shows a possible autocontrol system intended to control the pitch angle of an aircraft. Again each element is shown as a box without regard to its actual function; in particular the servomotor is shown as a single box in spite of having an internal feedback loop. The output pitch angle θ_o is fed back and is subtracted from the demanded pitch angle θ_i. To improve the damping in pitch the pitch rate is sensed and fed into the servomotor with the error in pitch angle signal. Optional compensators are shown at A and B to further process the

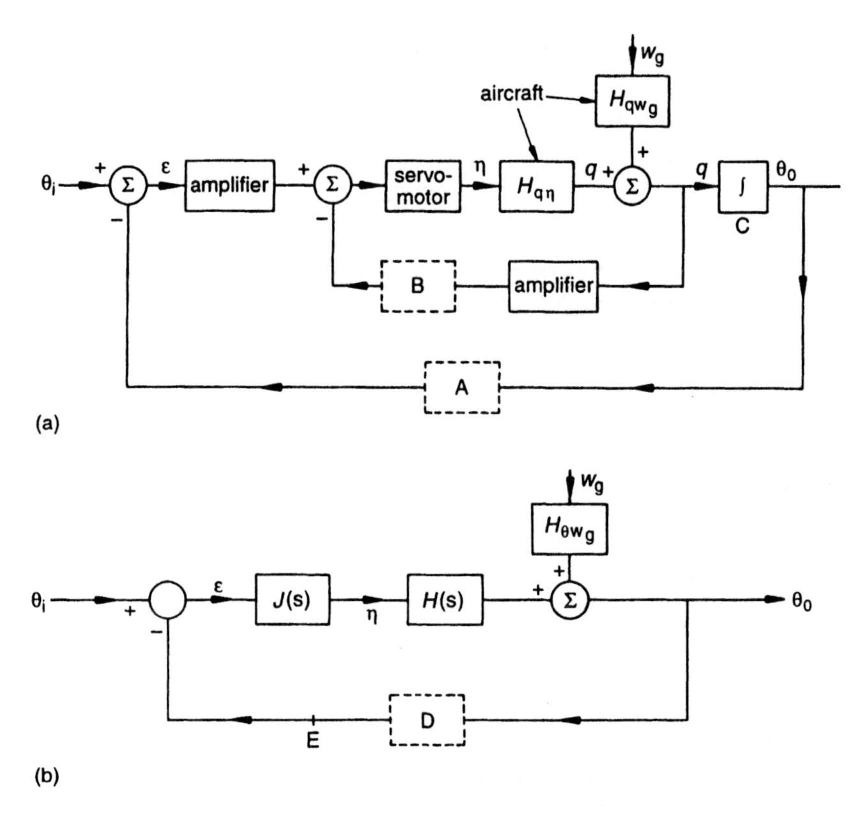

Fig. 13.4 Autocontrol system to control pitch angle of an aircraft

signals. The aircraft is also subject to disturbances such as gusts which add to the response to elevator angle deflection via the transfer function H_{qw_g}.

It would be equally possible to differentiate the output pitch angle signal to obtain a pitch rate signal, then factor it and add this to the output pitch angle to obtain an equivalent system to that of figure 13.4(a). If the demanded value of pitch angle is constant, differentiating the error signal would have the same effect. The process applied to the error signal can then be represented by the transfer function $J(s) = (k_0 + k_1 s)$. This box is often referred to as the 'control'; with minor changes it could also be placed at the position D in figure 13.4(b). By lumping together the transfer function of the servomotor, $H_{q\eta}$, and the integration process at C into $H(s)$, using the result (13.1), the system of figure 13.4(a) can be replaced by that of figure 13.4(b). The gust input now becomes $H_{\theta w_g} . w_g$. This system can be taken as the standard system to which all simple feedback systems can be reduced. Considering the input and output for each component in turn we have the relations

$$\bar{\theta}_o = H\bar{\eta} + H_{\theta w_g}\bar{w}_g \tag{13.2}$$

$$\bar{\varepsilon} = \bar{\theta}_i - \bar{\theta}_o \tag{13.3}$$

$$\bar{\eta} = J\bar{\varepsilon} \tag{13.4}$$

Substituting (13.3) into (13.4) and then substituting into (13.2) we find

$$\bar{\theta}_o(1 + HJ) = HJ\bar{\theta}_i + H_{\theta w_g}\bar{w}_g$$

The system transfer function in the absence of gust effects is then

$$\frac{\bar{\theta}_o}{\bar{\theta}_i} = \frac{H(s)J(s)}{1 + H(s)J(s)} \tag{13.5}$$

and the response to gusts in the absence of control input is

$$\frac{\bar{\theta}_o}{\bar{w}_g} = \frac{H_{\theta w_g}(s)}{1 + H(s)J(s)} \tag{13.6}$$

instead of $H_{\theta w_g}(s)$ for the aircraft without the control system.

A system which uses a feedback signal proportional to the controlled quantity is said to use 'proportional control'. If the signal going to the servomotor also contains a term proportional to the derivative of the controlled variable then the system is said to have 'rate control'. This anticipates future changes in the error; an alternative view is that it creates a phase lead correcting for lags elsewhere in the system. Two of the likely sources of lags are those due to the servomotor and the aerodynamic lag due to the time taken for the circulation around a wing surface to change after a change in control surface angle. If the aircraft flies through high frequency turbulence the term may become large and it may be necessary to reduce its effect by the use of a low pass filter. It can also be useful to include a term proportional to the integral of the controlled quantity; this for instance will prevent the build-up of error over a long time due to friction in components.

Now suppose the servomotor is a perfect device with unit transfer function; the elevator transfer function is in the form of the ratio of two polynomials, as in (10.6), that is $H_{\theta\eta}(s) = G(s)/F(s)$, say. The system transfer function can be written, using (13.5), as

$$\frac{\bar{\theta}_o}{\bar{\theta}_i} = \frac{(k_1 s + k_0)G(s)/F(s)}{1 + (k_1 s + k_0)G(s)/F(s)}$$

$$= \frac{(k_1 s + k_0)G(s)}{F(s) + (k_1 s + k_0)G(s)} \tag{13.7}$$

The characteristic equation for the combined system is therefore

$$F'(s) = F(s) + (k_1 s + k_0)G(s) \tag{13.8}$$

When s has a value equal to a root of (13.8) the system transfer function (13.7) becomes infinite; these roots (or zeros) are said to be the 'poles' of (13.7). The numerator of (13.7) can also become zero; these points are the zeros of (13.7). In a similar manner we can write the transfer function for gust velocity in the form $H_{\theta w_g} = G'(s)/F(s)$ leading to

$$\frac{\bar{\theta}_o}{\bar{w}_g} = \frac{G'(s)}{F(s) + (k_1 s + k_0)G(s)} \tag{13.9}$$

Worked example 13.1

A low aspect ratio aircraft has a value of $\hat{l}_p = 4.5$. Using the one degree of freedom model for this mode, check with Section 11.7 to see if the characteristic time t_c in the roll subsidence mode meets the handling requirements. If not, design a system to augment the damping in roll to bring the time to the required value. The normalized time unit τ has a value of 8 s and the characteristic time of the servomotor is 0.08 s.

Solution

This mode was discussed in detail in Section 11.4.3. To obtain the simplified response equation for this mode we use (11.20) with the second member of (11.6) and omit the terms in $\hat{v}$, ζ' and $\hat{r}$ to obtain

$$(\hat{l}_p + \hat{D})p = -l_\xi \xi \quad \text{(i)}$$

Putting the right-hand side to zero and solving as before, $\lambda = -\hat{l}_p$. The characteristic time t_c is then $\tau/\hat{l}_p = 8/4.5 = 1.78$. This value clearly violates the requirement $t_c \ngtr 1.5$ s.

The system we propose to use is that of figure 13.2 with the aircraft modelled by the transfer function for the roll subsidence mode combined with the transfer function for the servomotor. The aileron will be moved in proportion to the error in roll rate, p. Any problems can, of course, be solved in either dimensional or normalized terms, with identical results. It is, however, conventional to solve such problems in dimensional terms, probably because such systems are usually analyzed by control engineers with no training in aerodynamics. For the sake of demonstration we choose to use normalized variables.

A servomotor can be modelled by a simple first-order equation of the form

$$(t_s D + 1)\xi = f(t)$$

where $D = \mathrm{d}/\mathrm{d}t$ and t_s is the servomotor time constant or lag. Introducing the normalized time unit, τ, we have

$$\left(\frac{t_s}{\tau}\frac{\mathrm{d}}{\mathrm{d}t_s/\tau} + 1\right)\xi = \hat{f}(t/\tau)$$

We now write $\hat{t} = t/\tau$ as before and $\hat{t}_s = t_s/\tau$ to give

$$(\hat{t}_s \hat{D} + 1)\xi = \hat{f}(\hat{t})$$

where $\hat{D} = \mathrm{d}/\mathrm{d}\hat{t}$. We now take the Laplace transform as in Section 10.2.1, giving

$$(\hat{t}_s s + 1)\bar{\xi} = \bar{f}(s)$$

The transfer function for the servomotor is therefore

$$\frac{\bar{\xi}}{\bar{f}} = \frac{1}{\hat{t}_s s + 1}$$

Taking the Laplace transform of (i) above gives

$$\left(\hat{l}_p + s\right)\bar{p} = -l_\xi \bar{\xi}$$

The transfer function is therefore

$$\frac{\bar{p}}{\bar{\xi}} = -\frac{l_\xi}{\hat{l}_p + s}$$

Using proportional feedback we write $\xi = \hat{k}\varepsilon$, that is $J(s) = \hat{k}$, then from (13.8) the characteristic equation is $F + GJ$ or

$$\left(\hat{l}_p + s\right)\left(\hat{t}_s s + 1\right) - \hat{k} l_\xi = \hat{t}_s s^2 + \left(\hat{l}_p \hat{t}_s + 1\right)s + \hat{l}_p - \hat{k} l_\xi = 0$$

Substituting values gives

$$0.01s^2 + (4.5 \times 0.01 + 1)s + 4.5 - 50\hat{k} = 0$$

or

$$s^2 + 104.5s + 450 - 5000\hat{k} = 0 \tag{ii}$$

We require the characteristic time to be 1.5 s so converting this to normalized time gives $\hat{t}_c = 1.5/8 = 0.1875$. Then putting $s = -0.1875$ and solving (ii) we have

$$\hat{k} = \frac{(-0.1875)^2 + 104.5(-0.1875) + 450}{5000} = 0.086\,088$$

Substituting back into (ii) and solving checks this value and gives a second root of -104.31; this corresponds to a characteristic time of 0.009 586 which is associated with the servo. It is interesting to note that omitting the term in s^2 in (ii) makes almost no difference to the result; however, putting the servomotor characteristic time to zero results in a value for $\hat{k}$ of only 0.0167. Finally we convert the constant $\hat{k}$ to its dimensional equivalent k by noting that the aileron angle can be written as $\xi = kp$ or as $\xi = \hat{k}\hat{p}$ and that $\hat{p} = p\tau$ so that

$$kp = \hat{k}\hat{p} = \hat{k}p\tau \text{ or } k = \hat{k}\tau$$

then $k = 0.689$.

13.3.2 The effect of rate control

We now consider the longitudinal stability quartic characteristic equation (9.38) and now write the polynomial G in the form

$$G(s) = \alpha s^2 + \beta s + \gamma \tag{13.10}$$

We find that the characteristic equation is now

$$F'(s) = s^4 + (B_1 + k_1\alpha)s^3 + (C_1 + k_1\beta + k_0\alpha)s + (D_1 + k_1\gamma + k_0\beta)s + (E_1 + k_0\gamma) = 0 \tag{13.11}$$

Since the characteristic equation is still a quartic we can factorize it into a pair of quadratics, as in Section 9.3.4, that is

$$F'(s) = (\lambda^2 + a_1\lambda + b_1)(\lambda^2 + a_2\lambda + b_2)$$

Multiplying this out and equating the coefficients of s^3 we find

$$a_1 + a_2 = B_1 + k_1\alpha$$

Since the left-hand side of this expression is twice the sum of the damping constants of the SPPO and phugoid modes, it follows that unless rate control is used the total damping cannot be increased, only distributed differently between the modes. Since lags are introduced in real systems the total damping is actually reduced. This is a demonstration of a general observation, that unless the damping of a system is already adequate, rate control is required and proportional feedback on its own can worsen a system.

13.4 Methods of design

During the design of an aircraft it will become evident that an autocontrol system will be required when the aircraft is found not to satisfy the handling criteria of Sections 10.5.3 and 11.7; this will also define the desirable characteristics. In the design and development of a feedback system the difficulties arise over the choice of system and of the feedback constants which will give the aircraft desirable characteristics over the whole flight envelope. The steps in designing a system are then to decide which quantities to measure and whether inner loops are required to provide autostabilizers. The last step is to find the feedback constants; like any design process the whole process may have to be an iterative one. In the next section we discuss some simple methods for finding the feedback constants. We have already seen in Worked example 13.1 that it is possible simply to set up and solve an equation for them. If a test of the stability of the system is all that is required then the test functions of Section 9.3.3 can be used.

13.4.1 Frequency response methods

Referring to figure 13.4(b), if we cut the circuit at a point such as E in the feedback loop and treat the ends as an input and output we can determine the response. This is the 'open loop' response $H(i\omega)J(i\omega)$ and is normally a complex quantity. For simple feedback systems the open loop frequency response is a convenient quantity to use for investigating and improving the stability.

There are at least three ways of presenting the frequency response graphically as frequency is varied; these are the Nyquist diagram, the Bode diagram and the Nichols diagram. These are most easily determined by using standard computer software. The Nyquist diagram is a polar plot with the gain, $|HJ|$, as radius and $\arg(HJ)$ as polar angle and using frequency as parametric variable. Figure 13.5 shows a typical plot; subject to certain conditions a system is unstable if the plot encloses the point $(-1,0)$.

The Bode diagram is a logarithmic plot of gain and of phase angle against frequency; a typical plot is sketched in figure 13.6.

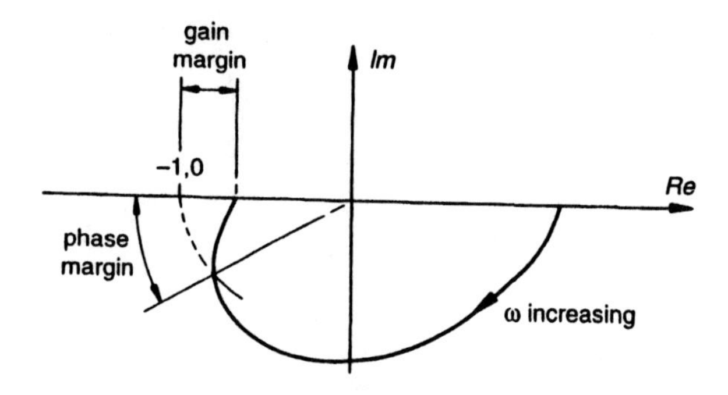

Fig. 13.5 Nyquist diagram

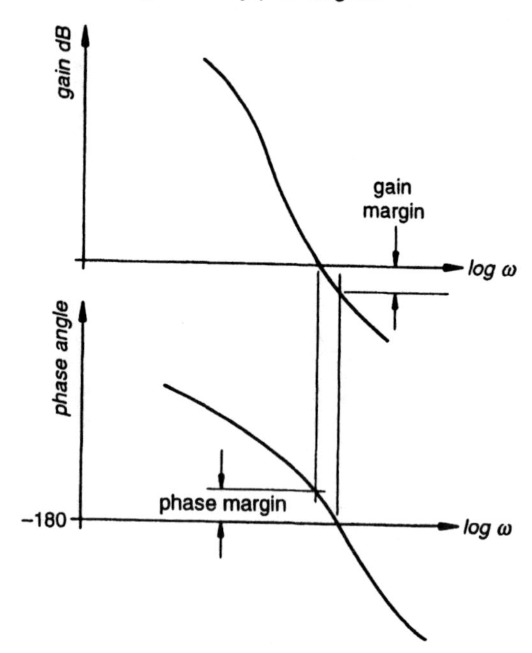

Fig. 13.6 Bode diagram

The Nichols diagram is a plot of the gain in decibels against the phase angle with frequency as parameter; a typical plot is shown in figure 13.7.

A Nichols chart is a specially printed chart on which to plot the gain and phase of the open loop transfer function. It also has contours on it from which the gain and phase for the closed loop system can be determined based on (13.4). The sensitivity of a system to variations in the characteristics of its components can be seen from these plots. A system having the response shown as the dotted line in figure 13.7 would be stable but it would only take a small error in the components or in estimating stability derivatives to produce a system with poor stability.

The adequacy of the stability of the system is judged by the 'gain margin' and the 'phase margin'; figures 13.5, 13.6 and 13.7 illustrate their determination. A gain margin of 6 dB and a phase margin of 60° are considered to provide adequate stability. For a system in which the modes are well separated in frequency, each oscillatory mode can be treated as if it were a separate second-order system. For a second-order system it can be shown that the phase margin is related to the relative damping by the approximate relation

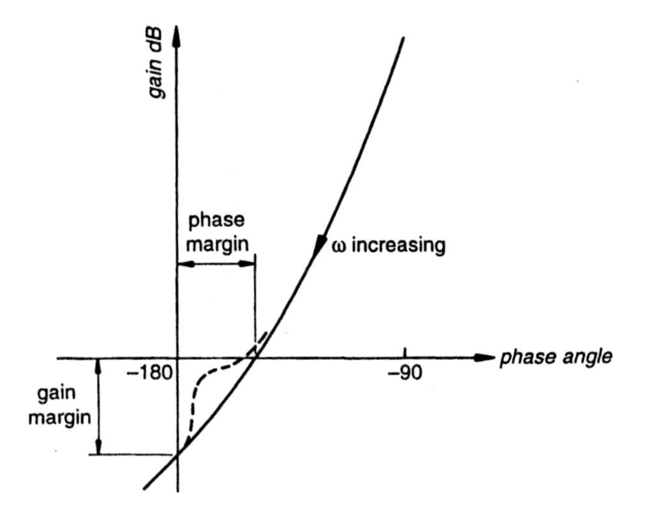

Fig. 13.7 Nichols diagram

$$\zeta = 0.011\varphi \text{ for } \zeta \leq 0.7 \tag{13.12}$$

where φ is the phase margin in degrees. This enables the additional damping required to be determined and can be used to design an autostabilizer.

13.4.2 Use of the root locus plot

The root locus plot has already been introduced in Section 9.4.3, where the locus of the roots of the longitudinal quartic as the cg margin varies was plotted. In this case the parameters used are the constants in $J(s)$. The roots of high-order polynomials are easily found using computer packages, without which the method could be rather tedious. From (7.34) we see that if a line drawn to a point in the left-hand half-plane from the origin makes an angle Φ to the negative axis then

$$\zeta = \cos \Phi \tag{13.13}$$

which is true whether we are using dimensional or normalized variables. We now consider an example.

The aircraft is the same one considered previously but with the cg margin reduced from 0.15 to 0.01, so that it is only just statically stable. At this value of the cg margin the roots of the SPPO mode are both real. It is intended to use pitch angle as the controlled variable and to use both proportional and rate feedback to give the phugoid mode a relative damping ζ of 0.7 with no deterioration in the damping of the SPPO. A perfect servomotor is assumed. Using (13.11) the resulting quartics were solved for a variety of cases and the results are shown in figure 13.8.

The locus of the phugoid roots is shown in figure 13.8(a). The point A represents the initial roots, variation of k_1 results in the locus AB while that of k_0 results in AC with the phugoid roots becoming a real pair. Also shown are lines at 45° to the negative real axis corresponding to a relative damping of 0.7. The corresponding variation of the SPPO roots is shown in figure 13.8(b); k_0 decreases the damping and k_1 increases it. It is evident that variation of k_0 increases the relative damping of the phugoid and reduces the damping of the SPPO in line with comments in Section 13.3.2. A value of k_0 of about -0.03 gives a value of ζ close to that

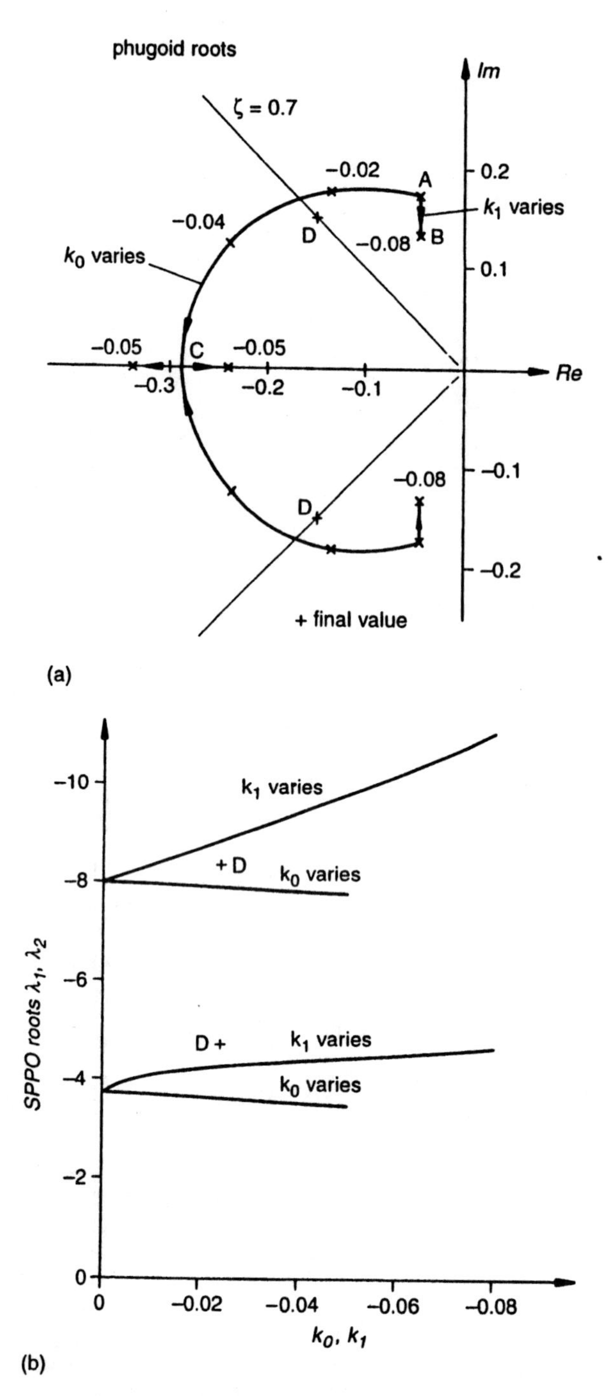

Fig. 13.8 Use of root locus: (a) root locus plot showing effects on phugoid roots, (b) plot showing effect on SPPO roots

required but reduces the damping of the SPPO; however, we can correct the latter using rate feedback. A little trial and error shows that $k_0 = -0.033$ and $k_1 = -0.03$ gives the required value of ζ for the phugoid and slightly improved damping for the SPPO, marked as the point D on the figures. Of course this solution is not unique.

13.5 Modern developments

The simply systems so far described cannot be used to improve all the flight characteristics, or produce suitable characteristics over whole flight envelopes due to the variations of speed, height and stability derivatives. Modern aircraft automatic control systems have therefore had to progress well beyond the simple systems described and frequently use multiple inputs and outputs. As a result more advanced analysis methods have been developed using state-space formulation of the equations. Some improvement has been obtained by developing methods to design systems which are optimal in some sense. To cope with the variations of speed and height through the flight envelope of the aircraft the feedback constants of the AFCS are made to vary with dynamic pressure; this is known as 'gain scheduling'.

In the most advanced aircraft autocontrol techniques are used to provide other functions; this is often referred to as 'active control'. We have already mentioned the possibility of stabilizing an unstable aircraft. A set of interrelated functions are those to improve the ride, alleviate the loads produced by flying through gusts, and alter the spanwise wing lift distribution to give lower root bending moments in manoeuvres by deflecting suitable control surfaces. Flutter is another possible area of application for these techniques, raising the flutter speed of a wing or saving wing weight by designing it to a reduced flutter speed.

For a number of years automatic control systems have operated on the analogue principle, that is voltages have represented the physical quantities being manipulated within the system. Current and new systems are digital, that is streams of digital pulses represent the physical quantities. This introduces new possibilities, new problems and techniques to solve them. In such systems the control law is realized by a digital computer. This has advantages such as making possible more complex laws and easy alteration of the laws by altering the software. Again three or four independent channels must be used to ensure reliability of the software used.

Beyond the advances described in this necessarily short account there are flight control systems which are designed to adapt to the current environment of the aircraft and the use of fuzzy logic and expert systems.

Student problems

13.1 Using the approximation for the dutch roll mode given in Section 11.4.4.2 find the proportional constant required for a yaw damper fitted to an aircraft with the following stability derivatives to have a damping constant of 0.7. The derivatives are $\hat{y}_v = 0.75$, $n_v = -61$, $n_r = 1.7$, $n_\zeta = 1.7$; assume that $y_\zeta = 0$. Also assume that the servomotor has a negligible time constant. (A)

13.2 Using the stability criteria in Section 9.3.3 (see Section 13.4) determine if the following characteristic equations represent stable or unstable systems and if unstable the number of roots with positive real parts. If possible indicate if there is an unstable oscillation. (A)

(a) $\lambda^3 - 9\lambda^2 + 2\lambda + 30$
(b) $\lambda^4 + 12\lambda^3 + 76\lambda^2 + 5\lambda + 15$
(c) $\lambda^4 - 8\lambda^3 + 11\lambda^2 + 50\lambda + 30$
(d) $\lambda^4 + 12\lambda^3 + 12\lambda^2 + 2\lambda - 1$
(e) $\lambda^4 + 10\lambda^3 + 46\lambda^2 + 432\lambda + 157$
(f) $\lambda^5 - 7\lambda^4 + 3\lambda^3 + 61\lambda^2 + 53\lambda + 30$
(g) $\lambda^5 - 31\lambda^4 - 19\lambda^3 + 39\lambda^2 + 20\lambda - 20$

Appendix A
Aircraft moments of inertia

The aerodynamic characteristics of an aircraft can be estimated fairly accurately given sufficient geometric information using the ESDU Data Sheets or other information sources. The same is not true for the aircraft inertias for which there is very little easily available information. This appendix presents the results of an attempt to correlate the inertias with geometric parameters. It can be used to estimate the inertias at an early stage of a design or for the purpose of setting student exercises. There is also some information on inertia in pitch in reference (A.1). Data for this correlation have been taken from references (A.2), (A.3) and (A.4) and cover a mass range from 800 to 210 000 kg. The inertias quoted in reference (A.3) were estimated using a method given in reference (1.1) and the aircraft types are not

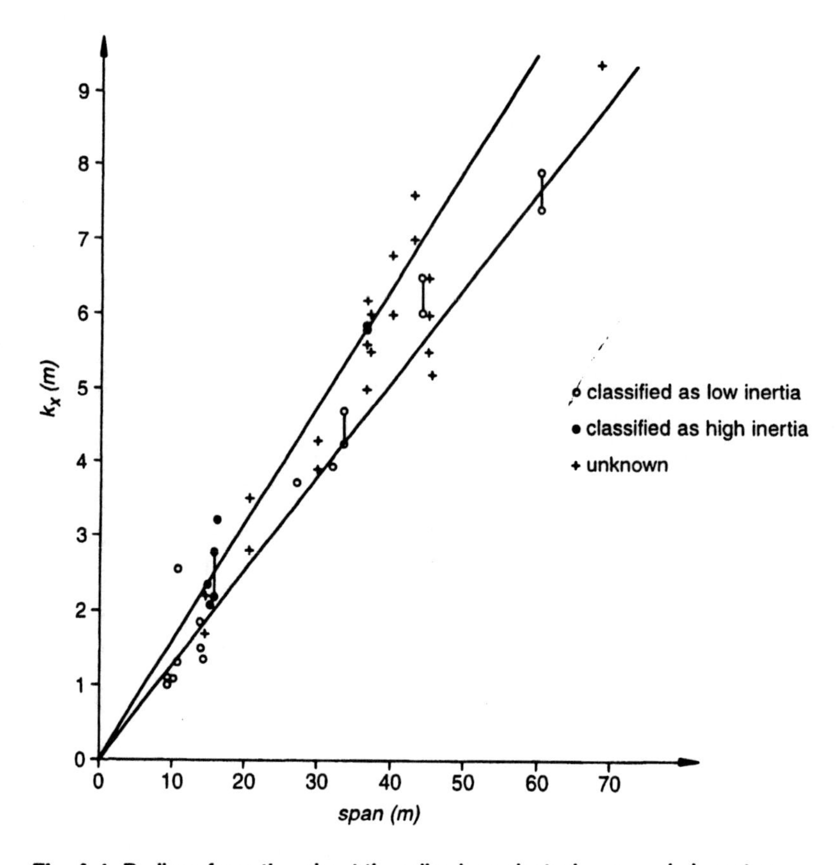

Fig. A.1 Radius of gyration about the roll axis against wing span, b, in metres

identified; the other inertias have been obtained by calculation from the known masses of the components and the aircraft types are given. Reference (A.3) correlated the inertias obtained by plotting the corresponding radii of gyration against an appropriate geometrical parameter; the same process and parameters have been used here. In a few cases data were available on an aircraft at both the maximum mass and the minimum; this is indicated on the plots by joining the points by a vertical line. It should be noted that these values of the radii of gyration are unlikely to be either the maximum or the minimum that can occur.

Figure A.1 shows a plot of radius of gyration about the roll axis against wing span, b, in metres. Fairly obviously, if the engines are located well out on the span of the aircraft we would expect the inertia to be larger than for a comparable aircraft with engines in or near the fuselage. For the data with known aircraft type, multi-engine propeller aircraft were classified, as expected, to have relatively high inertia; single-engined aircraft and jet-engined aircraft with engines close to the fuselage were classified as of low inertia. It proved possible to draw lines of reasonable fit to the two sets of data. The data reference (A.3) were then added and can be seen to fall close to the other data. The slopes of the lines are 0.16 for the multi-engined propeller aircraft and 0.128 for the other types; reference (A.3) found no effect of number of engines and a slope of 0.157.

Figure A.2 shows a plot of radius of gyration about the pitch axis against fuselage length, l_B, in metres. Again the aircraft were classified into low and high inertia, with aircraft with engines near a spanwise line through the cg classified, as expected, to have lower inertia. In this case the classification did not separate the data and a single relation between the parameters was assumed. Linear regression gives the radius of gyration in pitch as $k_y = 0.1816 l_B - 0.214$ m, with a standard deviation of the fractional error of 0.134.

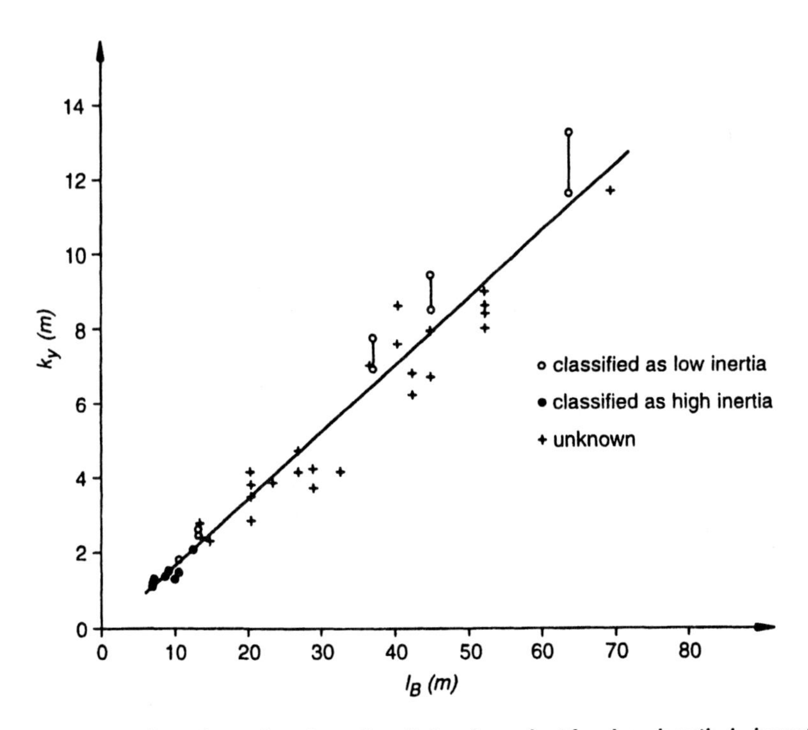

Fig. A.2 Radius of gyration about the pitch axis against fuselage length, l_B, in metres

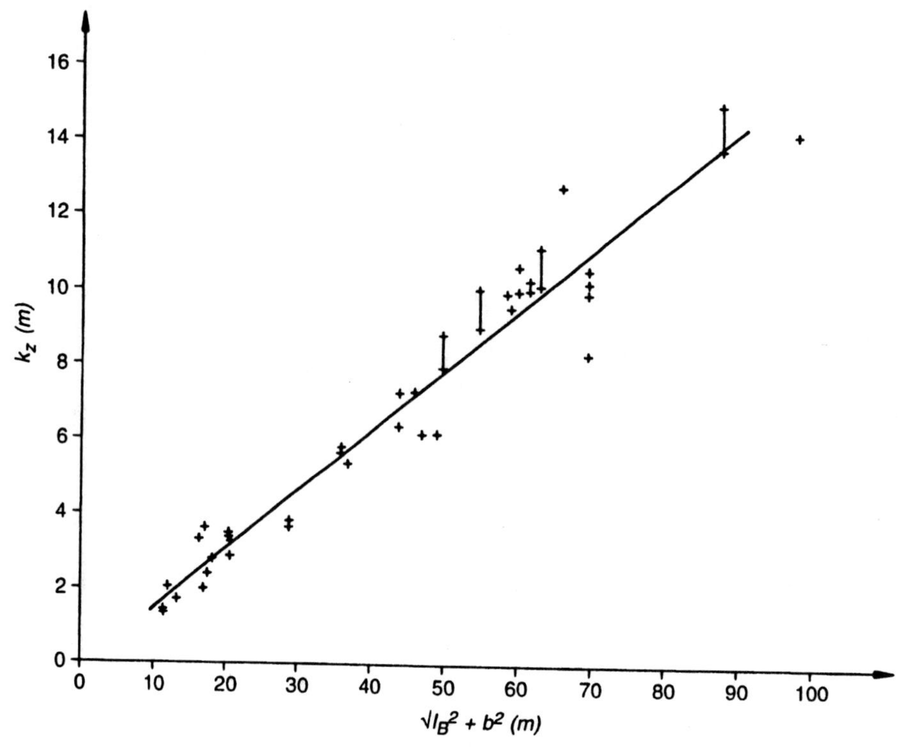

Fig. A.3 Radius of gyration about the yaw axis against $\sqrt{l_B^2 + b^2}$

Figure A.3 shows a plot of radius of gyration about the yaw axis against $\sqrt{l_B^2 + b^2}$. In this case no simple classification of the aircraft by engine arrangement was considered likely to be useful. Linear regression was used to obtain the relation $k_z = 0.1613\sqrt{l_B^2 + b^2} - 0.124$ m, with a standard deviation of the fractional error of 0.136.

Answers to problems

1.1 4.77
1.2 541 lb h^{-1} approximately
1.3 1.291, 1.007, 0.832, 0.334, 0.0766

2.1 0.5085, 0.024 34, 59.8 m s^{-1}
2.2 1.19, 54.7 m s^{-1}, 1310 kW; 0.687, 72.9 m s^{-1}, 17.93 kN
2.3 144 m s^{-1}
2.4 137 m s^{-1}
2.5 2.24
2.6 37.5 s
2.7 8.13 km using Simpson
2.8 2971 km; 2226 km; 98.4 m s^{-1}, 2910 km
2.9 1600 km; 126.7 m s^{-1}, 1014 km, 0.4184
2.10 5797 km
2.11 6.09 h

3.1 6.1°, 9.72 m s^{-1}; 6.3°, 108.8 m s^{-1}, 11.9 m s^{-1}; 161.3 m s^{-1}, 5.3°, 14.86 m s^{-1}
3.2 9.01°, 14.41 m s^{-1}; 82.6 m s^{-1}, 10.11°, 15.5 m s^{-1}
3.3 29.6°, 69.4 m s^{-1}
3.4 220 m s^{-1}, 18.35 m s^{-1}
3.5 0.008 552, 0.020 05, 0.574 m s^{-1}, 1.73°
3.6 8.6 km
3.7 9.57 m s^{-1}, 75.41 m s^{-1}
3.9 2.76 km, 8.6 km, 23.89 km, 67.4°
3.10 1320 m, 2.807
3.11 545 m
3.12 63.3°, 1138 m
3.14 236 m
3.16 263 m
3.17 383 m
3.18 217 m
3.19 488 m

4.1 Wing warping, all moving tips, auxiliary jets/rockets, dive brakes, reverse and deflected engine thrust, independent air-brakes

5.1 5.98 m, 4.83 m, 7.52 m, 43.3°
5.2 50 N, 0.0163 N m, 8.57°, 233 N
5.3 0.125
5.4 −1.82°, 0.125
5.5 −0.074
5.6 −5.92°, 84.5 N

5.7 0.36, 0.38
5.9 −9.2°
5.11 5.73°
5.13 −6.76°

6.1 93.4 m s^{-1}

7.1 113.2 s, 202.3 m, 7.94 m s^{-1}

9.1 −0.0326, 0.103, −0.42, −4.915, −0.455, −0.237, −1.3, 0.0, −0.4935, −1.29, −3.685, −0.671
9.4 58.8 s, 94 s
9.5 −0.0145 ± i0.1129, −0.0906 ± i0.7056
9.6 0.507 s, 8.098 s, 154.9 s, 1491 s, $k_2/k_3 = 1.1568 + i0.1096$, k_1/k_3 is left for the student

10.4 $n = -0.176[1 - e^{-0.66t}(\cos 0.718t + 0.92 \sin 0.718t]$

11.3 −9.804, −0.376, 0.0185 ± i6.517; comment: dutch roll mode is unstable, 0.4 s, 10.5 s, 213.5 s (to double), 5.595 s. Eigenvectors: $k_2/k_1 = 3.39 \pm i1.979$, k_3/k_1 is left for the student; $k_3/k_2 = 0.0253$, $k_1/k_2 = -0.0431$

13.1 −0.466
13.2 (a) one, (b) none, (c) one unstable oscillation, (d) one real, (e) one unstable oscillation, (f) one (oscillation) (g) two (one oscillation).

References

A reference in the form of a five-figure number in brackets, such as (710012), is a reference to the Data Item of the Engineering Sciences Data Unit with the given number.

1.1 *USAF Stability and Control Datcom*, Flight Control Division, Air Force Dynamics Laboratory, Wright Patterson Air Force Base, OH.

2.1 Bore, C. L. 1993: Some contributions to propulsion theory – fuel consumption and the general range equation. *The Aeronautical Journal* **97** (963), 118–20.

7.1 Thomson, W. T. 1983: *Theory of vibrations with applications*, second edition. London: George Allen & Unwin.

8.1 Bryan, G. H. 1911: *Stability in aviation*. London: Macmillan.

8.2 Hopkin, H. R. 1970: *A scheme of notation and nomenclature for aircraft dynamics and associated aerodynamics*. ARC Reports & Memoranda 3562.

8.3 Etkin, B. 1959: *Dynamics of flight – stability and control*. New York: John Wiley & Sons.

9.1 Hancock, G. J. and Lam, J. S. Y. 1987: On the application of axiomatic aerodynamic modelling to aircraft dynamics. Part 2 – Longitudinal aircraft motions with attached flow. *The Aeronautical Journal* **91** (901), 4–20.

9.2 Babister, A. W. 1980: *Aircraft dynamic stability and response*. Oxford: Pergamon International Library, Chapter 3.

10.1 Jain, M. K. 1979: *Numerical solution of differential equations*. New Delhi: Wiley Eastern, Chapter 2.

10.2 Lanczos, C. 1957: *Applied analysis*. London: Sir Isaac Pitman & Sons, Chapter 4.

11.1 Babister, A. W. 1980: *Aircraft dynamic stability and response*. Oxford: Pergamon International Library, Chapter 4.

11.2 Pinsker, W .J. G. 1961: *The lateral motion of aircraft and in particular of inertially slender configurations*. ARC Reports & Memoranda 3334.

13.1 Pallett, E. H. J. 1987: *Automatic flight control*, 3rd edition. Oxford: B S P Professional Books.

A.1 Torenbeek, E. 1982: *Synthesis of subsonic airplane design*. Delft/The Hague: Delft University Press and Martinus Nijhoff.

A.2 Stinton, D. 1983: *The design of the aeroplane*. Oxford: B S P Professional Books.

A.3 Mitchell, C. G. B. 1973: *A computer programme to predict stability and control characteristics of subsonic aircraft*. R A E Technical Report 73079.

A.4 Private communication, Prof. C. Leyman.

Further reading

Babister, A. W. 1980: *Aircraft dynamic stability and response*. Oxford: Pergamon International Library.
Duncan, W. J. 1952: *Control and stability of aircraft*. Cambridge: Cambridge University Press.
Etkin, B. 1972: *Dynamics of atmospheric flight*. New York: John Wiley & Sons.
Grantham, W. J. and Vincent, T. L. 1993: *Modern control systems analysis and design*. New York: John Wiley and Sons.
Hale, F. J. 1984: *Aircraft performance, selection and design*. New York: John Wiley & Sons.
Hancock, G. J. 1994: An introduction to the flight dynamics of rigid aeroplanes. Ellis Horwood.
Mair, W. A. and Birdsall, D. L. 1992: *Aircraft performance*. Cambridge: Cambridge University Press.
McLean, D. 1990: *Automatic flight control systems*. Hemel Hempstead: Prentice Hall.
Nelson, R. C. 1989: *Flight stability and automatic control*. New York: McGraw-Hill.
Pallett, E. J. H. 1987: *Automatic flight control*, 3rd edition. Oxford: B S P Professional Books.
Torenbeek, E. 1982: *Synthesis of subsonic airplane design*. Delft/The Hague: Delft University Press and Martinus Nijhoff.
Vinh, N. X. 1993: *Flight mechanics of high-performance aircraft*. Cambridge: Cambridge University Press.

Index

Printed in the United Kingdom
by Lightning Source UK Ltd.
109179UKS00002B/108

9 780340 631706